Forty Years of Scientific Instrument Studies

Scientific Instruments and Collections

STUDIES PUBLISHED UNDER THE AUSPICES OF THE SCIENTIFIC INSTRUMENT COMMISSION

VOLUME 11

The titles published in this series are listed at *brill.com/sico*

Forty Years of Scientific Instrument Studies

The Scientific Instrument Commission

Edited by

Sara J. Schechner
Richard L. Kremer

BRILL

LEIDEN | BOSTON

Cover illustration: SIC delegates studying instruments at the XXII Scientific Instrument Symposium, Stockholm, 2001. Photo: Klaus Staubermann

Library of Congress Cataloging-in-Publication Data

Names: Schechner, Sara, 1957- editor | Kremer, Richard L. (Richard Lynn) editor
Title: Forty years of scientific instrument studies : the scientific instrument commission / Edited by Sara J. Schechner, Richard L. Kremer.
Description: Leiden ; Boston : Brill, [2026] | Series: Scientific instruments and collections, 1877-7066 ; Volume 11 | Includes bibliographical references and index. |
Identifiers: LCCN 2025029257 (print) | LCCN 2025029258 (ebook) | ISBN 9789004737570 hardback | ISBN 9789004737587 ebook
Subjects: LCSH: Scientific apparatus and instruments—History | Scientific apparatus and instruments—Design and construction—History
Classification: LCC Q185 .F67 2026 (print) | LCC Q185 (ebook)
LC record available at https://lccn.loc.gov/2025029257
LC ebook record available at https://lccn.loc.gov/2025029258

Typeface for the Latin, Greek, and Cyrillic scripts: "Brill". See and download: brill.com/brill-typeface.

ISSN 1877-7066
ISBN 978-90-04-73757-0 (hardback)
ISBN 978-90-04-73758-7 (e-book)
DOI 10.1163/9789004737587

This book is printed on acid-free paper and produced in a sustainable manner.

In memory

of two outstanding and well-loved figures in the

first forty years

of symposia of

the Scientific Instrument Commission

Jim Bennett
1947–2023

and

Paolo Brenni
1954–2021

Contents

Figures and Tables

Figures

Tables

Notes on Contributors

Robert G.W. Anderson
spent practically his whole career working in museums, after having taken a doctorate in physical chemistry at Oxford University. First, he worked as a science curator, later becoming director of the National Museums of Scotland (Edinburgh), the British Museum (London), and the Science History Institute (Philadelphia), from which he retired in 2020. From 1982 he was President of the SIC for fifteen years, and at other times of the British Society for the History of Science and of the Society for History of Alchemy and Chemistry. He holds the Dexter Award (American Chemical Society) and the Bunge Prize (German Chemical Society). He is an Emeritus Fellow of Clare Hall, Cambridge and Honorary Fellow of St John's College, Oxford. When time allowed, his research was focused on early chemical instruments and apparatus, and on the Scottish Enlightenment chemist, Joseph Black (1728–99), whose correspondence he published in 2012. Throughout his life he has suffered from ICS (incurable collecting syndrome).

Jim Bennett (deceased)
was Keeper Emeritus at the Science Museum, London. He was formerly Director of the Museum of the History of Science and Professor of the History of Science in the University of Oxford. He was a previous President of the Scientific Instrument Commission.

Julia Bloemer
(ORCID: 0000-0002-0246-9684) is a historian of physics working at the European University of Flensburg, Germany. She received her PhD in Munich with a dissertation on experimental philosophy in southern German monasteries in the eighteenth century and spent a decade at the Deutsches Museum in Munich. Her research focuses on the history of experimentation, scientific instruments, and the relationship between science and religion. Together with three colleagues, she conducts an annual international seminar on material culture for graduate students and museum interns.

Marvin Bolt
was formerly Vice-President for Collections and Director of the Webster Institute for the History of Astronomy at the Adler Planetarium in Chicago and first Curator of the History and Technology of Glass at the Corning Museum of Glass. For more than twenty years he has teamed together with Michael Korey

to pull telescopes apart. He has most recently been extending and summarising this work while on research fellowships at the Technische Universität in Berlin and the Bibliotheca Hertziana in Rome. For his outreach and interpretation work he was awarded the Hazen Education Prize of the History of Science Society in 2017.

Randall C. Brooks
spent twenty years in a Canadian university astronomy department designing and servicing research instruments before turning his attention to the history of scientific instruments and technology. He then spent the next twenty years in a museum environment, beginning as the curator of physical sciences and space at the Canada Museum of Science and Technology in Ottawa and ending as the Vice President of Collections and Research of the Ingenium, the institution which operates three Canadian national museums related to science and technology, aviation, and agriculture. He also served as the editor of *Rittenhouse* and *eRittenhouse*.

Terje Brundtland
(ORCID: 0009-0008-4872-4549) is a former laboratory technician, specialised in modern vacuum technology. He has previously worked at the University of Tromsø, Norway. Later, turning from technology to history, he did his DPhil at Linacre College, Oxford, on the history of scientific instruments. He is author of "A Description of the Vacuum Chamber for the Blaamann Plasma Experiment," *Vacuum*, 1998; "After Boyle and the Leviathan: The Second Generation of British Air Pumps," *Annals of Science*, 2011; and "Francis Hauksbee and his Air Pump," *Notes and Records of the Royal Society of London*, 2012. A coming study is planned on Emanuel Swedenborg's air pump.

Ileana Chinnici
(ORCID: 0000-0002-3536-1709) is an astronomer at INAF-Osservatorio Astronomico di Palermo. Her main research field is the history of nineteenth-century astronomy and astrophysics. Her volume, *Decoding the Stars. A Biography of Angelo Secchi, Jesuit and Scientist* (Leiden: Brill, 2019) was awarded the 2021 Osterbrock Book Prize from the American Astronomical Society. Her interests include studies on historical scientific collections of instruments, books and archives. A member of the IAU C3 (History of Astronomy) Organising Committee in the last two triennia, she was elected incoming President of the Scientific Instrument Commission for the years 2025–2029.

Gloria Clifton
(ORCID: 0000-0002-1566-7911) is a PhD, and curator emeritus of Royal Museums Greenwich, after retiring as senior curator of the Royal Observatory. She published the *Directory of British Scientific Instrument Makers, c. 1550–1851* (1995) and continues to add material to the database of makers, extending to 1914. She is President of the Scientific Instrument Society and also of the Hakluyt Society, which publishes original historical accounts of travel, voyages and exploration.

Tiemen Cocquyt
(ORCID: 0000-0002-5426-4162) is curator of early modern natural sciences in Rijksmuseum Boerhaave, the Dutch national museum for the history of science and medicine in Leiden. He has held this position since 2010. Before, he worked at Utrecht University Museum. He primarily researches seventeenth-century optical instruments and cabinets of experimental philosophy. In this research, he has repeatedly applied forensic methods, such as interferometry, neutron tomography and XRF analysis, to query historic instruments.

Gaye Danışan
(ORCID: 0000-0002-8888-6533) is an assistant professor at Istanbul University, Department of the History of Science. Her research focuses on Ottoman history, particularly calendars, portable astronomical instruments, volvelles, navigation, and astrometeorology. She is currently working on the project *Portable Astronomical Instruments: The Processes of Adaptation and Diffusion of Medieval Islamic and European Examples in the Ottoman Geography* (1500–1700), funded by the Turkish Academy of Sciences – Outstanding Young Scientists Awards Program (TÜBA-GEBİP).

Richard Dunn
(ORCID: 0009-0006-2613-6477) is Keeper of Technologies and Engineering at the Science Museum, London. He was previously Senior Curator for the History of Science at Royal Museums Greenwich. His publications include *The Telescope: A Short History* (2009), *Finding Longitude* (2014, with Rebekah Higgitt), and *Navigational Instruments* (2016).

Jean-François Gauvin
(ORCID: 0000-0001-9426-8726) is a professor of history, scientific heritage, and museum studies at Université Laval in Québec City, Canada. He has published numerous articles on scientific instruments, two frequently-cited edited

volumes on globes and abbé Jean-Antoine Nollet (1700–1770), and curated and/or oversaw award-winning exhibitions in Canada and the United States. His recent book, *Instruments of Knowledge: Finding Meaning in Objects, Habits, and Museums* (Brill, 2023) explores the relationships between instruments (*organum*), scientific practices (*habitus*), and collections (*museum*). He is a member of the Centre de recherche Cultures-Arts-Sociétés (CELAT).

Neil Handley
is a PhD, AMA, FRSA, and is Curator of the British Optical Association Museum at the College of Optometrists, London. He is a Liveryman of the Worshipful Company of Spectacle Makers and current Chairman of the Scientific Instrument Society (SIS).

Rebekah Higgitt
(ORCID: 0000-0001-9387-4150) is Principal Curator of Science at National Museums Scotland. She did her PhD at Imperial College, London, and has previously worked at the University of Edinburgh, Royal Museums Greenwich, and the University of Kent. She is author or co-author of *Recreating Newton* (2007), *Finding Longitude* (2014), and *Metropolitan Science: London Sites and Cultures of Knowledge and Practice, c.1600–1800* (2024). In 2024 she was awarded the Paul Bunge Prize for her work on the history of scientific instruments. She is one of the editors of the Brill series, Knowledge Infrastructure and Knowledge Economy.

Boris Jardine
(ORCID: 0000-0002-9590-0102) is a rare book dealer, writer, and historian. Before entering the book trade full-time, he worked as a Research Associate at the University of Cambridge, and Curator of History of Science at the Science Museum, London.

Floor Anna Koeleman
(ORCID: 0000-0001-8176-4544) received the Excellent Thesis Award from the University of Luxembourg for her dissertation "Visualizing Visions: Re-Viewing the Seventeenth-Century Genre of Constcamer Paintings," which will be published as a book by Brill in the series Studies on Art, Art History, and Intellectual History. She is currently a postdoctoral researcher at the University of Lausanne and has been a member of the Scientific Instrument Commission since 2019. Her research focuses on the history and philosophy of art, science, and instrumentation.

Michael Korey
(ORCID: 0009-0009-7634-7872) When not tending to early-modern optics, including telescopes and burning lenses, Michael is busy making mischief with Renaissance planetary clocks and mathematical instruments as senior curator at the Mathematisch-Physikalischer Salon in Dresden, and was the Secretary of the Scientific Instrument Commission from 2013–2017. For his outreach and interpretation work he was awarded the Medienpreis of the Deutsche Mathematiker-Vereinigung in 2019.

Richard L. Kremer
(ORCID: 0000-0002-6235-7799) is Professor Emeritus of History at Dartmouth College where he taught the history of science and curated that institution's collection of historic scientific instruments. He served as President of the SIC from 2018–2021; co-edited SIC04, *Instruments on Display* (2014); co-authored *Study, Measure, Experiment: Stories of Instruments at Dartmouth College* (2005); and has published extensively on medieval astronomical instruments and tables.

Phil Loring
(ORCID: 0009-0002-2185-9386) is curator of the history of medicine at Norway's National Medical Museum, part of Norsk Teknisk Museum. In this role he was part of the team that delivered the major gallery "Life and Death" (2021) as well as the award-winning exhibition "FOLK – from Racial Types to DNA Sequences" (2018). He holds two master's degrees, in medical anthropology and history of science. Before coming to Norway, he worked for three years at Harvard University's Collection of Historical Scientific Instruments, and for five years as BPS Curator of Psychology at the Science Museum, London, where he curated the exhibition "Mind Maps: Stories from Psychology" (2013).

A.D. Morrison-Low
(ORCID: 0000-0003-4184-3090) is a Research Associate at National Museums Scotland, Edinburgh. She has been a member of the Editorial Board of Scientific Instruments and Collections since 2009, and General Editor since 2022. In retirement she has continued to write, publish and lecture in those subjects that engrossed her while still employed.

Jacqueline Marie Musacchio
(ORCID: 0009-0001-3209-0819) is Professor of Art History at Wellesley College, Massachusetts, United States. She publishes on Italian Renaissance domestic

art and on Americans in nineteenth-century Italy, including most recently *The Art and Life of Francesca Alexander 1837–1917* (2025).

Joshua Nall
(ORCID: 0009-0003-4942-2714) is Director of the Whipple Museum of the History of Science at the University of Cambridge. His book, *News from Mars: Mass Media and the Forging of a New Astronomy, 1860–1910* won the History of Science Society's 2020 Philip J. Pauly Prize for the best first book on the history of science in the Americas. Since 2021 he has been a member of the Editorial Board for the Scientific Instrument Commission's series Scientific Instruments and Collections.

David Pantalony
(ORCID: 0009-0007-1786-7187) is Curator of Science and Medicine at Ingenium – Canada's Museums of Science and Innovation and Adjunct Professor in the Department of History at the University of Ottawa, where he teaches collection-based approaches in the history of science. He won the Bunge Prize in 2012 for his book *Altered Sensations: Rudolph Koenig's Acoustical Workshop in Nineteenth-Century Paris*. Through a partnership between Ingenium and the University of Ottawa, he is currently developing a national inventory of precision instrument makers, instruments, and collections in Canada.

Richard A. Paselk
Emeritus Professor of Chemistry at California State Polytechnic University, Humboldt, is the founder, author, web-designer, creator and Emeritus Curator of the university's Scientific Instrument Museum. In addition to authoring numerous scholarly and educational articles during a four-decade teaching career, he is a maker of early scientific instrument recreations.

Pedro M.P. Raposo
(ORCID: 0000-0001-6724-2993), DPhil, is the Martha Hamilton and I. Wistar Morris III Executive Director of the Library and Archives at the Academy of Natural Sciences of Drexel University. Raposo holds a doctorate in history of science from the University of Oxford, where he was awarded the Magellan Prize for his doctoral project on the history of astronomical observatories in nineteenth-century Europe. Before joining the Academy, Raposo was a postdoctoral fellow and lecturer at the University of Lisbon, a visiting scholar at the Max Planck institute for the History of Science in Berlin, and Curator and Director of Collections at the Adler Planetarium in Chicago, where he was

responsible for the Planetarium's renowned collections of rare books, scientific instruments, and archives. The exhibition "What is a Planet?", in which he served as curatorial lead, was awarded the 2016 Great Exhibitions Prize of the British Society for the History of Science.

Elisabetta Rossi
(ORCID: 0000-0002-8975-6917) is a Research Fellow at the Department of Physics and Astronomy "A. Righi" (DIFA) and the University Museum Network (SMA), University of Bologna. She earned a PhD in Historical Studies from the University of Milan in 2025 and holds a degree in Astrophysics and Cosmology (2021). Her current research focuses on the invention of multi-mirror telescope technology pioneered in the mid-twentieth century. She also has experience in museum exhibition design, with a strong interest in scientific instruments and the material history of science.

Sara J. Schechner
(ORCID: 0000-0002-6335-1399) is Curator Emerita of the Collection of Historical Scientific Instruments at Harvard University, retiring in 2024 after twenty-four years of service. She is past Secretary of the Scientific Instrument Commission (2003–2013) and currently serves on the Editorial Board for the Scientific Instrument Commission's series Scientific Instruments and Collections. A founding member of the American Astronomical Society's Working Group for the Preservation of Astronomical Heritage, she is responsible for issues related to historical instruments. Her recent books include *Tangible Things: Making History through Objects* (2015, with Laurel Thatcher Ulrich *et al.*) and *Time of Our Lives: Sundials of the Adler Planetarium* (2019). Current research focuses on sundials, science, and social change; the representation of astronomers and their instruments in works of art and advertising; and the Pope Orrery, an enormous mechanical model of the solar system made in Boston during the American Revolution.

Robert W. Smith
(ORCID: 0009-0005-0759-6809) is a Professor of History at the University of Alberta. He has been awarded the History of Science Society's Watson-Davis Prize, the American Astronomical Society's Doggett Prize, and won the Bunge Prize in 2023. Robert's main research interests are in the history of astronomy and cosmology, particularly the material culture of space astronomy. Among his writings is a scholarly monograph, as well as three coauthored-popular

books, on the history of the Hubble Space Telescope. Currently he is at work on a history of the James Webb Space Telescope.

Giorgio Strano
(ORCID: 0000-0002-7587-0495) holds a PhD (2003) in History of Science, University of Florence, and is Head of the Collections at the Museo Galileo – Institute and Museum of the History of Science of Florence. He is actively involved in the study and popular dissemination of the history of astronomy. He has published extensively on the topic, and curated or collaborated in permanent and temporary exhibitions. He is a member of the Scientific Instrument Society and, from 2007 to 2020, has been the general editor of the series Scientific Instruments and Collections, published by Brill.

Sylvia Sumira
is an independent conservator specialising in globes. She worked in globe conservation at the National Maritime Museum, Greenwich, for several years before setting up her own studio. An accredited member of the UK Institute of Conservation, her clients include museums, libraries and other institutions in Britain and abroad, and also private owners of globes. Her book, *The Art and History of Globes*, was published by the British Library a few years ago.

Sofia Talas
(ORCID: 0000-0003-1971-8680) is the Curator of the Giovanni Poleni Museum, at the University of Padua. She carries out researches on the history of scientific instruments and the history of physics. From 2011 to 2017, she was President of Universeum, the European network aiming at the preservation, study, access, and promotion of academic heritage.

Simon Werrett
(ORCID: 0000-0002-4074-5264) is Professor of the History of Science in the Department of Science and Technology Studies at University College London. He is the author of *Fireworks: Pyrotechnic Arts and Sciences in European History* (Chicago, 2010) and *Thrifty Science: Making the Most of Materials in the History of Experiment* (Chicago, 2019) and co-editor, with Lissa Roberts, of *Compound Histories: Materials, Production, and Governance, 1760–1840* (Brill, 2017).

Ewa Wyka
(ORCID: 0000-0003-3822-7377) is a historian of science, with research interests in the history of mathematical and natural sciences, the evolution of scientific

instruments, aristocratic and academic scientific collections, making of scientific instruments, museology of science and technology, and the popularisation of science. Her latest book is: *Artefakty nauki: Historyczne przyrządy naukowe w zbiorach polskich muzeów* [Artifacts of Science: Historical Scientific Instruments in the Collections of Polish Museums] (Warsaw: PWN, 2023).

Zhao Ke

(ORCID: 0000-0002-7884-2879) is the director of the Electronic Science and Technology Museum and an associate professor in the University of Electronic Science and Technology of China. He earned his PhD in microelectronics and solid-state electronics. His research interests include the history of electronic science and technology and university museums. He currently serves as the Vice-Chair of Chinese Committee on the History of Scientific Instruments.

Introduction: Forty Years of Scientific Instrument Studies

Sara J. Schechner and Richard L. Kremer

The minutes of the plenary business meeting of the XXVIII Scientific Instrument Commission Symposium indicate that the group met on 31 July 2009 in Budapest. They do not reveal where the group met. Indeed, the only documentation for the location seems to be a slightly blurry photograph showing a group of grinning instrument curators and historians floating in the Gellért Bath, Budapest's iconic art nouveau spa overlooking the Danube River (*Fig. 0.1*). No blackboards, lecterns, vitrines, or historic scientific instruments are visible. Nonetheless, serious scholarly business was conducted that morning in the Gellért Bath. To explain how this could be true is the purpose of this book.

This collection commemorates the fortieth anniversary of the reconstitution of the Scientific Instrument Commission (SIC) of the International Union of History and Philosophy of Science (IUHPS), a step formalised in 1977 at the XV International Congress of the IUHPS meeting in Edinburgh and the establishment of its annual symposia, the first being held in Bucharest in 1981. In the 1950s the IUHPS had founded an earlier Scientific Instrument Commission to organise the preparation of national inventories of historic scientific instruments. The mission statement of the 1977 SIC also mentions inventories and catalogues but adds: "to facilitate the exchange of information between research workers in the field of historic scientific instruments."[1] The essays in this volume explore these facilitations over the past forty years and, in some

FIGURE 0.1 Conducting SIC business in the Gellért Bath, Budapest, 31 July 2009
PHOTO: JANE WESS

1 SIC, *Newsletter*, no. 1 (1980), 1.

 | DOI:10.1163/9789004737587_002

cases, speculate about how the SIC and instrument studies might evolve in a future of re-imagined museums, artificial intelligence, and the development of entirely new types of instrumentation for the practice of science.

The 1977 SIC launched its activity by circulating regular newsletters "to those people all over the world who are engaged in the study of the history of scientific instruments."[2] The Commission also began organising annual, week-long symposia, bringing together museum curators, historians of science and technology, collectors, dealers and independent scholars to present papers, visit museums and collections and exchange information. Meeting each year in a different city, the symposia by the mid-1980s were attracting between sixty and one hundred attendees and quickly became the lifeblood of the Commission.

Four chapters in this volume describe the activities of the SIC in a chronological arc, ordered by decade. Each author, Robert Anderson, Sara Schechner, Richard L. Kremer, and David Pantalony, had been heavily involved with the Commission during his or her decade so these are participants' accounts of the organisation, at times quite personal, at times more analytical. They tell stories of the SIC welcoming increasing numbers of students, young scholars, and more people from beyond Europe into its activities; embracing new media with the rise of the Internet; creating a culture of informality, even light-heartedness, and fostering not only professional but also personal relationships of exchange, criticism, and support. These chapters explain why the Gellért Bath might epitomise one essential facet of the SIC's first forty years.

Other chapters examine particular themes of instrument studies over the past four decades. Three chapters follow instrument people into libraries and museums. Boris Jardine reviews historiography, debates over research methods, and the SIC's long-standing insistence on materially-based interrogation of surviving artefacts and asks what art history might have to offer to instrument studies (*Fig. 0.2*). A "curator's meditation" by Giorgio Strano reflects on how museum architecture has shaped design possibilities for permanent exhibitions and the changing types of temporary exhibitions. A.D. Morrison-Low constructs a typology of instrument catalogues, defining five forms that have developed, or not, over the past forty years, ranging in scale from the national inventory to the private collector's labour of love. Topics such as these have been widely discussed at SIC symposia since 1981.

2 The first newsletter was circulated to eighty persons in eighteen nations. SIC, *Newsletter*, no. 1 (1980), 7–8.

FIGURE 0.2 In a scene reminiscent of the 1766 painting, *A Philosopher Lecturing on the Orrery* by Joseph Wright of Derby, SIC delegates examine and discuss rare scientific instruments in a royal collection otherwise inaccessible to the general public. This special visit to the Bernadotte Library in the Royal Palace, Gamla Stan, Stockholm was arranged as part of Symposium XX in October 2001.
PHOTO: KLAUS STAUBERMANN

The final three chapters move into wider settings. Jean-François Gauvin surveys how historic scientific instruments have been used to teach today's students about contemporary science, the "nature of science" or material culture more generally, and how efforts to replicate historical experiments or reconstruct historical instruments have enriched instrument studies. Sofia Talas asks about national variations in the study, collection, preservation, and exhibition of historic scientific instruments. Pedro M.P. Raposo and Floor Anna Koeleman describe how the SIC in the late 2010s began using social media platforms to introduce historic scientific instruments to wider audiences. And they report some interesting experiments that employed artificial intelligence and large language models to analyse images of historic instruments and their design and functional features.

These ten chapters offer description, analysis, and commentary on the SIC and instrument studies over the past four decades. But no commemoration of the Commission could be complete without examples of particular historic scientific instruments and the methodological challenges they present.

As noted by Jardine, the SIC has always exhibited a materialist streak, has always emphasised hands-on investigation of surviving historical artefacts. We thus have commissioned twenty "fingerposts" or short, illustrated essays pointing toward discrete problems interestingly encountered in instrument studies. We adopt the metaphor from Francis Bacon's "instances of the fingerpost" or those places where roads part and a "decisive and judicial" decision must be made about which path to follow. Such instances, wrote Bacon, "afford very great light and are of high authority, the course of interpretation sometimes ending in them and being completed."[3] The index fingers of our posts point, for example, to movement (from the Soviet Union to China), materials (gold), the quotidian (wine bottles), public display (massive globes), authenticity (forgers), patronage (cost-overruns), and much more. These vignettes show the SIC currently at work.

We dedicate this book to two outstanding practitioners of instrument studies and well-loved figures in the Scientific Instrument Commission. Jim Bennett, former President of the SIC, former President of the British Society for the History of Science, and long-time curator at the history of science museums in Cambridge, Oxford, and London, proposed the idea for this fortieth-anniversary commemorative volume back in 2021:

> It shouldn't just be a gather-up of articles like a Festschrift. It should have some theme or structure, like the development of instrument historiography over the period. An alternative to a temporal organisation could be a geographical one, emphasising our international character – views from different places and traditions. Perhaps both organising principles can be accommodated. There could be articles on events or trends – the influence of *Leviathan and the Air Pump* ... the material turn in the history of science ... who needs the Scientific Instrument Society? ... how we saw off the threat from Science Centres ... (only joking) ... trends in catalogues ... in museums. ... Anyway, to demonstrate that we are a vital organisation in a cogent discipline, there should be a structure or theme. We want scholars other than ourselves to be interested.[4]

The SIC Board enthusiastically supported Jim's proposal and he invited the current editors to form an editorial team. Together we defined the chapter topics but Jim came up with the "fingerpost" idea and took the lead in soliciting

3 Francis Bacon, *The New Organon* [1620], ed. Fulton H. Anderson (Indianapolis: Bobbs-Merrill, 1960), 191.

4 Bennett to Kremer, email, 12 January 2021.

contributors and editing their early drafts until his health started to decline in the spring of 2022. Unfortunately, we were unable to complete the volume before Jim's death.

We also dedicate this volume to Paolo Brenni, another former President of the SIC, intrepid explorer of instrument collections large and small located anywhere in the world, and theatrical demonstrator (and video-maker) of historic scientific instruments in action. In the SIC Paolo always exuded excitement, curiosity, and a delightful inability to keep his hands off historic apparatus. When in a museum he yelled "come and look at this," everyone came, looked, and learned. Paolo also introduced the buzzword contest into the SIC's annual symposia. At the opening session, that year's buzzword would be announced, with each speaker challenged to smoothly insert this word somewhere into his or her presentation.[5] Prizes were awarded, sessions enlivened, and the "exchange of information between research workers" enhanced.

5 The first buzzword, introduced in 2003 at the XXII SIC Symposium in Newport News, Virginia, was Toast. The prize was won by Admiral Welsh, speaking about the Enigma machine. Buzzwords were always arbitrary and humorous, sometimes related to the culture of the meeting locale. Other examples include Lawnmower, Borscht, Opossum, Coconut, Paprika, Pinocchio, Bratwurst, Samba, Platypus. Rhubarb, Fiat (automobile), Tulip, Rum, James Bond, Polka, and Oracle of Delphi.

Chapters

CHAPTER 1

Establishing the Scientific Instrument Commission in the 1980s

Robert G.W. Anderson

Having stood down from its Presidency some twenty-five years ago, I am delighted to see how the Scientific Instrument Commission (SIC) continues to flourish: to amass eighty-seven papers at this Athens meeting, not including the Paolo Brenni tributes and this session on SIC history, is a real achievement![1] This essay will attempt to give a flavour of the earliest days rather than a close analysis of the hundreds of papers that were delivered at them.[2]

There can be few people present in this lecture hall who were present at the 1981 International Union for the History and Philosophy of Science (IUHPS) meeting held in Bucharest, Romania, where the first of what has become the SIC annual series took place. Those who were there may remember the extremely long opening address to the Congress – I think that with its translation into English it lasted for two hours – given by Elena Ceaușescu, First Deputy Prime Minister, the Romanian President's wife, the woman who used to pretend she was a top research chemist.[3] Just as we thought that her lengthy diatribe was coming to an end, in flowed a troupe of folk dancers from the back of the stage, and we were subjected to an hour of music by Romania's most famous composer, Georges Enescu. That bit of the opening ceremony was engaging enough, but after more than three hours, some of us were desperate to discover the *toaletă*.

I am not going to give a listing of persons and papers given by delegates who made the pilgrimage to Bucharest, followed by further similar approaches to

1 Scientific Instrument Commission Symposium XLI: Meeting to mark the fortieth anniversary of annually held meetings. Athens, 19–23 September 2022.

2 R.G.W. Anderson and G.L'E. Turner, *An Apparatus of Instruments: The Role of the Scientific Instrument Commission* (London: Scientific Instrument Commission, 1993). ISBN 0-9522077-0-2.

3 When the spurious nature of Elena Ceaușescu's chemistry background and qualifications were revealed, various honours accorded to her were withdrawn, including a fellowship awarded by the Royal Institute of Chemistry in 1978. She and the President of Romania, her husband, were tried during the course of the Romanian Revolution and executed on 25 December 1989.

 | DOI:10.1163/9789004737587_003

the people and topics covered at nine subsequent meetings held at Greenwich, Munich, Amsterdam, Berkeley, Florence, Paris, London, Hamburg and Munich, and Cambridge, Massachusetts. That would be exceedingly dull, and without analysis, not of great interest. But as the first speaker in this sequence of four decadal papers today, I must give you an idea about what happened before these annual meetings took place. But even before that, I must emphasise that in presenting the history of the SIC there is no systematically preserved archive to refer to. One has to rely on the *Newsletters*, a few printed items, and sundry notes which happen to survive. What I shall be describing here is largely autobiographical and anecdotal. An approach which is purely descriptive of one meeting after another, one lecture after another, would scarcely stimulate the imagination and would lack the flavour of what went on behind the scenes.

So why did the SIC come into being when it did? From the early twentieth century, the Musée des Techniques in Paris and the Science Museum in London started to publish catalogues of their collections.[4] Additionally, Robert Gunther, founding curator of the Museum of the History of Science in Oxford, was particularly active and his *Astrolabes of the World*, of 1932, was a landmark publication.[5] The IUHPS, Division of the History of Science, decided that a number of Commissions should be established to deal with particular areas in the history of science to which more attention needed to be paid. In 1952, it created a Commission des Instruments Scientifiques, whose first major act in 1956 was to accept a proposal originally made in 1946, that a world inventory of historical scientific apparatus should be compiled. To achieve this aim, the SIC was established to promote the production of national printed inventories of instruments associated with the work of what were called "major scientists." This status clause, "major scientists," was later abandoned and emphasis was placed on the instruments themselves. It seems, however, that were it not for the coming of the Second World War, a Belgian initiative may have produced its early version independently of the IUHPS, for the Introduction to the Belgian inventory states, "Dès 1939, au moment où menaçait la guerre, le Comité Belge d'Histoire des Sciences avait pris l'intiative d'un recensement des instruments scientifiques de valeur historique, conservés dans les musées et les collections privées de notre pays. L'idée fut reprise et l'inventaire esquissé, en 1940, avec la

4 The number of newly published catalogues each decade by the Science Museum varied significantly. See Robert Anderson, "The Museum Catalogue – A Disappearing Art Form?" in *Museum Publications in Science, Technology and Medicine* (Oxford: Group for Scientific, Technological & Medical Collections, 1982), 24–30. See graph on p. 41, showing production by decade, 1860–1980.

5 Robert T. Gunther, *Astrolabes of the World, Based upon the Series of Instruments in the Lewis Evans Collection at Oxford*, 2 vols. (Oxford: Clarendon Press, 1932).

collaboration du personnel des Musées Royaux d'Art et d'Histoire." As it happened, the first inventory to appear was of Belgian collections, in two parts, in 1959 and 1960.[6] It was clear from the Belgian inventory that that country's scientific instruments were scattered amongst many museums and private collections, and if that pattern were repeated in inventories of other nations, the national inventories would provide an overview which, up to that time, had not been available. The next national inventories to be published were the five volumes of Italian instruments in 1963,[7] the French inventory in 1964,[8] the Soviet Union's in 1968,[9] Czechoslovakia's issued in microform in 1960,[10] and at about the same time came an undated microfilm of instruments in Polish collections.[11] A British and Irish inventory was started in 1967 by Mary Holbrook. She received grant funding from a Royal Society committee under the chairmanship of Sir Harold Hartley, but after five years she had not got very far with it; her grant had run out, and she opted to farm goats for cheese production outside Bath. An over-optimistic message in SIC *Newsletter* no. 3 states that the inventory would be published in 1982.[12] In fact at that stage it was very incomplete and quite unpublishable. Later, David Bryden (previously Curator of the Whipple Museum in Cambridge but who, by that time, was in charge of the Pictorial Collection in the Library of the Science Museum) and I picked up the pieces, made corrections, travelled to many venues, and expanded entries

6 Centre National d'Histoire des Sciences, *Inventaire des Instruments Scientifiques Historiques Conservés en Belgique*, two parts (Brussels: Centre National d'Histoire des Sciences, 1959 and 1960).

7 Union International d'Histoire et Philosophie des Sciences, Division d'Histoire des Sciences, Commission pour l'Inventaire Mondiale des Appareils Scientifiques d'Intérêt Historique, *Inventaire des Instruments Conservés en Italie*, 5 vols. (Milan: Museo Nazionale della Scienza e della Tecnica Leonardo da Vinci, 1963).

8 Comité National Français d'Histoire et de Philosophie des Sciences, *Inventaire des Instruments Scientifiques Historiques Conservé en France* (Paris: Centre de Documentation d'Histoire des Techniques, 1964).

9 Leonid Efimovich Maistrov, *Nauchnye Pribory. Pribory i Instrumenty Istorischeskogo Znacheniya* [Scientific instruments. Instruments of Historic Significance] (Moscow: Akademiya Nauk SSSR Institut Istorii Estestvoznaniya i Tekhniki, 1968).

10 Union Internationale d'Histoire et Philosophie des Sciences, Division d'Histoire des Sciences, Commission pour l'Inventaire Mondial des Appareils Scientifiques d'Intérêt Historique, *Contribution Tschechslovaque à l'Inventaire* (1960). Microform.

11 Union International d'Histoire et Philosophie des Sciences, Division d'Histoire des Sciences, Commission pour l'Inventaire Mondial des Appareils Scientifiques d'Intérêt Historique: Tadeusz Przypkowski, *Polish National Inventory of Scientific Instruments*, [1960], Science Museum London, Dana Research Centre and Library, microfilm B.28; see Ewa Wyka, "Les instruments historiques dans les musées polonais," *Artefact* 20 (2024): 283–309.

12 SIC, *Newsletter*, no. 3 (December 1981): 2.

FIGURE 1.1 The British National Inventory of Historical Scientific Instruments was published as *Science Preserved* in 1992. Included are a number of Irish collections
PHOTO: SARA SCHECHNER

considerably to increase the usefulness of the product. Eventually the inventory was published in 1992 as *Science Preserved*.[13] (*Fig. 1.1*)

The early inventories were made to very different standards and levels of comprehensiveness. Those appearing up to the late 1960s are of limited

13 Mary Holbrook with Additions and Revisions by R.G.W. Anderson and D.J. Bryden, *Science Preserved. A Directory of Scientific Instruments in Collections in the United Kingdom and Eire* (London: HMSO, 1992). This inventory includes instruments in 228 institutional and private locations, 18 of which are museums holding national status. A much-expanded inventory of instruments in Irish collections was later published by Charles Mollan, *Irish National Inventory of Historic Scientific Instruments* (Dublin: Royal Dublin Society, 1990), but this was not produced under the auspices of the IUHPS.

value today. Their production was essentially the only work the Commission undertook up until 1980. This did not mean that papers on historical instrument studies were not given at the then triennial IUHPS meetings. In 1977 at the Edinburgh meeting of the IUHPS, the broader purposes of the SIC were defined, which included the production of inventories, the encouragement of catalogue publication, and the exchange of information among research workers. For this, the aforementioned *Newsletter* would be produced. A Committee of the SIC was instituted with Lieutenant Commander Derek Howse of the National Maritime Museum, UK, as President, with Gerard L'Estrange Turner as Secretary, and with further members added from the USA, the Netherlands, East Germany, Poland, USSR, Denmark, France, and West Germany. The British Society for the History of Science held an influential meeting in Oxford in 1979 with the theme "Scientific Instruments and Scientific Theory" which turned out to be a sort of Invisible College event for the emerging new-version SIC and it led to the first of the *Newsletters*. Derek Howse was in charge of the SIC proceedings in Bucharest but he did not wish to continue as President beyond 1982. He pressed for the next SIC annual meeting to be held at his museum in Greenwich. At the conclusion of that meeting he would step down from the Presidency, 1982 being the year of his retirement from the museum.

From 1980, the annual *Newsletters* were produced. They are valuable records of the progress of the SIC and of its participants, though there are some inconsistencies in them, and facts need to be checked. Unfortunately, all except two of the *Newsletters* fail to list the papers presented at SIC sessions. The exceptions are the second, the Bucharest meeting, and twelfth, the Washington-Boston meeting. Even so, the former does not in fact list all the papers. The "western" papers are all given but only one of the five Romanian ones; the Czech paper by Otilie Skopova is also omitted.[14]

The Bucharest meeting, Symposium I, was not a particularly auspicious start for the SIC's annual series.[15] It bore the title, "The Role of Scientific Instruments in the Development of Science." Three half-days were allocated for the seventeen papers, including Gerard's offering and my own.[16] The programme was interspersed with papers by Romanian pseudo-historians, largely scientists drafted in to talk about the history of their specialist subjects (*Fig. 1.2*). Most

14 SIC, *Newsletter*, no. 2 (July 1981): 1.

15 This was the first meeting of the IUHPS to be held on a four-yearly rather than three-yearly cycle.

16 The SIC meeting was allocated session B.7, and outline details are found in the programme: *The 16th International Congress of the History of Science ... Scientific Programme Bucharest – Romania August 26–September 3* (Bucharest, 1981), 68–71, while paper summaries are given in *Proceedings of the 16th International Congress of the History of Science. B. Symposia* (Bucharest, 1981), 307–365.

FIGURE 1.2 Abstracts of the International Congress of the History of Science, Symposia, Bucharest 1981, includes the Scientific Instrument Commission's contribution
AUTHOR'S COLLECTION

of them could only offer their papers in Russian or broken French. Indeed, every section of the whole Bucharest meeting consisted of papers by scholars from abroad, with Romanian papers in-between. The intention was clearly to show that Romania had a leading history of science community.[17] It is important to make the point that at the IUHPS Congresses, scientific instrument

17 The printed List of Participants of the meeting indicates that 506 Romanians attended. This compares with forty-eight American and forty-seven British attendees.

papers might be given outside the SIC symposia programmes. In Bucharest, for example, Mary Robischon gave her paper, "Scientific Instrument Makers in London (17th–18th Centuries)" under the auspices of the International Council for Science Policy Studies, while Ferenc Szabadvary delivered "The History of Chemical Laboratory Equipment" at the Berkeley IUHPS meeting as a separate invited lecture.[18]

Gerard Turner and I decided beforehand that the history of instruments needed a more professional approach. True, there had been a few earlier, excellent publications, but a community of instrument scholars did not, at that stage, exist. We lived close-by to one another in the Oxford area, and we often discussed the future of instrument studies, with a mixture of optimism and pessimism; it didn't seem then that the subject was being taken very seriously by established science historians. Working as the senior curator in the Museum of the History of Science, Gerard had published his first history of science paper in 1966[19] and at an early stage in his career had become interested in the history of optical instruments. I had worked in the Royal Scottish Museum in Edinburgh as a history of science curator from 1970. In 1972, I published a brief exhibition catalogue of mariners' astrolabes and in 1978 produced a monograph describing a group of early chemical instruments and apparatus which had been used in teaching at the University of Edinburgh up to 1858.[20] We started to wonder how annual meetings concentrating on scientific instruments could be organised, thinking that it was not helpful to have to wait for four years until the next IUHPS congress came round. So, pre-Bucharest, we planned our strategy: I would stand for President of the Commission, and Gerard as Secretary, at the SIC business meeting in Bucharest. There turned out to be a rival nomination for President, Emmanuel Poulle of the Bibliothèque nationale in Paris (after 1988, Director of the École nationale des Chartes). Monsieur Poulle's speciality was medieval astronomical instruments, and he was a widely recognised scholar in his field. Exactly why it was me who was successful when the ballot took place I cannot say, except the feeling of the business meeting was strongly in favour of holding meetings annually, something which perhaps Gerard Turner and I were more likely to be able to achieve. Incidentally, *Newsletter* no. 3 omits to mention Emmanuel Poulle's candidacy.

18 This paper was later reissued as a separatum: *Periodica Polytechnica* 30 (Budapest, 1986).

19 Gerard L'E. Turner, "A Magic Mirror of Buddhist Significance," *Oriental Art* 12 (1966): 1–5.

20 R.G.W. Anderson, *The Mariner's Astrolabe* (Edinburgh: Royal Scottish Museum, 1972); R.G.W. Anderson, *The Playfair Collection and the Teaching of Chemistry at the University of Edinburgh 1713–1858* (Edinburgh: Royal Scottish Museum, 1978).

Holding an international congress in Bucharest, only several years after the 1977 massive earthquake, was challenging, to say the least. The economies of the Eastern European nations were not flourishing. In the coffee breaks there was scarcely enough coffee to meet half the demand, so delegates in-the-know slipped out of presentations before the speakers had finished. A conference buffet supper was held, but unless one had arrived thirty minutes before its stated start time, scarcely any food remained. There was clearly not a great deal of stock in shops, and the city did not seem to cater for international tourists in any significant way. I was anxious to take home some item of Romania's productions. This was difficult, and I succeeded only in finding a bookshop which had no scientific books, but which did have a pile of translations into Romanian of the Anglo-Saxon (Old English) epic poem, *Beowulf*.[21] (*Fig. 1.3*) I bought a number of copies, and later presented one of them to the British Library. Extramurally, there was a visit to Dracula's (supposed) Castle at Bran, which sadly did not include a dental museum, and to the medieval town of Brasov. On the way there we stopped by a monastery of the Romanian Orthodox Church, where a young supplicant demonstrated the use of the *crotalus*, wooden clappers which are used in place of church bells for the last three days of Lent. We also experienced the National Museum in Bucharest, the top floor of which was entirely dedicated as a shrine to Nicolae Ceaușescu (the display was dismantled after the 1989 Revolution but I believe that some components survive in storage).

One of the major points of conferences is to make contacts and discuss issues outside the formality of the papers. This cannot be done with Zoom. At the Bucharest conference I was staying in the Hotel Lido, as was the Marxist sinologist-historian, Joseph Needham. He was giving a paper on the origins of gunpowder in China.[22] We met over several breakfasts, and one day I boldly asked him what he thought of the communist regime of Romania. He wriggled out of addressing the question by telling me that the Ceauşescu regime was not a Communist one. I recall a problem in leaving Bucharest; right to the end there were difficulties. When the time came to return to London, I was the last to check-in at the airport, only to be told that my flight was overbooked by one person. I was offered a free additional (but undoubtedly solitary) night in a hotel. My complaint was such that Swissair agreed to take me (from the UK we had to travel via Zurich) if I squeezed-in alongside one of the stewardesses in her take-off seat. It didn't take me long to agree.

21 Dan Duțescu and Leon Levițchi, eds., *Beowulf* (Bucharest: Editura Pentru, 1969).

22 His co-author, who did not travel to the Conference, was Lu Gwei-Djen; the paper was delivered in the "Development of Science and Technology in the Far East" section.

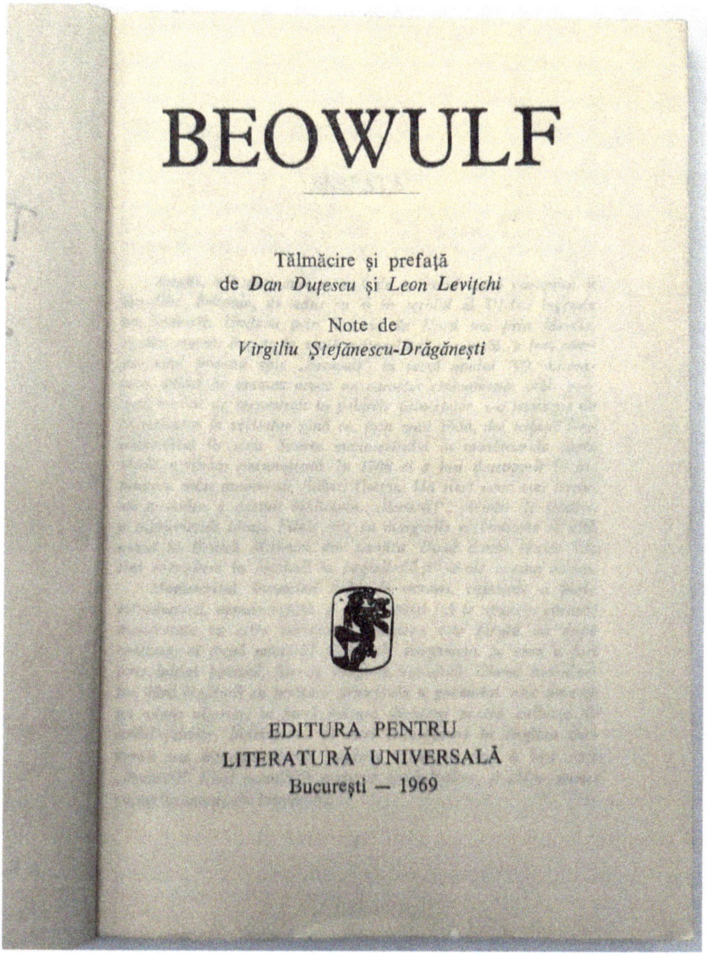

BEOWULF

Tălmăcire şi prefaţă
de *Dan Duţescu* şi *Leon Leviţchi*
Note de
Virgiliu Ştefănescu-Drăgăneşti

EDITURA PENTRU
LITERATURĂ UNIVERSALĂ
Bucureşti — 1969

FIGURE 1.3 A Romanian translation of the tenth- to eleventh-century Anglo-Saxon (Old English) epic poem, Beowulf, purchased during Symposium I in Bucharest
AUTHOR'S COLLECTION

In contrast to Bucharest 1981, the second annual SIC Symposium at Greenwich in 1982 was held in an institution not recovering from an earthquake and experienced in the art of holding conferences (*Fig. 1.4*). Derek Howse discussed, retrospectively, how his monumental world list of astronomical observatories, published in three volumes in 1975, had been compiled. Prospectively, Gerard Turner talked on the subject "Humphrey Cole and the Origins of Instrument Making in Britain" which would develop into his last book, *Elizabethan Instrument Makers*, not published until 2000. David Bryden gave a most stimulating lecture, which asked the central question: Is the

FIGURE 1.4 Delegates attending Scientific Instrument Commission Symposium II at the National Maritime Museum, Greenwich, 1982
AUTHOR'S COLLECTION

relative unimportance of studies of scientific instruments within the community of historians of science inherent in the nature of science, or is it merely a contemporary expression of the inability of such historians to tap the potential of material evidence? Although a vitally important aspect of instrument curation is conservation, it is a rarely discussed subject. But at Greenwich, two members of the National Maritime Museum's Conservation Department gave dire warnings in their paper, "The Dangers of Do-it-Yourself Conservation." It had been hoped that Derek de Solla Price of Yale University, who had published the seminal paper on the Antikythera Mechanism in 1974, would attend as he had played an important part in a meeting on scientific instrument fakes and facsimiles, including legal aspects, at Greenwich in the previous year;[23] but his health was poor and sadly he died not so long afterwards at a relatively young age, in September 1983. At the SIC business meeting, it was reported that

23 The Greenwich meeting on fakes was not part of the SIC sequence. However, the Whipple Museum in Cambridge (Price had been its curator in the 1950s) was much involved in studying the involvement of Anton Mensing of Amsterdam in the early twentieth century, who sold many fake instruments to museums. The SIC XXXIV Symposium at Turin in 2015 had a session, "Learning from Forgeries."

agreement had been reached that responsibility for globes would transfer from the Bibliographic Commission to the Scientific Instrument Commission.[24]

Symposium III was held in Munich at the Deutsches Museum in 1983. Its Director was, unusually, a recognised historian of science and technology, Otto Mayr, who had been poached from the Smithsonian Institution. One of the organisers was the head of physics, Alto Brachner, and he had mounted an exhibition for the meeting on the works of the Bavarian instrument maker Georg Friedrich Brander.[25] Twenty-five papers were presented and abstracts of them survive; there were eighty-two delegates present in Munich, showing that the SIC had established a strong following by only its third annual meeting, although all delegates save one came from Europe. There was a heavily Bavarian streak to several of the papers. H.R. Bachmann gave a paper on "Die mathematisch-physicalische Sammlung der Bayerischen Akademie der Wissenschaften," G. Hartl spoke on Brander, and Max Seeberger on Joseph von Fraunhofer. There were intriguing visits to Fraunhofer's Glassworks in Benediktbeuern, to the Kremsmünster Benedictine Monastery cabinet of instruments (in Austria, about 250 km east of Munich), and to the Oktoberfest to study liquid dynamics and the physiological effects of dilute ethanol. Subsequent to this meeting, the first of the SIC bibliographies of printed works on the history of scientific instruments was issued. Gerard and I created the very earliest one, but David Bryden took my place subsequently. To make the 1983 to 1995 bibliographies more useful, a special classification system was devised. The thirteen issues up till then were published together as a single volume in 1997.[26]

The fourth meeting in 1984 was held in Amsterdam at the Nederlands Scheepvaart Museum, the local organiser being senior curator Willem Mörzer Bruyns (*Fig. 1.5*). For the first time, there were more than one hundred

24 In 1958, the IUHPS's Commission on Bibliography reported that its "sub-commission for international catalogues of ancient globes and other scientific instruments" would be transferred to the newly created "Commission for Ancient Instruments" but that responsibility for cataloguing globes would remain with the Bibliography Commission. See "Documents officiels, Commission de Bibliographie," *Archives internationales d'histoire des sciences* 11 (1958): 391–393.

25 The exhibition and the book were done explicitly for the Symposium, and probably neither would have happened without all those foreign delegates: A. Brachner *et al.*, *G.F. Brander 1713–1783 Wissenschaftliche Instrument aus seiner Werkstatt* (Munich: Deutsches Museum, 1983).

26 These bibliographies were republished in a single volume by the SIC as: G.L'E. Turner and D.J. Bryden, *A Classified Bibliography on the History of Scientific Instruments* (Oxford: Scientific Instrument Commission, 1997).

FIGURE 1.5 Delegates at Symposium IV, held at the Scheepvaart Museum, Amsterdam, 1984
AUTHOR'S COLLECTION

delegates.[27] In *Figure 1.6*, I am seen giving my paper. During the business meeting, a special plea was made that a publication of instrument makers, listing them country-by-country, should be a priority. It was pointed out that listings already existed for the UK and for the Netherlands, but not as yet for Germany, France, or Italy.

Symposium V was held away from Europe for the first time, in conjunction with the 17th International Congress held at the University of California, Berkeley. The book of abstracts lists the sixteen SIC papers. Amongst these, there were contributions by historians from four continents. The fifth, Australia, would have been represented if Rod Home had not given his in another Section. It is perhaps worth observing that SIC papers are to this day overwhelmingly given by historians from Europe and America. The SIC symposium was titled "The Role of Scientific Instruments in the Development of Science" (conference titles have to be general enough to allow for inclusion of any worthy proposal!). Three half-days were allocated for the papers. A key paper for everyone at Berkeley was the plenary lecture given by Thomas Kuhn, author of the seminal book, *The Structure of Scientific Revolutions* (1962).

27 Selected material from the fourth symposium was published as: Peter R. de Clercq, ed., *Nineteenth-Century Scientific Instruments and Their Makers* (Leiden: Museum Boerhaave and Rudopi, 1985).

FIGURE 1.6 Robert Anderson delivering the Symposium IV presidential address at the Scheepvaart Museum, Amsterdam, 1984 AUTHOR'S COLLECTION

Symposium VI was held in Florence and Modena in 1986, organised locally by our late lamented colleague, Paolo Brenni, and by Mara Miniati (*Fig. 1.7*). This was held in the Casa di Risparmio in Florence and the Palazzo Ducale in Modena. A rather finely designed broadsheet giving the programme on the verso of the general symposium notice was produced, and a rival sheet of equal refinement was developed for the seventh meeting the following year in Paris (*Fig. 1.8*). The main visit in Florence, unsurprisingly, was to Palazzo Castellani (at the back of the Uffizi Gallery), housing the Istituto e Museo di Storia della Scienza, now generally known as the Museo Galileo. The broadsheet lists thirty-six papers, fifteen of which dealt with Italian instruments to a greater or lesser extent, including those in the Adler Planetarium in Chicago, a presentation given by its benefactor, Roderick Webster.

Symposium VII was held in Paris in 1986 at the then new Cité des Sciences et de l'Industrie which had opened only a year earlier in the Parc de la Villette. Some of the papers were published in 1989 at Anthony Turner's initiative in a

FIGURE 1.7 Notice for Symposium VI, held at the Istituto e Museo di Storia della Scienza in Florence in 1986
AUTHOR'S COLLECTION

FIGURE 1.8 Notice for Symposium VII, held at the Cité des Sciences et de l'Industrie (la Villette) and at CNAM (Musée des Techniques), Paris, in 1987
AUTHOR'S COLLECTION

volume called *Studies in the History of Scientific Instruments*.[28] Though selective and doubtless edited, it remains – along with the publication of part of the Amsterdam symposium – one of the few records of the actual presentations made at the early SIC meetings. Poster sessions, now commonplace, were held for the first time in Paris. Two exhibitions were prepared for the meeting at the Observatoire de Paris, entitled *Newton* and *La Mesure du Ciel*. Further visits were arranged to the laboratory of Eduard Branly and to the Departement du Cartes et Plans at the Bibliothèque nationale. A sub-meeting was held to discuss computer cataloguing of instruments as well as continuing efforts to produce national inventories by traditional techniques. In the UK, the Museums Association had somewhat earlier developed its IRGMA system (Information Retrieval Group of the Museums Association) for museum artefacts, and it was Willem Hackmann of Oxford who applied this to scientific instruments and who participated in this sub-meeting.[29] This approach to record-keeping, now commonplace, was then only just starting to challenge curators.

The eighth SIC Symposium was in London, mainly based at the Society of Antiquaries (founded in 1707) with its own historical collection, although a whole day was spent at Greenwich at the Royal Naval College and the Old Royal Observatory. Gerard and Helen Turner took on the whole burden of the organisation. There were seventeen papers and two posters, one of which reported on the production of the catalogue of the King George III Collection in the Science Museum.[30] Prior to the meeting we were told: "As was intimated at our Paris meeting, the Eighth Symposium will be different in character from the seven which preceded it. The time has now come to assess the progress which has been made over the past decade, to consider the position which instrument studies occupy in the broader spectrum of history of science, and to contemplate new directions for the future. ... A number of eminent 'outsiders' will be invited to comment on how they view the scientific instrument scene." This may have been the ambition, but it scarcely seems to have been achieved if judged by the symposium programme. However, the first paper of the second day, 13 September, was given by the eminent art historian Martin Kemp on the topic "Basic Research and Public Performance."[31]

---

28 Christine Blondel, François Parot, Anthony Turner, and Mari Williams, *Studies in the History of Scientific Instruments. Papers presented at the 7th Symposium of the Scientific Instruments Commission of the Union Internationale d'Histoire et de Philosophie des Sciences (Paris 15–19 September 1987)* (London: Rogers Turner Books, 1989).

29 Willem D. Hackmann, "Ten Years of IRGMA," *Museums Journal* 77 (1977): 11–14.

30 This was not published until 1993: A.Q. Morton and J.A. Wess, *Public and Private Science: The King George III Collection* (Oxford: Oxford University Press, 1993).

31 Martin Kemp, though by profession an art historian of the Renaissance, has frequently been intrigued by the connections between art and science. Such works as *Science of Art*.

Symposium IX, organised by Ludolf von Mackensen, was to have been held at Kassel in West Germany but few had committed themselves by the deadline and so it had to be cancelled. For that reason, SIC Symposium IX was added on to the coat-tails of the 1989 IUHPS Congress series with the general title "Science and Political Order" (the Congress with this rubric planned for Berlin in 1940 had, not surprisingly, been cancelled) (*Fig. 1.9*). The introductory statement made the point that research in the history of science was strongly Eurocentric but that this Congress would include sessions on science in Islam, Asia, the Pacific, and Central and South America.[32] Scientific instrument studies, with one or two exceptions, were far from approaching this ambition, and the question arises as to whether ghettoising papers culturally is better than attempting to integrate them. The full meeting was bifurcated, being held first in Hamburg followed by Munich, with a delightful day's train ride in the middle. There were just eight SIC papers in Hamburg, although these were attended by a decent-sized audience of up to forty delegates. The Kassel-lesson was learnt; one can't expect conference-goers to pay registration before the meeting is declared definitely to be taking place.

My final report, of Symposium X, is also of a dual-venue meeting, that held in Washington, DC at the Smithsonian and in Cambridge, Massachusetts, at Harvard University. A good description of the meeting is to be found in *Newsletter* no. 12.[33] The meeting began with a reception held by Silvio Bedini in his home and the *Newsletter* waxes eloquently on the extent of Silvio's publishing and on the richness and orderliness of his study. Silvio had briefly been director of the National Museum of History and Technology, as it then was known, which had included substantial instrument displays. In 1980 it was wrenched away from its focus on the history of science, technology, and medicine by a director, Roger Kennedy, who turned it into the National Museum of American History, a social history museum. Not everybody was then, or has since been, happy with this transformation. I have often made the point that, very surprisingly, there is no comprehensive science museum in America on the lines of the Science Museum or Deutsches Museum. At Boston we visited the Computer Museum, a fine, well-curated collection which quite soon afterwards was dispersed. This can be a danger in America: private collections,

---

Optical Themes (New Haven: Yale University Press, 1992); *Spectacular Bodies* (London: Hayward Gallery Publishing, 1997); *Visualisations: the Nature Book of Art and Science* (Oxford: Oxford University Press, 2000); *Seen | Unseen: Art, Science and Intuition from Leonardo to the Hubble Telescope* (Oxford University Press, 2006) explore these themes.

32 Fritz Krafft and Christoph J. Scriba, eds., *Science and Political Order/Wissenschaft und Staat … Abstracts* (Hamburg and Munich: ICHS, 1989).

33 SIC, *Newsletter*, no. 12 (January 1991).

FIGURE 1.9 Abstracts of the XVIII International Congress of History of Science held in Hamburg and Munich in 1989, which incorporated Symposium IX
AUTHOR'S COLLECTION

masquerading as public museums, can evaporate. This also happened to the Time Museum at Rockford, near Chicago, in 1999.

To strike a brighter note, the Harvard Collection of Historical Scientific Instruments in the Science Center was full of interest, the President and Secretary being seen in *Figure 1.10* at the entrance to a prep room across from its stores (the Collection staff were not responsible for the title above the door but enjoyed the joke).[34] Together with the Adler Planetarium in

34 The preparation room for the Science Center's lecture demonstrators was across the hall from the rooms storing the historical apparatus, and the demonstrators were having some fun in distinguishing themselves.

FIGURE 1.10 Gerard Turner (*left*) and Robert Anderson come across the Collection of Hysterical Scientific Instruments in the basement of the Science Center at Harvard University, Symposium X, 1990
AUTHOR'S COLLECTION

Chicago, it is as close as one gets in America to a traditional scientific instrument museum, although it is relatively limited in scope. At the meeting, the Adler Planetarium's curator, Sara Schechner, gave a paper "Scientific Theory Made Manifest in Instrument Design." Owen Gingerich, the sometime senior astronomer of the Harvard Smithsonian Center for Astrophysics and professor of astronomy and the history of science at Harvard, gave us a fine closing lecture at the Boston Museum of Science's planetarium on "The Dance of the Planets." He was a leading Copernicus scholar and was frequently seen at European meetings, including SIC Symposia. He died very recently.[35]

Having run through my ten Symposia at breakneck speed, I need to try to draw some conclusions. I have not discussed whether the content of the SIC lectures show a meaningful change of topic or approach over the ten years. My feeling is that 1981 and 1990 are not fundamentally different. It could be argued that the SIC culture was dragging its feet behind the changing history

35 Owen Gingerich obituary, *New York Times*, 11 June 2023, https://www.nytimes.com/2023/06/11/science/space/owen-gingerich-dead.html.

of science scene in universities. Did the publication of Steven Shapin and Simon Schaffer's historiographically sophisticated *Leviathan and the Air Pump* of 1986 have any impact on approaches taken by speakers at SIC meetings? I judge little or not at all. In that same year, 1986, our delegates were publishing works such as *The Great Age of the Microscope* (Gerard Turner) and *Brass and Glass: Scientific Instrument Making in Scotland* (Tristram Clarke, Alison Morrison-Low, and Allen Simpson). Both are splendid in their traditional antiquarian way.

Gerard Turner's recently unearthed 1988 paper, "Recent Advances in the History of Scientific Instruments," was given at Symposium VIII in London.[36] It provides an excellent survey of instrument studies at the time. On the other hand, it is an entirely inward-looking, antiquarian piece, as though such studies are ends in themselves and relate little to broader questions in the history of science.

In that same year, 1988, a piece in the *Daily Telegraph* on 26 October, a highly conservative British daily newspaper, quoted Jim Bennett, then still at Cambridge, talking about a collection of pocket calculators that the Whipple Museum had just acquired: "They are worthless in a sense because no one wants them," it is claimed that he had said, to which the blustering, narrow-minded journalist responds sarcastically, "I have good news for the Whipple, it can branch out from mere science and become a major tourist attraction for its peerless collection of obsolete gadgets of all kinds." And who was that journalist writing thirty-four years ago? Why, none other than the twenty-four-year-old Boris Johnson (today, a former UK Prime Minister)! Does he have a point, though? A relevant question is: were the history of scientific instrument studies and science history topics pursued in universities in the 1980s adequately connected? Are things different now? To what extent do fashionable feminism, decolonialism, and internationalism generally concern our scientific instrument agenda today? I would challenge anyone to read Gerard's paper, survey the topics presented at the 1988 meeting, and only then make a judgement. Certainly, subject trends are difficult to discern in the first decade of annual meetings. Very frequently historians and curators of scientific instruments agonised about the reluctance to study and publish on modern or contemporary themes, yet these concerns are not reflected in the topics of papers which were offered at SIC symposia.

Both in terms of the number of papers offered and in the willingness of local organisers to arrange meetings in their institutions and find the resources to hold them, the SIC has been able to thrive. Small grants to our Commission

36 Gerard L.E. Turner, "Recent Advances in the History of Scientific Instruments," *Bulletin of the Scientific Instrument Society*, no. 19 (1988): 3–6.

from the IUHPS in the 1980s paid for the *Newsletters* and for their distribution. The SIC was not a very democratic body, even though there were yearly business meetings, but the near-autocracy seems to have been accepted with equanimity. There was no formal review process for proposed symposium papers. Some by younger curators were naïve, but our feeling was that having to present a paper at an international meeting was an excellent training exercise. Nearly all papers offered were accepted, which sometimes led to programme overload. Gerard Turner and I ran the show for fifteen years without being challenged. We were buoyed up by being told that we were the most active of all the Commissions in the IUHPS; no other at the time held annual meetings.

Those who have known Gerard will know that he had a fierce, at times bullying, manner, and he made quick decisions as to whether he approved of you or not. If the former, you would enjoy a fruitful relationship and be asked to read papers at meetings. If not, you were out in the cold. He enjoyed seeming to appear eccentric, such as never taking the ferry to the Continent without his jar of Cooper's Oxford marmalade, without which his breakfasts would not be complete (*Fig. 1.11*). His inner coterie was numerically quite small. But it cannot be denied that his *modus operandi* was in its way effective, and he worked extremely hard to achieve his ends. After all, we are still here today … and thriving.

FIGURE 1.11
Cooper's Oxford Marmalade, without which Gerard Turner would never travel abroad
PHOTO: BY AUTHOR

The 1970s–1980s was a time when museum and university employment in the history of science was quite buoyant, at least in the UK. When I joined the Science Museum in 1975, sixty staff could be defined as curators. They were distributed around seven subject departments such as physics, chemistry, mathematics, medicine, and so on. From the 1990s, this structure was destroyed by successive directors and at one stage afterwards only an estimated fifteen curatorial staff remained. Hence the SIC's annual symposia had been started at a good time. Some science museums were even able to re-present themselves. In *Figure 1.12*, Gerard can be seen being anatomised in a reconstructed anatomy theatre (originally of 1593); this scene is taking place at the 1991 reopening of the Dutch national science museum, the Rijksmuseum Boerhaave, in Leiden. At this time, other germane academic and social bodies were developing devoted to the study of artefactual history. In Britain, the Group for Science, Technology and Medical Collections started in the mid-1970s (I acted as its chairman for a year). The Scientific Instrument Society, in which Gerard Turner played a major role (though I never joined), was founded in 1983,

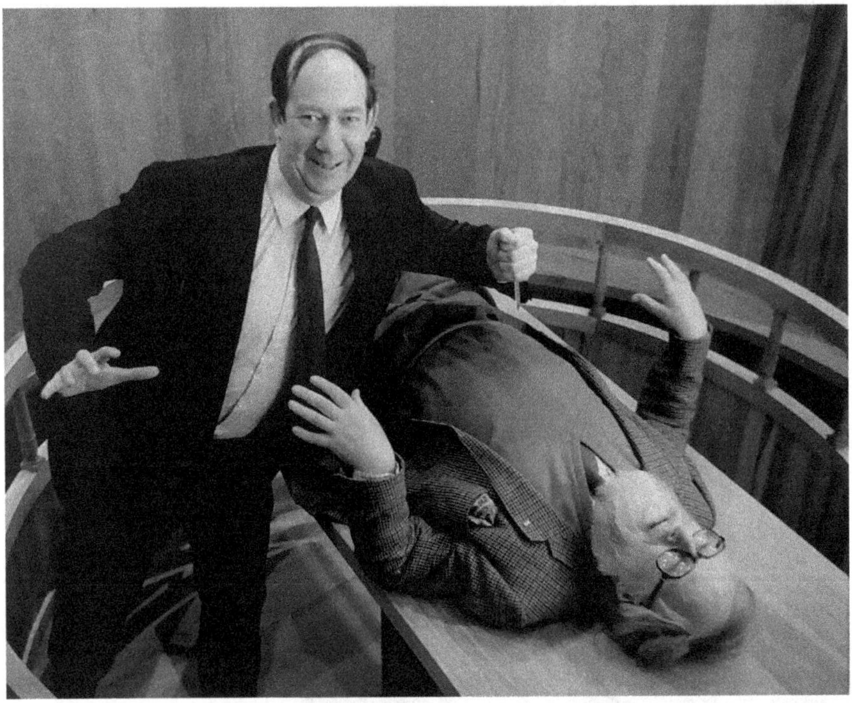

FIGURE 1.12 Robert Anderson (*left*) anatomising Gerard Turner in the reconstructed anatomy theatre (originally 1594) built into the Museum Boerhaave in Leiden, in 1991
AUTHOR'S COLLECTION

and it brought together a wider circle of instrumentalists than those simply working in museums. It developed as an international group in spite of the fact that it was UK-based and it included dealers, auctioneers, scientists who were interested in their subject's history, collectors, and amateur enthusiasts. It still thrives and produces the excellent *Bulletin of the Scientific Instrument Society*. In 1979 Gerard instituted his private dining club for those interested in scientific instrument collecting, study, and eating well. He called it the Equinoctial Club (*Fig. 1.13*); it had no formal connection to the SIC, although most of its members were regular attendees at its symposia. It met in the Oxford and Cambridge Club at Pall Mall in London. Carefully selected males were invited by Gerard to dine – but they had to pay for themselves. Most of them were British but Gerard had a particular fondness for the Dutch, and his friends from the Netherlands often travelled over to the dinners. Very occasionally women came as guests. After Gerard's declining health made him incapable of continuing with the Equinoctial Club's organisation, Jim Bennett, Will Ryan, and I took it over and changed its form fundamentally. It is still in existence, still without formal membership or subscriptions. Its meetings were interrupted by COVID-19 in 2000, but it has revived, and the most recent dinner, held on 27 September 2024, was its eighty-third.[37] It now welcomes men and women

FIGURE 1.13 An Equinoctial Club dinner held in the Oxford and Cambridge Club, London, at the Vernal Equinox, March 1984
AUTHOR'S COLLECTION

37 Records of attendances are found in two record books, the first running from the establishment of the Equinoctial Club in March 1979 up to 2008, the second from its reorganisation in 2008, continuing to the present day (2024). Currently, both of them are in my care.

FIGURE 1.14 Three early Presidents of the Scientific Instrument Commission together at MIT (*from left*): Robert Anderson (1982–1997), Paolo Brenni (2003-2013), and Jim Bennett (1997–2003)
AUTHOR'S COLLECTION

equally, holding a semi-annual social meeting at the equinoxes for instrumentalists, cognoscenti, and young curators, alike. It meets in pubs rather than in posh Pall Mall clubs, and dinners are nearly always preceded by the examination of an exhibition or collection that has scientific instrument interest.

I did not step down from the SIC until 1997, after fifteen years. *Figure 1.14* shows the first three SIC Presidents at an MIT meeting in Cambridge, Massachusetts: Jim Bennett who took my place in 1997, and then after Jim, in 2003, Paolo Brenni. But here I must end, and now it's for others to pick up the story, post-1990.

Bibliography

Anderson, R.G.W., and Gerard L'E. Turner. *An Apparatus of Instruments: The Role of the Scientific Instrument Commission*. London: Scientific Instrument Commission, 1993.

Blondel, C., F. Parot, A.J. Turner, and M. Williams, eds. *Studies in the History of Scientific Instruments. Papers Presented at the 7th Symposium of the Scientific Instruments Commission*. London: Rogers Turner Books, 1989.

Clarke, T.N., A.D. Morrison-Low, and A.D.C. Simpson. *Brass & Glass. Scientific Instrument Making Workshops in Scotland.* Edinburgh: National Museums of Scotland, 1989.

Hackmann, W.D. "Ten Years of IRGMA, 1967–1977." *Museums Journal* 77 (1977): 11–14.

Hackmann, W.D. *The Evaluation of a Museum Communication Format. Part 1. Collection of Input Data. OSTI Report no. 5154.* London: Office of Scientific and Technical Information, 1973.

Holbrook, Mary, with Additions and Revisions by R.G.W. Anderson and D.J. Bryden. *Science Preserved: A Directory of Scientific Instruments in Collections in the United Kingdom and Eire.* London: HMSO, 1992.

Scientific Instrument Commission. *Newsletters*, nos. 1–12 (May 1980–January 1991).

Shapin, Steven and Simon Schaffer. *Leviathan and the Air Pump.* Princeton: Princeton University Press, 1989.

The 16th International Congress of the History of Science. Proceedings B. Symposia. Bucharest: Publishing House of the Academy of the Socialist Republic of Romania, 1981.

Turner, Gerard L'E. "Recent Advances in the History of Scientific Instruments." *Bulletin of the Scientific Instrument Society*, no. 19 (1988): 3–6.

Turner, Gerard L'E. *The Great Age of the Microscope. The Collection of the Royal Microscopical Society Through 150 Years.* Bristol: Adam Hilger, 1989.

Turner, G. L'E., and D.J. Bryden. *A Classified Bibliography on the History of Scientific Instruments. Based on the SIC Annual Bibliographies of Books, Pamphlets, Catalogues and Articles ... 1983 to 1995.* Oxford: Scientific Instrument Commission, 1997.

CHAPTER 2

The Scientific Instrument Commission in the 1990s

Sara J. Schechner

I have been a member of the Scientific Instrument Commission (SIC) and a participant in its annual symposia from the time that I was hired as the Curator of the Adler Planetarium in 1983. I shared a desk in the curatorial office (between the outer and inner shells of the Planetarium's dome!) with voluntary curators Roderick S. and Marjorie K. Webster, who were Trustees of the Adler, generous benefactors, and collectors. Rod and Madge were already engaged with the programmes and people remembered by Robert Anderson in his account of the 1980s and earlier (see Chapter 1). They travelled to Europe every year to visit collections, meet curators, attend conferences, and buy scientific instruments and related rare books. They also brought back amazing stories, and extended hospitality in Chicago to the personalities they met abroad. As Madge would say, "Now, Sara dear. You really must meet so-and-so." And so, Rod and Madge introduced me to the SIC and took me on their jaunts to SIC symposia.

This preface is by way of saying that having been present in the formative decade, I can speak with some authority on developments in the SIC's second decade of symposia. Several major themes stand out in the expansion of the SIC in the 1990s: (1) There was a shift in paper topics at symposia. (2) Members showed interest in the conservation and authenticity of scientific instruments. (3) The big projects continued to be national inventories. (4) Computers transformed our lives. And (5) there was a change in the way the SIC was administered. This essay will examine each of these in turn.

Table 2.1 contains a list of the SIC International Scientific Instrument Symposia in this decade, and gives the themes of each symposium, estimated attendance, and numbers of papers. The table also includes three significant workshops sponsored by the SIC but held independently of the Scientific Instrument Symposia. This period is bookended by my two most memorable meetings – Scientific Instrument Symposium X (in Washington, DC and Boston) and Symposium XVIII (in Moscow and St. Petersburg).

 | DOI:10.1163/9789004737587_004

TABLE 2.1 SIC Symposia and major meetings in the 1990s

Year	Primary location	Symposium number	Host	Programme themes	Attendees	Papers + posters
1990	Washington, DC / Boston	SIS X	Smithsonian Institution: National Museum of American History / Harvard Collection of Historical Scientific Instruments	The American Instrument Enterprise The Role of instruments in the Social Construction of Knowledge	46	22
1991	Bologna	SIS XI	University of Bologna	[Cataloguing, research uses, international trade of instruments]	68	65
1992	Edinburgh	SIS XII	National Museums of Scotland	Historiography of Scientific Instruments Curation of Collections, including uses, conservation, and documentation.	100+	39
1993	[Zaragoza]		[XIX IUHPS Congress]	[No SIC sessions]		
1994	Leiden		Museum Boerhaave	Origins and Evolution of Collecting Scientific Instruments	115	15
1994	Leiden	SIS XIII	Museum Boerhaave	What more can research on scientific instruments offer us than a mere increase of our knowledge of these instruments?	115	39
1995	Prague	SIS XIV	National Technical Museum	[No theme]	71*	31
1996	Ottawa	SIS XV	National Museum of Science and Technology	Transformation of Canada Canadian Context	53	31

TABLE 2.1 SIC Symposia and major meetings in the 1990s (*cont.*)

Year	Primary location	Symposium number	Host	Programme themes	Attendees	Papers + posters
1997	Liège	SIS XVI	XX IUHPS Congress	National Inventories of Historical Scientific Instruments	61*	15
1998	Sorø	SIS XVII	Sorø Academy	Instruments and Makers in Scandinavia Cabinets of Physics and Chemistry	100+	64
1998	Florence		Istituto e Museo di Storia della Scienza and the Opificio delle Pietre Dure	Conservation and Restoration of Scientific Instruments: Issues of Principle and Practice		10
1999	Moscow / St. Petersburg	SIS XVIII	Institute of the History of Science and Technology, Russian Academy of Science	Historical Relations between Russia and Other Countries in the Field of Sentific Instruments	53**	31
1999	Leiden		Museum Boerhaave	Scientific Instruments: Originals and Imitations. The Mensing Connection		9

NOTES

* Estimate is based on the headcount recorded for the AGM, which is always lower than overall.

** Published list of attendees does not correlate to speakers or those present, especially in the case of Russians. Estimate based on latter.

1 Symposia

The first symposium of the decade was the first to be held in North or South America. This was Scientific Instrument Symposium X, which took place 20–25 September 1990 and was divided between Washington, DC and the Boston area. The hosts were the Smithsonian Institution and Harvard University.

The meeting began with Silvio Bedini (retired Deputy Director of the National Museum of History and Technology) and his wife welcoming guests into their home. The SIC *Newsletter* described the occasion as:

> a most delightful 'Get Together' .... Silvio has by now published more books and more papers on the history of scientific instruments than anyone else, and to see the extent of his study and its orderliness was an experience that can never fade from memory.[1]

In Washington, talks were held at the National Museum of American History, the National Air and Space Museum, and the S. Dillon Ripley Center, with various tours of collections, exhibitions, and libraries. Excursions were also made to the original "foggy bottom" site and current "observatory hill" site of the United States Naval Observatory, and to the National Museum of Health and Medicine, which housed the Billings Microscope Collection. Special note deserves to be made of the fabulous, costumed demonstration of eighteenth-century philosophical apparatus by John McKnight and Hans C. Von Baeyer of the College of William and Mary. This performance took place at the National Museum of American History.[2]

In Boston, the meeting kicked off with a tour and dinner at the Computer Museum. The next day, talks took place at the Harvard College Observatory, where visitors could check out the Great Refractor, Plate Stacks, and library. Formal excursions were made to the Harvard Collection of Historical Scientific Instruments and Houghton Library (for an exhibition on sundials and rare dialling books), leaving participants time to visit other Harvard museums at their will. The symposium concluded with delegates taking a train to Salem, Massachusetts to see navigational instruments at the Peabody Museum (today known as the Peabody Essex Museum) as special guests of David P. Wheatland, the founding curator of Harvard's Collection of Historical Scientific Instruments. We also visited the Salem Maritime Historic Site.

1 SIC, *Newsletter*, no. 12 (January 1991): 2. Silvio Bedini (1917–2007) began his Smithsonian career in 1961 as Curator in the Department of Mechanical and Civil Engineering at the National Museum of History and Technology (today known as the National Museum of American History). He served as Deputy Director from 1972 until 1978, at which time he became Keeper of Rare Books at the Dibner Library until his retirement in 1987. He was then appointed Historian Emeritus. Anthony J. Turner, "Silvio A. Bedini," *Nuncius* 23, no. 2 (2008): 353–371; Robert C. Post, "Silvio A. Bedini, 1917–2007," *Technology and Culture* 49, no. 2 (2008): 522–529.

2 National Museum of American History, *Program: 10th Scientific Instrument Symposium. September 20–September 25, 1990. Washington, D.C.* [*and*] *Boston Area* (Washington, DC: Smithsonian Institution, 1990).

Back in Boston that evening, we were treated to a lecture by Owen Gingerich (Harvard-Smithsonian Center for Astrophysics) on "The Dance of the Planets," in the Charles Hayden Planetarium of the Museum of Science.[3]

The symposium had two themes: "The American Instrument Enterprise" and "The Role of Instruments in the Social Construction of Knowledge."[4] The first theme was defined as including "the manufacture, marketing, and use of instruments in both American Continents." Addressing the first theme, speakers included David Bryden, Deborah Jean Warner, and Randall Brooks, who were not strangers to the SIC, as well as those coming newly and briefly into our orbit, such as Robert Smith (who spoke about the Hubble Space Telescope), Stephen Dick (astronomical instruments in the service of navigation at the US Naval Observatory), Rand Evans (scientific patronage and the American psychological instrument trade), and Carlene Stephens (instruments of empire, the Harvard College Observatory, and US topographical engineers).[5]

The second theme was spun off into a "Workshop on Scientific Instruments and the Social Construction of Knowledge" held on September 20, the day prior to the first official papers of Scientific Instrument Symposium X. It was organised by Nathan Reingold, senior historian of American science, and had three papers.[6] This left only one paper on the social construction of knowledge in the official Symposium X, an outlier in the Boston portion of the meeting by a young Sara Schechner Genuth (Adler Planetarium, Chicago): "Scientific Theory Made Manifest in Instrument Design." It was about air pumps turned into electrical machines in the early eighteenth-century work of Francis Hauksbee and the Royal Society, and the theory-laden nature of scientific instruments. "This provoked some excited discussion, in which some members of the audience felt that theory was instrument-laden rather than vice-versa," the *Bulletin of the Scientific Instrument Society* reported in what was an understatement.[7] In particular, Gerard Turner (Imperial College), the imposing SIC Secretary, would

3 *Program: 10th Scientific Instrument Symposium.*

4 SIC, *Newsletter*, no. 11 (January 1990): 1–2.

5 The speakers represented these institutions: David Bryden, National Museums Scotland, Edinburgh; Deborah Jean Warner and Carlene Stephens, National Museum of American History, Smithsonian Institution, Washington, DC; Randall Brooks, St Mary's University, Halifax, Nova Scotia; Robert Smith, National Air and Space Museum, Smithsonian Institution, Washington, DC, Steven Dick, US Naval Observatory, Washington, DC; Rand Evans, University of Baltimore, Baltimore, MD.

6 SIC, *Newsletter*, no. 12 (January 1991): 3.

7 A.D. Morrison-Low, "10th Scientific Instrument Symposium, 20th–25th September 1990, Washington, D.C. and the Boston Area," *Bulletin of the Scientific Instrument Society*, no. 27 (December 1990): 26–27.

have none of this. He railed against my thesis, implying that it was wrongheaded to analyse the development of instruments from the perspective of philosophy or the social creation of science. Unfortunately, I was awkwardly tasked with chairing the session that contained my own talk. When I noted that the allotted time for questions had expired, Gerard exclaimed that he was in charge and if he wanted to ask more questions, he would do it as long as he liked! No one in the room dared to contradict him. I stood trembling and answered as best I could, but firm in my conviction that as an academically-trained historian and philosopher of science and a museum curator, I was bringing a new way of looking at scientific instruments to the table – one vetted in fact by Jim Bennett and Simon Schaffer, under whom I had written a Cambridge thesis on this very topic.[8] But I knew my approach was not going to be everyone's cup of tea. Needless to say, the session went way over time until the audience demanded its lunch.

Scientific Instrument Symposium XI was hosted by the Department of Physics of the University of Bologna, Italy on 9–14 September 1991. The meeting had no particular theme, but papers were invited on "historical scientific instruments and their significance, cataloguing, research uses, theoretical and methodological aspects, [and] the international trade in instruments."[9] Over sixty-five papers and posters were delivered, including the following related to international trade: Jim Bennett, "The English Quadrant in Europe;" Gloria Clifton, "Growth of the British Scientific Instrument Trade, 1600–1850;" and Anita McConnell, "Instruments for South America." Papers under the theoretical and methodological category included: A. Drago, "Towards a General Theory of Measurement Instruments;" Bruno Jacomy, "Scientific Instruments as Technical Objects for Science;" D.L. Sepper, "Showing, Doing, and the Ontology of Using Scientific Instruments;" and Gerard Turner, "Reconstruction of a Sixteenth-Century Florentine Workshop" by means of "fingerprinting" the letter and number punches of Johannes Battista Giusti.[10] Proceedings of the Bologna meeting were later published in 1994.[11]

8 Sara J. Schechner, "The Influence of Theory on Instrument Design: The Case of Hauksbee's Friction Electric Machine" (MPhil thesis, University of Cambridge, 1980).

9 SIC, *Newsletter*, no. 12 (January 1991): 1.

10 SIC, *Newsletter*, no. 13 (January 1992): 2; A.D. Morrison-Low *et al.*, "XI International Scientific Instrument Symposium, Bologna, Italy, 9–14 September 1991," *Bulletin of the Scientific Instrument Society*, no. 31 (December 1991): 11–20.

11 Giorgio Dragoni, Anita McConnell, and Gerard L'E. Turner, eds., *Proceedings of the 11th International Scientific Instrument Symposium: Bologna University, Italy, 9–14 September 1991* (Bologna: Grafis Edizioni, 1994).

Excursions were made to Bologna's physics and zoology museums, the anatomical theatre, the botanic garden, the original observatory (completed in 1726) in the Palazzo Poggi, and Cassini's meridian line (1656) in the Basilica di San Petronio. The group also spent a day in Ferrara and visited the fourteenth-century dam at Casalecchio and the Museo Marconi at the Villa Griffone, the country house of the Marconi family from which Guglielmo Marconi transmitted his first radio signal in 1895.[12]

A notable feature of the meeting was the presence of delegates from countries formerly behind the Iron Curtain, which was "falling" in 1990–1991.[13] As President of the SIC, Robert Anderson (British Museum) observed:

> It was delightful to have an Estonian, a Czech, a Romanian, and a (former) East German present. They were in Bologna because they had made strenuous efforts to come, not because they had toed the line in the past and were being rewarded. … One said to me that it would be pleasant not to have to report to his security service on his return. … We must ensure that the contacts we have now made will be nurtured, and this will probably mean that we shall have to look for grants to help our Central European friends to travel to future meetings.[14]

This friendship and collaboration with scientific-instrument scholars in the USSR and its former satellites would be solidified further in the 1995 and 1999 SIC symposia, which were held in the Czech Republic and Russia, and in later meetings in the former East Germany (2004), Poland (2006), Hungary (2009), and Estonia (2014).

Edinburgh was the destination for the SIC's Scientific Instrument Symposium XII, which was held 7–11 September 1992. Hosted by the National Museums Scotland, it had an overflow crowd of more than 100 participants.[15]

---

12 SIC, *Newsletter*, no. 12, p. 1; and Morrison-Low, *et al.*, "XI International Scientific Instrument Symposium."

13 After the Second World War, the Soviet Union had gradually erected the Iron Curtain – a figurative and physical barrier – to split the communist East under its control from the allied, democratic European countries in the West. The Berlin Wall separating East and West Germany was the most notable manifestation. It fell on 9 November 1989 as part of reform movements sweeping through Poland, Hungary, Czechoslovakia, Estonia, Latvia, Lithuania, Romania, and East Germany. By October 1990, Germany was reunified, and the constituent Soviet republics were becoming independent. On 25 December 1991, the Union of Soviet Socialist Republics collapsed.

14 Robert Anderson, "The XI Scientific Instrument Symposium: Summing Up," p. 20 in Morrison-Low *et al.*, "XI International Scientific Instrument Symposium."

15 SIC, *Newsletter*, no. 12 (January 1992): 1; SIC, *Newsletter*, no. 14 (January 1993).

The theme was the "historiography of scientific instruments with a sub-theme on the curation of collections, including discussions on the ethics of conservation, the uses of collections, and the documentation of collections."[16] A day devoted to the "Industrial Applications of Scientific Instruments" included papers related to naval architecture, the electrical industry, steel manufacturing, aeronautics, and the military. It culminated with a talk on the "History and Characteristics of Malt Whisky," by chemist Peter Dryburgh, which was followed by a whisky tasting. Another day's session focused on the "Historiography of Scientific Instruments," which had reports on the preparation of an encyclopaedia on the history of scientific instruments,[17] Project SIMON's latest revelations concerning the London Guilds and optical instrument making, the establishment of the new Observatoriemuseet in Stockholm, and works related to documenting instruments in Ireland and Britain. A session on "Collections and Curation of Scientific Instruments" will be discussed below.[18]

In Edinburgh, symposium participants explored the collections of the Royal Museum of Scotland, the Royal College of Surgeons of Edinburgh, the Royal Observatory, and the Crawford Collection of early astronomical books. Beyond Edinburgh, delegates could visit collections of the Universities of St. Andrews, Glasgow, Strathclyde, and Aberdeen (this last brought south to Edinburgh for a special display). An excursion to a Scoonie Hill farm cottage near St. Andrews brought to light Gregory's Pillar, a meridian marker for James Gregory's observatory, originally installed circa 1673.[19] In Glasgow, highlights were the Hunterian Museum to see Lord Kelvin's apparatus and other philosophical instruments, and the Burrell Collection housing the extensive art collection of industrialist

16 David Bryden and A.D. Morrison Low, "First Circular, XII Scientific Instrument Symposium," Edinburgh, 1991; quoted text is a paraphrase published on the SIC website, "XII Scientific Instrument Symposium," SIC, accessed 6 December 2024, https://scientific-instrument-commission.org/event/xii-scientific-instrument-symposium/.

17 Later published as Robert Bud and Deborah Jean Warner, eds. *Instruments of Science: An Historical Encyclopedia* (New York: Science Museum, London, and National Museum of American History, Smithsonian Institution, in association with Garland Pub., 1998).

18 National Museums Scotland, *XII Scientific Instrument Symposium, Edinburgh, September 1992: Programme for Main Meeting* (Edinburgh: National Museums Scotland, 1992); and *Programme and Abstracts: Scientific Instrument Commission, IUHPS: XII Scientific Instrument Symposium, National Museums of Scotland, Edinburgh, 7–11 September 1992* (Edinburgh: National Museums Scotland, 1992); both can be found at https://scientific-instrument-commission.org/event/xii-scientific-instrument-symposium/.

19 The original marker was wood and rotted away. It was replaced by a stone pillar in 1757 or 1775. See John Ceres Amson, "Gregory's Meridian Line of 1673–74: a St Andrews Detective Story," *BSHM Bulletin* [today: *British Journal for the History of Mathematics*] 23, no. 2 (2008): 58–72.

Sir William Burrell. At the University of Strathclyde, delegates examined the books and instruments of eighteenth-century natural philosopher John Anderson, and the alchemical library of James "Paraffin" Young.[20]

In 1993 the International Union for the History and Philosophy of Science, Division of History of Science held its XIX International Congress of the History of Science in Zaragoza, Spain from August 22 to 29. It contained a free-standing symposium on "The Oldest Latin Astrolabe: Historical Context" – referring to the Carolingian astrolabe at the l'Institute de monde arabe in Paris. Although papers were read by some SIC regulars, this session was not sponsored by the SIC.[21] Indeed, the leadership of the Scientific Instrument Commission felt it could not adequately arrange for a Scientific Instrument Symposium within the Congress, and "decided to produce a publication on the topic of the work of the Commission over the past decade, and how the study of scientific instruments brings historians and museums together." Entitled *An Apparatus of Instruments: The Role of the Scientific Instrument Commission*, 700 copies of the booklet were distributed not only at plenary sessions in Zaragoza, but also at the "International Conference on Technological Change," held in Oxford from 8–11 September 1993. The four authors of the chapters were Robert G.W. Anderson ("Introduction"), Gerard L'E. Turner ("A Decade in the Study of Scientific Instruments"), James A. Bennett ("Science Museums and Science History"), and Robert Fox ("Collections and Research in the History of Science").[22]

The Scientific Instrument Commission's next outing was a triple-header hosted by the Museum Boerhaave in Leiden. From Wednesday to Friday, 7–9 September 1994, the museum held a scholarly conference on the "Origins and Evolution of Collecting Scientific Instruments." It included fifteen papers and an excursion to the castle-museum Sypesteyn in Nieuw Loosdrecht.[23]

20 "XII Scientific Instrument Symposium ... Programme;" and "Some Highlights from the XII Symposium of the Scientific Instrument Commission, Edinburgh, 7–11 September 1992," *Bulletin of the Scientific Instrument Society*, no. 35 (December 1992): 29–32.

21 The symposium was organised by Wesley Stevens, and had these speakers in addition to himself: A.J. Turner, E. Poulle, A. Duhoux-Tihon, D.A. King, B.S. Eastwood, M. Viladrich, A.M. Mundo, G.L'E. Turner, G. Beaujouan, M. Barrandon, and J. Samsó.

22 SIC, *Newsletter*, no. 15 (January 1994): 2; and *Newsletter*, no. 16 (January 1995): 2; R.G.W. Anderson and G.L'E. Turner, eds., *An Apparatus of Instruments: The Role of the Scientific Instrument Commission* (London: Scientific Instrument Commission, 1993).

23 The papers are described in Anita McConnell *et al.*, "The Three Dutch Conferences at Leyden, 7–16 September 1994," *Bulletin of the Scientific Instrument Society*, no. 43 (December 1994): 3–11. The proceedings were published in Peter de Clercq and Anthony Turner, eds., "Origins and Evolution of Collecting Scientific Instruments," special issue, *Journal of the History of Collections* 7, no. 2 (1995).

The meeting accompanied the opening of a new exhibition, *The Brass Cabinet: Treasure Houses of Science, 1550–1950*,[24] and both examined how monarchs, institutions, and private individuals have since the Renaissance acquired scientific instruments for different motives – e.g., as current research and teaching tools, as curiosities, for display, and as antiques. On Saturday, September 10, a one-day meeting was held for the general public on the practice and problems of collecting in modern museums ("Verzamelen Nu – Collecting Now"). The next day saw the start of Scientific Instrument Symposium XIII, which ran 11–16 September 1994. As the museum director, G.A.C. Veeneman, explained in the introduction to the combined programme booklet:

> Here we meet the people, who do the collecting that the above mentioned talks will be about. And, for that matter, those who criticize the collecting policies of museums for the history of science will be present as well. Thus, lively discussions are inevitable. And part of the debate will be about the central theme of the symposium: *What more can research on scientific instruments offer us than a mere increase of our knowledge of these instruments?*[25]

Some speakers accepted this challenge. Jim Bennett urged instrument historians and curators to bring instruments from the sidelines into the mainstream by showing historians of science "*how instruments have worked to determine and, perhaps, even to define the methods and content of science*" ("Instruments and the History of Science"). Paolo Brenni argued that "a careful analysis of the instruments presented at the international exhibitions between 1851 and 1900 gives us not only a detailed view of the developments of precision industry but also a more general image concerning the state of science and technology in different countries" ("Scientific Instruments Go to the Fair: Precision Industry

24 *Het koperen cabinet: Schatkamers van de wetenschap, 1550–1950* [The brass cabinet. Treasure houses of science, 1550–1950]. The exhibition was divided into four sections: "Courtly Collections," assembled in the sixteenth and seventeenth centuries by the Habsburgs, the Electors of Saxony and the Dutch princes of Orange; "Science at Home," featuring eighteenth-century apparatus from the fourth Earl of Orrery and Ernestus Ebeling; "Physics for the Public," showcasing nineteenth-century instruments from a regional scientific society (in Deventer) and a secondary school (in Tilburg); and "Old Wine in New Bags," featuring the collections of Dutch antiquarians. A sixty-four page catalogue was published in Dutch: *Het koperen kabinet: schatkamers van de wetenschap, 1550–1950* (Leiden: Museum Boerhaave, 1994). A review by Marijn van Hoorn can be found in McConnell *et al.*, "The Three Dutch Conferences," 8–11.

25 Extract from p. 2 of the 60-page programme booklet for all three conferences, simply entitled, *Museum Boerhaave Leiden* (Leiden, 1994), italics added.

at the Great Exhibitions 1851–1900"). For the benefit of economic historians, A.D.C. Simpson brought evidence from surviving metrological standards "to illuminate otherwise inaccessible aspects of trading custom and practice in the early modern period" ("Objects and Evidence: The Use of Measurement Standards in Trading Practice"), and Alison Morrison-Low ("The Role of the English Subcontractor in Instrument Manufacture") used artefacts to build an economic picture of the Midlands as a place of production and London as a marketplace for goods, thereby revising our understanding of trade during the early Industrial Revolution. Sara Schechner analysed pocket sundials as evidence of the precision wanted to run lives and businesses, the places users travelled or dreamed of going, the fashion and fads found in different cities and times, the religious and political orientations of users, spending habits and conspicuous consumption, and marketing practices by makers ("Scientific Instruments as Evidence of Consumer Culture"). And Françoise Le Guet Tully ("Information to Be Gained from Astronomical Instruments") offered case studies of astronomical instruments that provided information "dealing with the human, social or political context" and told us about "scientific controversies, technological progress, … financial support, scientific policies, etc."[26]

Excursions included a visit to the oldest, still operational radio telescope in the world, the Dwingeloo Radio Observatory, the University Museum of Groningen, the Eise Eisinga Planetarium (completed 1781 in the living room ceiling of Eisinga's house) and the Museum 't Coopmanshûs in Franeker, the Lorentz locks on the Afsluitdijk, and the Zaandam Horological Museum.

Thus, a participant could spend nearly two weeks in the vicinity of Leiden seeing wonderful things and hearing the most stimulating papers, but I recall the weather being continuously grey, gloomy, and rainy. As someone accustomed to more sunlight, I descended into a seasonal affective disorder whose depression a steady diet of *pannenkoeken* and a follow-on trip to Tuscany could not fix (indeed, the episode contributed to the dissolution of my first marriage). On a brighter note, this meeting was the first symposium to have a hands-on session for participants. This has become a staple of all SIC meetings since and truly sets the Commission apart from other societies.

Scientific Instrument Symposium XIV was held in Prague during 23–28 September 1995 at the National Technical Museum (Národní technické

26 Quotations taken from abstracts in the programme booklet. The speakers represented these institutions: Jim Bennett, Whipple Museum, Cambridge; Paolo Brenni, IMSS, Florence; A.D.C. Simpson and Alison Morrison-Low, National Museums Scotland, Edinburgh; Sara Schechner Genuth, University of Maryland, College Park; Françoise Le Guet Tully, Observatoire de la Côte d'Azur, Nice.

museum) and with the support of the Society for the History of Science and Technology of the Czech Republic, the Mathematical/Physical Faculty of the Charles University, the Astronomical Institute of the Academy of Sciences, and the Council of Learned Societies of the Czech Republic. The symposium had no special theme, but the speaker list was notable in that forty-five percent hailed from the former Soviet Union and its satellites.[27] In addition to seeing Prague's famous sites – e.g., the astronomical clock in the Old Town Square, or the Astronomical Tower and Baroque Library inside the Klementinum – the participants visited the old silver mines of Kuntná Hora (*Fig. 2.1*).

From Prague, the Scientific Instrument Commission ventured across the Atlantic and up the St. Lawrence River to Ottawa and Montreal for Scientific Instrument Symposium XV on 9–12 September 1996. This was only the second time that the annual conference had taken place in the Americas, and the two themes – the "Transformation of Canada" and "Canadian Context" – encouraged a good participation by Canadian scholars. The organiser was the National Museum of Science and Technology (today known as the Canada

FIGURE 2.1 All dressed up for a visit to the medieval silver mines of Kuntná Hora, an excursion during Scientific Instrument Symposium XIV in Prague, 1995
AUTHOR'S COLLECTION

27 SIC, *Newsletter*, no. 17 (January 1996): 3.

Science and Technology Museum) but the events were divided between Ottawa and Montreal, with a side trip to Quebec City.[28]

In Ottawa, delegates explored the exhibits of the National Museum, its main storage warehouse and library, as well as the Canadian Museum of Civilization (home to the controversial "Champlain" astrolabe) and the internationally renowned Canadian Conservation Institute. Highlights in Montreal were the Musée Stewart and the Rutherford and McPherson collections at McGill University. Housed at the time in an eighteenth-century fort on Île Sainte-Hélène in the St Lawrence River, the Musée Stewart showcased the apparatus of Abbé Nollet, splendid globes by Blaeu and Coronelli, and other fine instruments.[29] In McGill's physics department, delegates saw the instruments used by Ernest Rutherford for his early research into radioactivity, as well as the McPherson Collection of nineteenth- and early twentieth-century apparatus. An optional excursion to Quebec City gave participants the chance to study the teaching apparatus of the Roman Catholic Séminaire de Québec in the storage of the Musée de l'Amérique française and to explore the Musée de la civilisation. Attendance by Europeans at this symposium was low, no doubt due to the travel costs, but those that came were well treated. Peter de Clercq (Museum Boerhaave, Leiden) spoke for the group, "Once again our warmest thanks to our Canadian friends for the effort they put into this SIC meeting. We saw no grizzlies, but will we ever finish the maple syrup?"[30]

The SIC's Scientific Instrument Symposium XVI was held as part of the XX International Congress of the History of Science in Liège, Belgium on 21–22 July 1997. The two-day gathering took place in the Maison de la métallurgie et de l'industrie de Liège. The theme was "National Inventories of Historical Scientific Instruments," with fifteen papers addressing inventories, catalogues, and problems of method. A report in the SIC *Newsletter* observed that the meeting:

> became a critical review of different genres of documentation. There was a mood of introspection and assessment; it was an opportunity to take stock, to consider what the Commission and the wider world of

28 See the bilingual programme booklet, *XVth Scientific Instrument Commission Symposium, September 9–10–11–12, 1996* (Ottawa: National Museum of Science and Technology, 1996), which includes abstracts and a list of participants. SIC, *Newsletter*, no. 18 (April 1997): 2. Peter de Clercq *et al.*, "The XVth Scientific Instrument Symposium in Canada: A Report," *Bulletin of the Scientific Instrument Society*, no. 51 (December 1996), 17–21.

29 The site permanently closed in 2021, and the holdings have been relocated to the McCord Stewart Museum.

30 de Clercq *et al.*, "The XVth Scientific Instrument Symposium," 21.

> instrument history had achieved in recent decades and to consider to what extent the current direction was appropriate for the future. This seemed a particularly suitable time for reassessment as new technologies offer fresh possibilities for information management. Some papers addressed these challenges, while others reviewed the achievements and continuing value of conventional catalogues and inventories.[31]

Sorø, Denmark was the next stop for the SIC, when the Scientific Instrument Symposium XVII, 20–25 July 1998, was hosted by the Sorø Academy. More than 100 people participated in this meeting, with many taking advantage of student housing on the academy grounds and eating in its dining hall. Along with opportunities for walks in the gardens, boating on the lake, and musical concerts, this resulted in a summer-course atmosphere. The themes of the conference were "Instruments and Instrument Makers in Scandinavia" and "Cabinets of Physics and Chemistry." The latter was prompted by the Hauchs Physiske Cabinet – one of Europe's finest collections – which was assembled between 1787 and 1815 by a Danish military figure, lord chamberlain, and natural philosopher Adam Wilhelm Hauch, who sold it to King Frederik VI in 1815, who in turn donated the collection to the Sorø Academy in 1827. The first day was devoted to papers on Scandinavian instruments; the second day to cabinets of physics in Italy, the Netherlands, Estonia, and Denmark; and a third day to miscellaneous topics. The number of papers offered was so plentiful that the third day was the first occasion of a symposium having parallel sessions.[32]

During a full-day trip to Copenhagen, there were visits to see instruments at the Rosenborg Castle, the Rundetårn (built as an observatory in 1636), the Medicinsk Museion, and the Nationalmuseet. A post-conference tour to Aarhus offered a visit to the Steno Museet (the Danish National Museum for the History of Science and Medicine) and the Jelling Stones (tenth-century carved runestones).

The Scientific Instrument Symposium XVIII took place in Moscow and St. Petersburg, Russia from 20 to 25 September 1999. We were guests of the Russian Academy of Sciences in recognition of the 275th anniversary of its founding. The organisers included the Institute of the History of Science and Technology

31 SIC, *Newsletter*, no. 19 (June 1998): 3. The Liège papers were later published in Maurice Dorikens, ed., *Scientific Instruments and Museums* (Turnhout: Brepols, 2002).

32 SIC, *Newsletter*, no. 19 (June 1998): 3, and no. 20 (June 1999): 3–4. Jan Tapdrup, ed. *Conference Programme for the XVII International Scientific Instrument Symposium, 20–25th July 1998, Sorø Academy* (Sorø: Sorø Academy, 1998). Howard Dawes, "The XVIIth Scientific Instrument Symposium in Denmark: A Report," *Bulletin of the Scientific Instrument Society*, no. 59 (December 1998): 33.

of the Russian Academy of Sciences and the Union of Scientific and Engineering Associations. The main theme was "Historical relations between Russia and other countries in the field of scientific instruments."[33] Accommodations in Moscow were at the Hotel Sputnik on Leninsky Prospekt, near the 141-foot (43 m) tall, titanium monument of Yuri Gagarin, the Soviet cosmonaut. Talks were held in the Presidium of the Russian Academy of Sciences. Only about twenty of the SIC regulars took part, but the numbers were more than made up by Russian academicians, who entertained us with celebratory dinners during which they continually made short Russian speeches followed by vodka toasts.

This meeting was my most memorable ever. The opening ceremony included an amazing performance on the theremin by the virtuoso Lydia Kavina, who had studied with her cousin the inventor Lev Sergeyevich Termen (better known as Leon Theremin). In Moscow, excursions were made to the Museum of the Faculty of Physics at Moscow State University, Red Square (*Fig. 2.2*), the State Politechnical Museum, the Kremlin complex of museums and its Armoury Chamber, and the Museum of the Moscow State University of Geodesy and Cartography.

FIGURE 2.2 SIC group photograph in Red Square. Scientific Instrument Symposium XVIII held in Moscow and St Petersburg, 1999
AUTHOR'S COLLECTION

33 SIC, *Newsletter*, no. 20 (June 1999): 4. *XVIII International Scientific Instrument Symposium: Abstracts and Materials* (Moscow: Publishing House "Engineer," 1999).

An overnight train was to take participants from Moscow to St. Petersburg for a two-day excursion, and we were settled into nice sleeping compartments by 11:00 PM. I shared a compartment with Sofia Talas (Museum of the History of Physics, Padua), and the Russian academicians in the adjoining compartment invited us and Jim Bennett (Museum of the History of Science [MHS], Oxford) to pass the time with them by eating ryebread-and-cucumber sandwiches, drinking vodka, and singing rowdy Russian labour songs. The plan had been to arrive in St. Petersburg at 8:00 AM, have breakfast at our hotel, and then begin the day with a visit to the State Hermitage. Unfortunately, the train was delayed ten hours en route, the restaurant car ran out of food, and we arrived after dark the next day.

But all was not lost. We were whisked off directly to the Hermitage, which was held open for us, and revived with tea and biscuits in the sumptuous personal theatre of Catherine the Great. The Hermitage's Vice Director, Dr V. Matveev, gave us a brief lecture on "Scientific Instruments and Watches at the Hermitage Collections" and then raced us through art galleries formally closed to the public until we reached a gallery with seventeenth- and eighteenth-century clocks, automata, and royal furniture. Matveev gleefully wound up and started various automata, including the celebrated Peacock Clock by James Cox, commissioned for Catherine in 1777. He also "sprung" the mechanical desks and cabinets created by David Roentgen for the Tsarina. When secret mechanisms were activated, sections of the furniture automatically opened up to reveal hidden compartments, writing implements, musical clocks, and even the desk chair. Now here were symbols of power. In another area of the Hermitage complex, we were introduced to Peter the Great's fascination with natural philosophy, medicine, and trades. We saw his workshop, including his lathe, dental tools (which he famously used to extract courtiers' teeth), and other apparatus.

By now it was close to 10 PM, but dinner awaited us at the Nikolaevsky Palace with wait staff robed in the attire and wigs of Tsarist Russia. Around midnight, we exhaustedly boarded our coach for a ride to the hotel. Along the way, the police pulled our coach over, arrested the driver, and hauled him off! We sat in stunned abandonment. We had no clue where we were or where we were going. Fortunately, the Russian academicians travelling with us managed to free the luggage from the locked compartment underneath the coach and ordered taxis. Upon arriving at the hotel, we presented our passports to check in. One registered guest in our group, Christine Blondel (Cité des sciences et de l'industrie, Paris), had forgotten hers back in Moscow. Despite entreaties from the academicians, the management refused to let her go up to a room or sit in the hotel lobby overnight, even though the lobby bar was filled with hookers who were freely going up and down the elevators with their clients. It was unacceptable for Christine to be put out on the street. So I waited for my floor lady or *dezhurnaya* – one of the stern stolid sentries who used "to guard the keys and the morals of nearly every hotel floor in the country"[34] – to leave her desk on a break. I then boldly snuck Christine up to my room.

The next day we visited the summer residence of the Tsars, the Yekaterininsky Dvorets (Catherine Palace) in Pushkin, followed by the Mikhail Lomonosov Museum, the Dmitry Mendeleev Museum and Archives, and the Museum of Metrology. The Lomonosov Museum houses a remarkable collection of scientific instruments and books dating from the fourteenth to the nineteenth century, including the Gottorp Globe (1651–1664). It is situated in the Kunstkamera – the original home of the Russian Academy of Sciences (founded in 1724), complete with an anatomy theatre and observatory in its tower.[35] The latter was under renovation, but many daringly stepped out onto the scaffolding to get a better view (*Fig. 2.3*).

FIGURE 2.3 Intrepid SIC members ventured out onto scaffolding surrounding the old observatory atop the Kunstkamera – the original home of the Russian Academy of Sciences – in St Petersburg, 1999
PHOTO: BY AUTHOR

34 Michael Specter, "Vladivostok Journal: Beware the Floor Lady (She Barks and Can Bite)," *New York Times*, August 24, 1995, section A, p. 4.

35 Sara J. Schechner, "Cover Story. Art in the St Petersburg Observatory: Putti with Scientific Instruments," *Bulletin of the Scientific Instrument Society*, no. 64 (March 2000): 2.

2 What We Talked About

When we examine the papers presented across all the meetings during the decade, we notice a shift in topics and types of analysis.

Many of the papers delivered at symposia in the 1980s were "getting to know you" papers. Speakers were delighted to find companions with similar interests, and they introduced their institutional or private collections to each other in slideshows. These introductions continued into the early 1990s. For instance, the 1991 and 1992 symposia introduced delegates to collections in Cambridge (Massachusetts), Dresden, Tartu, La Plata, Madrid, Prague, Sydney, and various academies and colleges, which were unrelated to the host city or country. Such talks petered out in later meetings, except for an uptick in 1996 where participants in Ottawa heard about the Jena Optical Museum, the Urbino Ducal Palace, and Soviet aerospace museums. By the mid-decade, papers only introduced a collection if it were related to the history or political geography of the meeting site – e.g., in Sorø, there were papers about Danish museums, or in Prague, on collections in the Czech Republic, the Former Soviet Union, and Central Europe.

Earlier meetings in the decade also had papers introducing particular classes of instruments – e.g., the cross staff, slide rule, diptych sundial, stethoscope, air pump, or surveyor's compass – to listeners who were presumed to be unfamiliar with them. In later meetings, such papers were rarely given. One reason was that SIC members were becoming more informed by instruments outside of their specialties. Another reason was that an encyclopaedic volume was under way, under the editorship of Robert Bud and Deborah Jean Warner. Begun in 1992, this reference work would be in print by 1998.[36]

Therefore, papers devoted to a class of instrument were replaced by more sophisticated reports in which the instruments played larger roles in some context. The themes of symposia encouraged this advance, and many examples of such papers have already been mentioned in the survey of symposia above. But it should be known that papers setting scientific instruments into the social, cultural, geographical, methodological, and philosophical contexts of history of science were already present in the SIC symposia from the very beginning of the decade. See for instance two papers given at the Bologna symposium in 1991: A. Rupert Hall's talk on the telescopes of Huygens, Newton, Gregory, and Dollond ("The Contributions of Science and Technology to the Design of Optical Elements") or Alison Morrison-Low's paper which turned

36 Bud and Warner, *Instruments of Science.*

attention away from the success stories of makers like the Dollonds toward enterprises that failed economically ("The Road to Ruin? Bankruptcy and Some Legal Consequences for the Instrument Maker in Nineteenth-Century Britain").[37] Indeed, in a closing address to the XIIIth Symposium in Leiden in 1994, Jim Bennett pointed to the evolution of the historiography of scientific instruments and how much the historical instrument community had to offer the discipline of the history of science. Until recently, historians of science had been primarily focused on the history of ideas and had put instruments on the sidelines, whereas the historians and curators of instruments were comfortable with material culture and were adept at interpreting it as evidence of social, cultural, and spatial forces influencing the practice of science.[38]

As curators in the SIC got to know and trust each other, "best practices" became a major focus of the symposia. In 1990, Gerard Turner (Imperial College) and Jon Eklund (National Museum of American History) spoke generally on "The Purpose and Practice of Cataloguing," and cataloguing methods were central themes during the Edinburgh and Liège symposia (1992 and 1997 respectively). The use of video documentation was raised in 1990 and 1992 (in Washington and Edinburgh), and computerised registration and collection databases in 1992 and 1994 (in Edinburgh and Leiden).[39] Today, it is hard to imagine a meeting in which the importance of cataloguing would have centre stage. The practice is something we all now take for granted.

37 A.R. Hall's and A.D. Morrison-Low's papers in the Bologna 1991 symposium; *Bulletin of the Scientific Instrument Society*, no. 31 (December 1991): 12–13.

38 J.A. Bennett, "Instruments and the History of Science," paper delivered at Scientific Instrument Symposium XIII, Leiden, 11–16 September 1994. See the programme booklet, *Museum Boerhaave Leiden* (Leiden, 1994), for the abstract.

39 On video, see for instance, 1990: David DeVorkin, "Using Video in Documenting the Astronomical Collections" of the Smithsonian's National Air and Space Museum; 1992: William Andrewes and Andrew McKinney, "Uses of Multi-Media in the Dissemination of Knowledge." On computerisation, 1992: John Burnett, "Information and Instruments in Britain and Ireland," a paper on computerisation in the National Museums Scotland; Mara Scheelings, "The Museum Boerhaave's Approach to Documentation;" 1994: Mara Miniati, "SIC [Scientific Instrument Catalogue]: A Newly Prepared Exchange Format," a paper on a computerised catalogue system based on ISIS, a software programme available free from UNESCO; and G. van den Dool, "Computerizing Registration in Museum Boerhaave."

3 Conservation and Authenticity

Conservation and restoration were also significant topics. At the Ottawa meeting in 1996, Paolo Brenni asked, "How Far Can We Go? The Restoration of Scientific Instruments: Problems and Suggestions." An entire workshop on the topic – Conservation and Restoration of Scientific Instruments: Issues of Principle – was sponsored by the SIC and organised by the Istituto e Museo di Storia della Scienza (IMSS) and the Opificio della Pietra Dure in Florence for 14–15 December 1998. Speakers came from Italy, Great Britain, Germany, The Netherlands, and France, and the proceedings were published.[40] Curators at meetings also learned about the latest techniques in conservation science to analyse materials and check the authenticity of objects. For instance, the use of X-ray fluorescence to investigate brass instruments for clues to their dates and geographical origins was described by Willem Hackmann in Bologna's symposium (1991), and "provoked a lively debate, centring on … the re-use of metals, the use of different alloys in the same instrument, and the longevity of identical techniques."[41]

Indeed, the recognition that there was fakery in our midst led to the Anton Mensing Scientific Instrument Project and a two-day meeting hosted at the Museum Boerhaave in Leiden on 15–16 October 1999: "Scientific Instruments: Originals and Imitations. The Mensing Collection." The Anton Mensing Scientific Instrument Project was a joint venture of the Museum Boerhaave, Leiden, the Nederlands Scheepvaartmuseum, Amsterdam, the Utrecht University Museum, and the Adler Planetarium, Chicago. Its purpose was to trace and document the scientific instruments that had passed through the hands of Anton W.M. Mensing (1866–1936), the director and owner of the Frederik Muller & Co. auction house in Amsterdam, and had ended up in the aforementioned collections. Some of these instruments had been proven fake or to have imitation components.[42] Conferences such as the Mensing meeting and the now routine hands-on sessions at the SIC symposia advanced the connoisseurship of the SIC community.

---

40 *Bulletin of the Scientific Instrument Society*, no. 60 (1999): 27. Istituto e Museo di Storia della Scienza and Opificio delle Pietre Dure, *The Restoration of Scientific Instruments: Proceedings of the Workshop Held in Florence, December 14–15, 1998* (Florence: Le Lettere, 2000).

41 *Bulletin of the Scientific Instrument Society*, no. 31 (December 1991): 17.

42 SIC, *Newsletter*, no. 20 (May 1999): 5, 7. Peter de Clercq, ed. *Scientific Instruments: Originals and Imitations. The Mensing Collection. Proceedings of a Symposium held at the Museum Boerhaave, Leiden, 15–16 October 1999* (Leiden: Museum Boerhaave, 2000).

4 Big Projects – National Inventories

National, collective inventories were the focus of the SIC throughout the 1990s, as witnessed by reports on them in the SIC *Newsletters* and in the minutes of the annual general meetings (AGMs) at the symposia. We wanted to know where the good stuff was. *Table 2.2* lists the national inventory projects of 1990s by the date they were started, the key individuals involved, the date completed, and any capstone publications. Another chapter in this volume by Alison Morrison-Low will delve into the history of these inventories and museum catalogues, so little needs to be said here.

TABLE 2.2 National inventories completed in the 1990s

Year started	Completed	National inventory	Key people	Publications
1968	1992	Great Britain and Ireland	Mary Holbrook, R.G.W. Anderson, and D.J. Bryden	Mary Holbrook, R.G.W. Anderson, and D.J. Bryden, *Science Preserved: A Directory of Scientific Instruments in Collections in the United Kingdom and Eire* (*London, 1992*)
1984	1995	SIMON index of London makers	G.L'E. Turner, Michael Crawforth, and Gloria Clifton	Gloria C. Clifton, *Directory of British Scientific Instrument Makers, 1550–1851* (London, 1995)
1989	1995	Irish	Charles Mollan, Royal Dublin Society, and Royal Irish Academy	Charles Mollan, *Irish Interim Inventory* (1990) Charles Mollan, *Irish National Inventory of Historic Scientific Instruments* (Blackrock, Co. Dublin, 1995)
1990	1995	Danish	Hemming Anderson	Hemming Anderson, *Historic Scientific Instruments in Denmark* (Copenhagen, 1995)

TABLE 2.2 National inventories completed in the 1990s

Year started	Completed	National inventory	Key people	Publications
1991		Italian	Paolo Brenni, Istituto e Museo di Storia della Scienza	
1992	1996	Belgian	Victor Rasquin, Robert Halleux, Maurice Dorikens,	V.A. Rasquin, *Dictionnaire des constructeurs belges d'instruments scientifiques* (Brussels, 1996)
1993		Hesse, Germany	Karl-Heinz Schaldach	

A few projects, however, are worthy of a brief note. In 1990, Charles Mollan of the Royal Dublin Society published an interim report on the status of the Irish National Inventory, which at the time had reached 809 pages. The SIC *Newsletter* praised it for including a chapter on methodology (confirming the general interest in best practices). It was the first of such inventories to make use of a word processor and database, the IBM Filing Assistant software.[43]

At the Edinburgh meeting in 1992, physicist Desmond King-Hele launched the publication of *Science Preserved*, a directory of instruments in the UK and Ireland.[44] More than twenty-five years in the making, this came as a relief to the SIC, whose exasperated members had expressed during the AGM of Scientific Instrument Symposium XIX (Hamburg 1989), "a fervent wish … for the publication of the long-completed British National Inventory, which seemed to have been captured by a bureaucratic black hole."[45]

And 1995 was the year of Gloria Clifton's landmark publication, *Directory of British Scientific Instrument Makers, 1550–1851*. It was the culmination of Project

43 SIC, *Newsletter*, no. 12 (January 1991): 4.

44 Mary Holbrook, *et al.*, *Science Preserved: Directory of Scientific Instruments in Collections in the United Kingdom and Eire* (London: HMSO Books, 1992).

45 SIC, *Newsletter*, no. 11 (January 1990): 4. The British and Irish National Inventory had begun with a grant from the Royal Society but was really stuck in institutional limbo. It was not an institutional priority for the Science Museum, London. The SIC only became involved because Robert Anderson and David Bryden were in positions in both the SIC and their institutions to push it to completion – twenty years after it had begun. See Robert Anderson's chapter on the 1980s in this volume.

SIMON (Scientific Instrument Makers, Observations, and Notes) whose origins dated back to 1984.

5 Computers Changed Our Lives

The second half of the 1980s saw the introduction of personal computers (PCs) into people's homes and businesses. The Adler Planetarium's History of Astronomy Collection had its own computer and primitive database before the rest of the museum did, even before the financial office and sky shows. This early adoption was due to the foresight of Rod and Madge Webster and their ability to personally finance it. It was not until the 1990s that computers could be expected in the back-of-the-house operations of many museum collections. As previously mentioned, talks and posters in SIC symposia – especially at the Edinburgh 1992 meeting – drew attention to the possibilities of computerised databases. But we must remember that these databases were for a long time without images and very primitive.

It is hard to recall the world without email and the World Wide Web. In 1990 some of us attached to universities were privileged to have a klutzy, UNIX-based email system, which used a dial-up service via telephone line. The current email standards with which we are familiar did not come into being until 1995. The first website was created by Tim Berners-Lee at CERN in 1991, but the Web did not enter regular use until 1993–1994. I first heard the abbreviation "HTML" for hypertext markup language from Jim Bennett around this time.

Jim was SIC's pioneer in bringing us into the Internet age. In 1995 Jim had the inspiration to launch "an electronic bulletin board" that would be administered by the Museum of the History of Science (MHS), Oxford. Endorsed during the AGM in Prague that year, Rete was born! The planning was done by Jim (then MHS director) along with Stephen Johnston and Giles Hudson. The listserv went into operation in early 1997. The name was eminently appropriate for the Commission's discussion network. As the MHS's newsletter, *Sphaera*, explained: "The bulletin board has been christened 'RETE', reflecting not only the most prominent component of the astrolabe but also, in the sense of the original Latin, an actual 'net' or, indeed, the Web itself."[46]

The Museum of History of Science at Oxford had the earliest website that I recall seeing devoted to our field. *Figure 2.4* has a screenshot of the museum's site as it appeared in 1996, courtesy of the Internet Archive's Wayback

46 "RETE: An Electronic Bulletin Board," *Sphaera*, no. 3 (Spring 1996); SIC, *Newsletter*, no. 17 (January 1996): 4; announcement of "Rete operational," *Sphaera*, no. 5 (Spring 1997); Stephen Johnston, email to author, 24 December 2024.

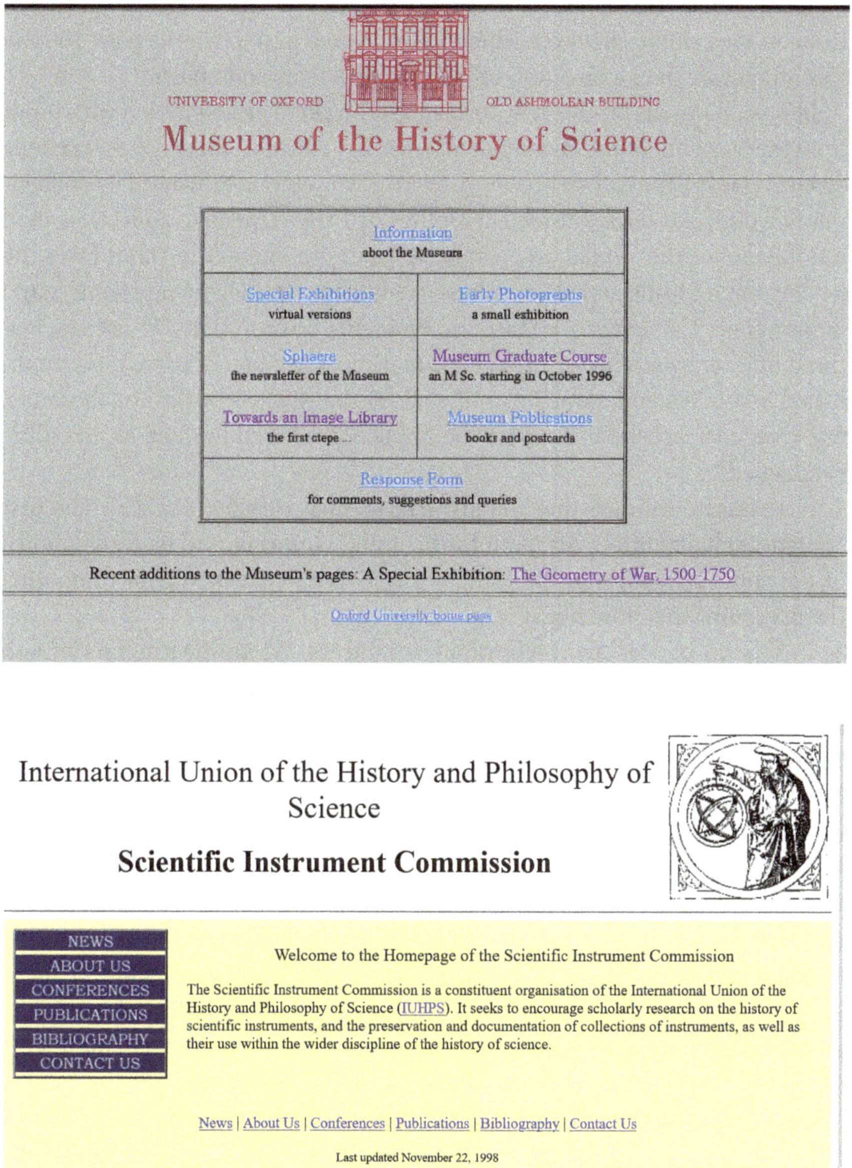

FIGURE 2.4 Earliest websites: (*above*) Museum of the History of Science, Oxford, 1996. (*below*) Scientific Instrument Commission, 1998
SCREENSHOTS COURTESY OF THE INTERNET ARCHIVE'S WAYBACK MACHINE

Machine. It had useful information about the museum, its activities, and a link to an online exhibit. However, when one navigated past the home page, the site was little more than a directory of links to short texts and images.

In 1996 materials for the web portal *Epact* began to be assembled with input from the Museum of the History of Science, the Istituto e Museo di Storia della Scienza (IMSS), the British Museum, and the Museum Boerhaave. The collaborators would not have known each other or their respective collections had it not been for the annual SIC symposia,[47] but Jim Bennett in Oxford was the leading force. He discussed *Epact* in some detail at the Liège symposium (1997) in his paper, "Cataloguing a Museum: Problems and Solutions."[48] The site featured objects by scientific instrument makers active before 1600. It was published first in a stand-alone form in 1998 for the sole use of the collaborators, and then as a website in 2001 for the public. *Epact* is still valued as an online catalogue.[49]

It was not until 1998 that the SIC had its own website (*Fig. 2.4*). The first content included information on forthcoming symposia, and featured photos of the Sorø Academy and the Hauchs Physiske Cabinet by way of promoting the upcoming meeting there.[50]

By 1994 as part of the celebrations around the Whipple Museum's fiftieth anniversary, Jim (at the time still Keeper there) had begun a "virtual teaching collection," which he described as a "computerized visual database aimed at giving those who teach history of science without ready access to a collection a resource for integrating scientific instruments into teaching and learning."[51] And in 1999, Jim also launched the "Online Register of Scientific Instruments" (http://www.isin.org) – an international, shared database of instruments accessible via the Internet – which had been developed at Oxford and endorsed by the SIC at the Sorø meeting (1998).[52] It was modelled on the "Computerized Checklist of Astrolabes" in that a common system of numbering scientific instruments in historical collections would be adopted (the so-called International Scientific Instrument Numbers, or ISIN numbers),

47 See A.D. Morrison-Low's chapter on catalogues in this volume.

48 J.A. Bennett, "Cataloguing a Museum: Problems and Solutions," pp. 165–190 in *Scientific Instruments and Museums*, ed. Dorikens (2002).

49 *Epact: Scientific Instruments of Medieval and Renaissance Europe*, last modified 2006, https://www.mhs.ox.ac.uk/epact/.

50 SIC, *Newsletter*, no. 19 (June 1998): 4. The original address was http://www.sic.iuhps.org.

51 J.A. Bennett, "The Whipple at 50: the Sphere of the World, the New Age, the Virtual Collection," *Bulletin of the Scientific Instrument Society*, no. 42 (September 1994): 25–26.

52 The "Online Register" was the brainchild of Giles Hudson and Jim Bennett. See SIC, *Newsletter*, no. 20 (May 1999), 6; and Stephen Johnston, email to author, 24 December 2024.

but differed in that institutions would be responsible for entering their own records and for answering queries. All that the centralised service provided was a core set of fields, technical and procedural instructions, and the maintenance of the unique ISIN numbers. According to Stephen Johnston, MHS curator, the idea was "ahead of its time:"

> The idea of a decentralised federation of institutions cooperating on the creation of a global database was a really clear-sighted and innovative response to the potential of the web. But it didn't really catch on, in a world where institutions didn't have their own websites and online records.[53]

Although the "Online Register" is now inactive, the point remains that we embraced new technologies in our studies of antique ones.

6 Administrative Matters

On 1 January 1998 Robert Anderson (British Museum) and Gerard L'E. Turner (Imperial College) stepped down as President and Secretary of the Scientific Instrument Commission. Robert had served since 1982 and Gerard since 1977. Jim Bennett (Museum of the History of Science, Oxford) was appointed the new President, and he invited Peter de Clercq (of the Museum Boerhaave) to be the Secretary. A new position of Vice President was created for Paolo Brenni (IMSS) in order "to broaden the Commission's geographical horizon." When Peter de Clercq left the Museum Boerhaave and moved to London a year later, it was considered important to maintain a connection to the Boerhaave. Therefore, Marian Fournier, its assistant director, was invited to become the SIC Treasurer (a newly created position).[54]

During the twenty years of Robert Anderson's and Gerard Turner's leadership, the Commission had grown in stature, hosting annual international symposia and conferences, publishing a newsletter and regular bibliographies, and

53 Stephen Johnston, email to author, 24 December 2024.

54 SIC, *Newsletter*, no. 19 (June 1998): 1, and no. 20 (May 1999): 1. During this period of the SIC's history, there were neither elections of officers nor ratification of the appointments by the AGMs. The introduction of democratic elections of the SIC officers was my idea during my term as SIC Secretary, 2003–2013. Then President Paolo Brenni and I created a system for nominating and electing SIC officers, with the turnover of leadership to be ratified by and take place at the General Assemblies of the IUHPS. The first such election was in 2013.

encouraging instrumental scholarship, catalogues, websites, and exhibitions. By the 1990s, the SIC was recognised as the most active commission in the International Union of the History and Philosophy of Science.

The 1990s also saw the solidification of programming that distinguished the Scientific Instrument Commission from other societies. Oral and poster papers were certainly an important part of meetings, but not the be all and end all. Fifty percent of the time was spent visiting regional collections and historic sites related to scientific instruments, going behind the scenes in venues closed to the public, and having the opportunity to handle and examine rare scientific instruments. The Scientific Instrument Commission was celebrated for its collegiality and friendliness. We were learners together with a common passion. In the 1990s, the annual meetings were cultural events and truly international in participation, welcoming new members from the Former Soviet Union and its satellites, as well as those from North America, India, and even Japan.

Bibliography

Amson, John Ceres. "Gregory's Meridian Line of 1673–74: A St Andrews Detective Story." *BSHM Bulletin* [today: *British Journal for the History of Mathematics*] 23, no. 2 (2008): 58–72.

Anderson, R.G.W. "The XI Scientific Instrument Symposium: Summing Up." P. 20 in "XI International Scientific Instrument Symposium," by A.D. Morrison-Low *et al. Bulletin of the Scientific Instrument Society*, no. 27 (December 1990).

Anderson, R.G.W., and G.L'E. Turner, eds. *An Apparatus of Instruments: The Role of the Scientific Instrument Commission*. London: Scientific Instrument Commission, 1993.

Bennett, J.A. "Cataloguing a Museum: Problems and Solutions." Pp. 165–190 in *Scientific Instruments and Museums*, edited by Maurice Dorikens. Turnhout: Brepols, 2002.

Bennett, J.A. "The Whipple at 50: the Sphere of the World, the New Age, the Virtual Collection." *Bulletin of the Scientific Instrument Society*, no. 42 (September 1994): 25–26.

Bud, Robert, and Deborah Jean Warner, eds. *Instruments of Science: An Historical Encyclopedia* New York: Science Museum, London, and National Museum of American History, Smithsonian Institution, in association with Garland Pub., 1998.

Dawes, Howard. "The XVIIth Scientific Instrument Symposium in Denmark: A Report." *Bulletin of the Scientific Instrument Society*, no. 59 (December 1998): 33.

De Clercq, Peter, ed. *Scientific Instruments: Originals and Imitations. The Mensing Collection. Proceedings of a Symposium held at the Museum Boerhaave, Leiden, 15–16 October 1999*. Leiden: Museum Boerhaave, 2000.

De Clercq, Peter, and Anthony Turner, eds. "Origins and Evolution of Collecting Scientific Instruments. Papers from the conference held in the Museum Boerhaave, Leiden, September 1994." Special issue, *Journal of the History of Collections* 7, no. 2 (1995).

De Clercq, Peter, *et al.* "The XVth Scientific Instrument Symposium in Canada: A Report." *Bulletin of the Scientific Instrument Society*, no. 51 (December 1996), 17–21.

Dorikens, Maurice, ed. *Scientific Instruments and Museums*. Turnhout: Brepols, 2002.

Dragoni, Giorgio, Anita McConnell, and Gerard L'E. Turner, eds. *Proceedings of the 11th International Scientific Instrument Symposium: Bologna University, Italy, 9–14 September 1991*. Bologna: Grafis Edizioni, 1994.

Epact: Scientific Instruments of Medieval and Renaissance Europe. Last modified 2006. https://www.mhs.ox.ac.uk/epact/.

Holbrook, Mary, with additions and revisions by R.G.W. Anderson and D.J. Bryden. *Science Preserved: A Directory of Scientific Instruments in Collections in the United Kingdom and Eire*. London: HMSO, 1992.

Istituto e Museo di Storia della Scienza and Opificio delle Pietre Dure. *The Restoration of Scientific Instruments: Proceedings of the Workshop Held in Florence, December 14–15, 1998*. Florence: Le Lettere, 2000.

McConnell, Anita, *et al.* "The Three Dutch Conferences at Leyden, 7–16 September 1994." *Bulletin of the Scientific Instrument Society*, no. 43 (December 1994): 3–11.

Morrison-Low, A.D. "10th Scientific Instrument Symposium, 20th–25th September 1990, Washington, D.C. and the Boston Area." *Bulletin of the Scientific Instrument Society*, no. 27 (December 1990): 26–27.

Morrison-Low, A.D., *et al.* "XI International Scientific Instrument Symposium, Bologna, Italy, 9–14 September 1991." *Bulletin of the Scientific Instrument Society*, no. 31 (December 1991): 11–20.

Museum Boerhaave. *Het koperen kabinet: schatkamers van de wetenschap, 1550–1950*. Leiden: Museum Boerhaave, 1994.

Museum Boerhaave. *Museum Boerhaave Leiden* [programme booklet for "Origins and Evolution of Collecting Scientific Instruments," "Collecting Now," and XIII Scientific Instrument Symposium]. Leiden: Museum Boerhaave, 1994.

Museum of the History of Science, University of Oxford. *Sphaera*, no. 3 (Spring 1996) and no. 5 (Spring 1997).

National Museum of American History. *Program: 10th Scientific Instrument Symposium. September 20–September 25, 1990. Washington, D.C. [and] Boston Area*. Washington, DC: Smithsonian Institution, 1990.

National Museum of Science and Technology (Ottawa). *XVth Scientific Instrument Commission Symposium, September 9–10–11–12, 1996*. Ottawa: National Museum of Science and Technology, 1996.

National Museums Scotland. *Programme and Abstracts: Scientific Instrument Commission, IUHPS: XII Scientific Instrument Symposium, National Museums of Scotland, Edinburgh, 7–11 September 1992*. Edinburgh: National Museums Scotland, 1992.

National Museums Scotland. *XII Scientific Instrument Symposium, Edinburgh, September 1992: Programme for Main Meeting*. Edinburgh: National Museums Scotland, 1992.

Post, Robert C. "Silvio A. Bedini, 1917–2007." *Technology and Culture* 49, no. 2 (2008): 522–529.

Russian Academy of Sciences. *XVIII International Scientific Instrument Symposium: Abstracts and Materials*. Moscow: Publishing House "Engineer," 1999.

Schechner, Sara J. "Cover Story. Art in the St Petersburg Observatory: Putti with Scientific Instruments." *Bulletin of the Scientific Instrument Society*, no. 64 (March 2000): 2.

Schechner, Sara J. "The Influence of Theory on Instrument Design: The Case of Hauksbee's Friction Electric Machine." MPhil thesis, University of Cambridge, 1980.

Scientific Instrument Commission. *Newsletters*, nos. 11–20 (January 1990–May 1999).

Specter, Michael. "Vladivostok Journal: Beware the Floor Lady (She Barks and Can Bite)." *New York Times*, August 24, 1995, section A, p. 4.

Tapdrup, Jan, ed. *Conference Programme for the XVII International Scientific Instrument Symposium, 20–25th July 1998, Sorø Academy*. Sorø: Sorø Academy, 1998.

Turner, Anthony J. "Silvio A. Bedini." *Nuncius* 23, no. 2 (2008): 353–371.

CHAPTER 3

The Scientific Instrument Commission in Its Third Decade, 2000–2009

Richard L. Kremer

My experience with the Scientific Instrument Commission (SIC) began in 2004 when I attended the symposium in Dresden. Trained as an historian of science and teaching the history of science since 1985, I had paid little attention to instruments (even if my first published paper had considered a fifteenth-century armillary sphere). But then in 2001 I inherited, rather by accident, the curatorship of my university's collection of about 4000 historic scientific instruments. Realising I had much to learn, I made my way to Dresden and opened a new page for my career, encountering for the first time museum curators, dealers, collectors, material culture specialists, and gloved hands-on examination of museum objects. My learning continued at subsequent SIC symposia and this chapter will draw on those experiences, as well as on the traditional sources for which I was trained to interrogate. To remain with the constraints of a few pages, I will limit my discussion of the SIC's third decade (roughly 2000–2009) to four facets: the institutional framework of the Commission; the contents of its annual symposia; its media and informal networks of communication; and its scholarly publications. I will suggest that in its third decade the SIC consolidated its life as a mature international organisation, a lively scholarly community, and a venue for focused scholarly publication.

1 Institutional Framework

The SIC had begun with a distinctly British orientation. Early symposia were held at Greenwich (1977, 1982) and the founding leadership was British.[1]

1 Derek Howse, keeper, National Maritime Museum, Greenwich; Gerard L'E Turner, senior curator, Museum of the History of Science, Oxford, Robert Anderson, keeper of the chemistry department, Science Museum, London, subsequently director of National Museums Scotland, Edinburgh, and from 1992–2002 director of the British Museum. Of the thirty speakers at the two Greenwich symposia, twenty-five were British or worked at British institutions. See SIC, *International Symposium on Scientific Instruments, Greenwich 1977, Summaries of Papers* and SIC, *2nd Greenwich Scientific Instruments Symposium, 1982, Summaries of Papers.*

 | DOI:10.1163/9789004737587_005

Although during the first two decades SIC symposia locations would cross Europe (with excursions to North America in 1990 and 1996), British curators, historians, and collectors comprised a significant proportion of the participants. British collections, British instrument-makers, and the history of British science and technology featured prominently in the early years, although less so by the 1990s.[2]

Attendance at the Annual General Meetings (AGMs), held during each symposia, reached forty in Bucharest (1981), fifty-two in Munich (1983), and remained fairly constant at that level through the 1990s. The officers on the SIC Board remained unchanged from 1982 until 1998. And in its first two decades, the SIC maintained a respectful distance from its parent organisation, the Division of History of Science and Technology (DHST) of the International Union of History of Philosophy of Science and Technology (IUHPST), which is under a UNESCO umbrella. Despite receiving a small annual subsidy to help cover costs of the symposia and mailing its newsletters, the SIC organised its own affairs and its officers and participants generally did not assume positions of responsibility within the DHST. The SIC cultivated an independent life within the DHST and its international congresses.

By 2000 the SIC had consolidated its shift from an initially Anglo-centric to a European focus, with participation stretching from the UK across Europe, eastward to Russia, southward to Greece. In 1998, the founding officers passed the SIC Board into new hands: Jim Bennett (Museum of the History of Science, Oxford), Peter de Clercq (Museum Boerhaave, Leiden) and Paolo Brenni (Museo Galileo, Florence), soon to be joined by Marian Fournier (Museum Boerhaave, Leiden) after de Clercq moved to London. From 2003 until 2013, Brenni, Sara Schechner (Collection of Historical Scientific Instruments, Harvard), and Hans Hooijmaijers (Museum Boerhaave) served as SIC officers. Attendance at the annual symposia increased, reaching an average of about eighty to ninety participants per year; at the symposia in Oxford (2000) and Lisbon (2008), attendance topped 150 (see *Table 3.1*).

As can be seen in *Table 3.1*, the SIC during the 2000s held three of its symposia outside Europe, twice in the United States and once in China (during an IUHPST International Congress). In 2001, the International Congress was held in Mexico, but it had "proved impossible" to organise a SIC symposium within

2 See the chapter by Robert Anderson in this volume. According to my rough count, one-third of the SIC symposia papers during the 1980s were presented by scholars working in the UK or Ireland; during the 1990s, this proportion dropped to one-quarter. For characterising the geographical distribution of SIC participants, I consider their home institutions (not personal nationality) and have included Russia as part of Europe.

TABLE 3.1 SIC Symposia from 2000–2009
Locations in red indicate symposia within the International Congress for the History of Science and Technology organised by the IUHPST

Date	Location	Attendance	AGM attendance	Presen-tations	Non-European participants
2000	Oxford	150	80	86	Algeria, India, Canada (2), USA (7)
2001	Stockholm		50	55	Japan, India, Canada, USA (4), Israel
2002	Athens		50	65	Japan, India (2), USA (6), Israel
2003	Newport News	66		40	Japan, Canada (2), USA (13)
2004	Dresden	92		59	India, China, USA (8), Israel
2005	Beijing		25	14	China (2), Syria, USA (3)
2006	Cracow	97	84	65	Israel (2), Canada, USA (8)
2007	Cambridge, MA	100		66	Mexico (3), Israel (2), USA (27)
2008	Lisbon	148		70	Brazil (14), Australia, Japan (3), Canada, USA (13)
2009	Budapest	27	25	39	Brazil (4), Japan, USA (3)

Data retrieved from surviving SIC newsletters, programmes, abstracts, AGM minutes, and (when all else failed) group photos (2003, 2004, 2006, 2009). Symposium attendance at the International Congresses is difficult to reconstruct since attendance at the individual sessions was not recorded. For example, SIC attendance at the 2009 Budapest Congress is documented by two group photos, one before a university building (20 persons), another in the Gellért Baths (15 persons), both in possession of the author.

that event. Rather, the SIC met in Stockholm in October of that year.[3] In its third decade, the SIC remained anchored in Europe. Its impact beyond Europe was scattered and inconsistent; over the decade roughly one-quarter of the SIC presentations were by authors whose home institutions were located outside Europe.

During its third decade, the profile of the SIC within the DHST expanded. The SIC was already recognised as the largest and most active of the twenty-plus content-specific commissions of the DHST due to its annual meetings (often over 100 participants) outside of the quadrennial congresses and its robust communications network (see below). Eftymios Nicolaidis, who hosted the 2002 SIC Symposium in Athens, served as Treasurer of the DHST from 2001–2005 and as Secretary General from 2005–2013. In 2009, SIC President Paolo Brenni was elected a First Vice-President of the parent organisation and served one term until 2013.[4] In those years, the influence of the SIC within the institutional structures of the DHST increased. Nonetheless, the SIC's annual grant from the DHST remained fixed at around $1000, on par with other commissions. And the SIC continued to preserve its distance from the organisational, political, and financial intricacies of the DHST and the IUHPST. Indeed, with its lean institutional framework and an expanding base of museum curators, historians, students, and collectors, the SIC during the 2000s became more than an organisation; it matured into a lively international (if mostly Euro-American) scholarly community, meeting in annual symposia and embracing, as we shall see below, newly emerging digital tools of communication.

2 Symposia Content

The heart of the SIC has always been located in the programmes of its annual symposia. Traditionally these gatherings have featured three types of content: short, formal presentations by individuals (or occasionally by groups) followed by questions and discussion; visits to local museums, collections, libraries, observatories, or historic laboratories; and a symposium dinner. Certain aspects of this content would shift during the SIC's third decade.

---

3 SIC, *Minutes of the 20th Plenary Session of the SIC held at the Royal Academy of Sciences, Stockholm, on 19 October 2001*. Interestingly, at the 2001 AGM in Stockholm, an invitation was received from the Algiers Astronomical Observatory to hold the 2006 SIC symposium in Algeria. But for reasons I have not been able to ascertain, this invitation subsequently disappeared from the documentary record. To date, the SIC has not met in Africa.

4 "Councils of DHS and DHST since 1947," Division for the History of Science and Technology (IUHPST), accessed 15 June 2024, https://dhstweb.org/history-past-councils-of-dhst.

I begin with the presentations. Although symposia attendance grew from 2000 to 2009, the number of presentations (papers and posters) remained roughly constant at about sixty per year. SIC presentations had always featured visual material (initially, 35mm slides and transparencies); with, however, the ubiquitous availability of presentation software by 2000, SIC talks became ever more richly illustrated. Adventuresome participants even presented movies of historic instruments in action or of re-performed historical experiments. However, since most of these presentations remained unpublished and surviving abstracts are scattered and brief, I must limit my analysis to the titles of the presentations.[5]

The majority of topics for the presentations of the 2000s were discrete, focused on a particular instrument or type of instrument, a collection, maker, or experiment, each considered as a fixed historiographical entity. An SIC talk might begin with this sextant in that collection or that telescope, now lost, used for transit of Venus observations in 1761. Hence, symposia during the third decade included titles such as:

- "The Marseille Observatory and its Instruments"
- "The Koenig Sound Analyzer"
- "Nineteenth-century Stellar Photometers"
- "The Technology Museum of Thessaloniki"
- "The Black Body as a Revolutionary Instrument"
- "Diagonal Scales for Measuring Instruments"
- "Eighteenth-century Research with Air Pumps"
- "Women as Designers Rather than Merely Users of Apparatus for Radioactivity Research"
- "Dangerous Experiments for Boys"

In the broader world of the history and philosophy of science and technology, important works on what might be called the "theory" of instrument studies appeared during the first decade of the twenty-first century.[6] For the most

5 Likewise, the personal notes I recorded during sessions I attended are too incomplete and incoherent to use in this chapter.

6 A selection might include Frederic L. Holmes and Trevor H. Levere, eds., *Instruments and Experimentation in the History of Chemistry* (Cambridge, MA: MIT Press, 2000); Lorraine Daston, ed. *Biographies of Scientific Objects* (Chicago: University of Chicago Press, 2000); Bernward Joerges and Terry Shinn, eds., *Instrumentation between Science, State and Industry* (Dordrecht: Kluwer, 2001); Patrick Carroll-Burke, "Tools, Instruments and Engines," *Social Studies of Science*, 21 (2001): 593–625; Hans Radder, ed. *The Philosophy of Scientific Experimentation* (Pittsburgh: University of Pittsburgh Press, 2003); Nelly Oudshoorn and Trevor Pinch, eds., *How Users Matter: The Co-Construction of Users and Technologies* (Cambridge, MA: MIT Press, 2003); Graeme Gooday, *The Morals of Measurement: Accuracy, Irony and Trust in Late Victorian Electrical Practice* (Cambridge: Cambridge University Press, 2004); Davis Baird, *Thing Knowledge: A Philosophy of Scientific Instruments* (Berkeley:

part, such works did not significantly impact the SIC presentations or conversations in that decade. Yet a few SIC titles do point toward interest in theory or self-reflexive historiographical reflection:

- "On Two Symbols of Power"
- "Instruments as Indicators of Social Context"
- "Writing about Instruments in Twentieth-Century India"
- "Instruments as Interfaces"
- "On the Social and Political Origins of the Cabinet of Physics"
- "On Craft Knowledge in Understanding Scientific Instruments"
- "Meaning and Reality of Scientific Instruments"
- "Material and Mathematical Instruments as Extensions and Extensors of the Cognitive Actions of the Subject, Their Epistemological Meaning in the Framework of a New Experience Notion"

Yet such papers remained isolated and apparently gained little traction; I find no sustained discussion of "theory" in the SIC presentations from 2000–2009.

One significant shift did emerge, however, as SIC presentations began to introduce more movement into their analyses, as attention shifted from local to global contexts, from technical functions of the single artefact to the circulation of instruments, experimental practices, commodities and material culture within Europe and within regions beyond Europe. Combining archival studies with hands-on artefact interrogation, SIC members began to consider the roles of instruments in colonialism, global exploration and trade, industrialisation, wars and nation-building.[7] Early examples of this shift appear in SIC presentations examining movement within Europe:

- "British Microscopes in the Zeiss-Jena Collection"
- "Donors Sending Instruments to Nineteenth-Century Greece"
- "Adam Clark and the Chain Bridge in Budapest"
- "Interactions between English-French Makers after Napoleon"
- "St. Petersburg Academy's Prizes and European Inventors"

Increasingly during the early 2000s, SIC scholars widened their gaze to global dynamics:

University of California Press, 2004); Soraya de Chadarevian and Nick Hopwood, eds., *Models: The Third Dimension of Science* (Stanford: Stanford University Press, 2004); Christoph Hoffmann, *Unter Beobachtung; Naturforschung in der Zeit der Sinnesapparate* (Göttingen: Wallstein, 2006); Henning Schmidgen, *Die Helmholtz-Kurven: Auf der Spur der verlorenen Zeit* (Berlin: Merve-Verlag, 2009). See the chapter on historiography by Boris Jardine in this volume.

7 For an early example of such scholarship, see Marie-Noëlle Bourguet, Christian Licoppe, and H. Otto Sibum, eds., *Instruments, Travel and Science: Itineraries of Precision from the Seventeenth to the Twentieth Century* (London: Routledge, 2002), which includes a chapter by Jim Bennett, former President of the SIC.

- "American Purchasing Trips to Europe"
- "Reception of Newtonian Corpuscularism in Japan"
- "Introduction and Diffusion of the Atwood's Machine in Japan"
- "Japanese and Chinese Instruments at the National Maritime Museum"
- "The Polariscope in Berlin and Barbados"
- "The Skinner Box in Brazilian Psychology Laboratories"
- "The Superconducting Supply Chain from Africa to Canada"
- "The 1876 Special Loan Exhibition as a Global Snapshot of Astronomy"[8]
- "Polar Expeditions and Electron Optics at AEG in the 1930s"

The 2006 SIC Symposium in Cracow aimed its theme at cultural exchange: "East and West, the Common European Heritage." The 2007 symposium in Cambridge, Massachusetts included two presenters from Mexico; the 2008 symposium in Lisbon had fourteen presenters from Brazil.[9] These shifts, from the static to the dynamic and from the European to the global, mark one of the most significant developments of the SIC's third decade.

Another novelty in those years was the addition of practical workshops to the programmes of the symposia. Some SIC museum curators had long offered hands-on experiences for the public at their home institutions. But as the SIC participant base widened, especially as more students and young curators and scholars joined the community, planners realised that formal avenues for sharing hands-on experience would be useful additions to the SIC symposia. Hence, the 2007 Symposium in Cambridge, Massachusetts featured a three-hour optional workshop on:

> Tools and Techniques of Traditional Instrument Makers: Participants in this workshop will make and take home a small brass navigator's quadrant using traditional hand tools and techniques for layout, scribing, and punching. Tools and materials will be provided.[10]

8 This is perhaps the first appearance of the term "global" in a SIC title, presented in 2006 in Cracow.

9 Ewa Wyka, Maciej Kluza and Anna Karolina Zawada, eds., *Proceedings of the XXV Scientific Instrument Symposium: East and West, the Common European Heritage* (Cracow: Jagiellonian University Museum, 2006); SIC, *General Schedule of the XXVI SIC Symposium 2007*; SIC, *Program of the XXVII Symposium of the SIC 2008*. In 2012, the SIC went to Rio de Janeiro for its first symposium in Latin America. Marcus Granato and Marta Lourenço, eds., *Scientific Instruments in the History of Science: Studies in the Transfer, Use and Presentation* (Rio de Janeiro: Museum de Astronomie e Ciéncias Afine, 2014).

10 SIC, *General Schedule of the XXVI SIC Symposium 2007*. SIC member Richard Paselk organised and led this workshop, assisted by the author.

This workshop nicely illustrated the challenges of a classic task in instrument-making, viz., dividing a circle on metal without using a protractor, and revealed, to the amusement of some, that not all senior curators or historians are accomplished with hand tools. At the 2008 Symposium in Lisbon a retired officer of the Portuguese Navy, Comandante José Manuel Malhão Pereira, offered a workshop on navigation using wooden replicas of eighteenth-century quadrants, sextants, and backstaffs (*Fig. 3.1*). Once again, participants learned that taking a solar altitude, even while standing on the stationary roof of a university building, is more difficult than might be imagined after reading eighteenth-century navigation handbooks.

Unchanged, in the third decade, was the long-established SIC practice of visiting local collections and museums during the symposia. Extant programmes reveal that a total of fifty-four visits or excursions were offered in the ten symposia from 2000 to 2009. Like the hands-on workshops, these communal experiences enabled students and younger scholars to learn from senior scholars as the group discussed how exhibits work, how to write labels, how given instruments might have been made or used, and, most importantly, pulled on gloves and handled historic instruments in storage rooms. Indeed, in my opinion, the

FIGURE 3.1 SIC navigators measuring angles with the Jacob's staff on the roof of the Escola Politécnica in Lisbon, September 2008
PHOTO: BY AUTHOR

SIC "visits to collections" have contributed, more than any other feature of the organisation, to its growth, intellectual vibrancy, and openness to persons with all levels of experience in dealing with historical objects.

3 SIC Communications and Media

Over its first two decades, the SIC officers communicated to the participants primarily via printed newsletters of two-to-six pages. Twenty-two newsletters, dating from 1980 to 2001, provided information on the annual symposia, minutes of the AGM, museum news, reports of other workshops or conferences concerned with instrument studies, and obituaries of persons prominent in instrument studies. In the second half of the 1990s, the SIC slowly entered the world of digital communication. In 1996, Jim Bennett, Director of the Museum of the History of Science established "an electronic bulletin board for the instruments community" to be hosted by that museum and soon to be called "Rete" (meaning net).[11] In 1998, three SIC members created the first website for the organisation.

By the 2000s, the internet began to powerfully impact instrument studies. First larger museums and soon thereafter smaller collections began featuring not only news of exhibitions but also searchable electronic databases of their collections on newly created web pages. The early SIC website provided links to the *Websters' Instrument Makers Database* and to earlier printed installments of the SIC Bibliography that the Commission had initiated in 1983.[12] The SIC website also linked to the first digital, multi-museum catalogue of early scientific instruments, opened in 2001. Drawing on the collections of the Museo Galileo, Museum Boerhaave, British Museum, and the Museum of the History of Science at Oxford, *Epact* catalogued more than 500 late medieval and Renaissance mathematical instruments, providing photographs, short and longer technical descriptions, and a glossary. Offering search tools for makers, instrument types, and provenance, *Epact* quickly became a model database

11 SIC, *Newsletter*, no. 17 (January 1996): 4.

12 Created by Roderick and Marjorie Webster at the Adler Planetarium in Chicago, this massive database documents more than 15,000 makers whose signatures are found on extant instruments and continues to serve the scholarly community. Twenty-four annual installments of the SIC Bibliography were prepared, first by Gerard L'E. Turner from 1983 to 1995, and then from 1995 to 2004 by librarians at the Museum Boerhaave. See *The Webster Signature Database*, accessed 15 June 2024, http://historydb.adlerplanetarium.org/signatures/; and "Bibliography," Scientific Instrument Commission, accessed 15 June 2024, https://scientific-instrument-commission.org/resources/bibliography/.

for museum curators seeking to create simple online catalogues of scientific instrument collections.[13]

Given the SIC's increased web and publication profile (see below), members at the 2006 AGM in Cracow asked whether the organisation needed a distinctive graphic logo.[14] The SIC Board launched a design contest and members started submitting ideas. A proposal by the well-known Italian graphic designer Valentino Szemere and SIC President Paolo Brenni eventually won the day. Featuring the "meter" as an iconic scientific instrument, the new logo gave a nineteenth-century flavour to the Commission and reflected a change in its centre of gravity from astrolabes and the Museum of the History of Science to more recent, industrially produced instruments found in hundreds of regional, university and school collections across the globe.[15] On the other hand, the three typefaces of the logo span the centuries: the "I" recalls the Roman style popularised by the Venetian printer Aldus Manutius around 1500; the serif font of the "S" with its strong contrast between thick and thin strokes looks like Giambattista Bodoni's late eighteenth-century typeface; and the sans-serif "C" is not unlike Paul Renner's Bauhaus-inspired geometric Futura font of the 1920s.[16] This clever, multivalent logo first appeared in 2009 on the inaugural SIC volume published by Brill (*Fig. 3.2*).[17]

FIGURE 3.2
SIC logo created by Valentino Szemere and Paolo Brenni, circa 2007
SCIENTIFIC INSTRUMENT COMMISSION

13 *Epact* continues to be available, more than twenty years after its creation. See *Epact*, accessed 15 June 2024, https://www.mhs.ox.ac.uk/epact/.

14 SIC, *Minutes of the XXV International Scientific Instrument Symposium, 10–14 September 2006*, p. 3.

15 For a useful typography of scientific instruments as "meters," "graphs," "scopes" or "chambers," see Carroll-Burke, "Tools, Instruments and Engines."

16 I thank Sara Schechner, then SIC Secretary who participated in the logo selection, for contributing to this interpretation of the Szemere/Brenni design.

17 Giorgio Strano *et al.*, eds., *European Collections of Scientific Instruments, 1550–1750*, Scientific Instruments and Collections, 1 (Leiden: Brill, 2009), rear cover.

During its third decade, the SIC website and the Rete on-line listserv significantly expanded the Commission's accessibility to new audiences. Increased numbers of PhD students, postdocs, and young curators began participating in the symposia. Joint papers and collaborative research projects became ever more common. Wide-ranging public discussions on Rete prompted curators to reconsider instruments in their collections, historians to move beyond paper sources, and the SIC community to spirited (even competitive!) debates about the identities of so-called "mystery objects." Perhaps not surprisingly, the digital age helped to widen and enliven the SIC.

4 Scholarly Publications

The third decade also saw the SIC regularise and expand its publishing programme. In its first two decades, the Commission had sporadically published seven volumes of proceedings, issued by various presses and irregularly edited.[18] In 2000 Paolo Galluzzi, Director of the Instituto e Museo di Storia della Scienza in Florence, wrote to the SIC, offering to devote one section per year of that institution's scholarly journal, *Nuncius*, to instrument studies. President Jim Bennett noted: "[T]he fact that Nuncius has a global distribution in academic libraries will ensure that SIC-material published in this journal reaches a wide readership." At the 2001 AGM in Stockholm, members agreed to accept Galluzzi's proposal.[19] Three SIC-related sections would appear in *Nuncius*. In 2001, the Parisian instrument dealer Anthony J. Turner edited eight papers delivered at a session of the 2000 Oxford symposium on *Archives of Scholars, Collectors and Dealers*. In 2002–2003, *Nuncius* included twelve papers edited by nineteenth-century historians Charlotte Bigg and Klaus Staubermann in a section entitled *Spectroscope Histories*, researches presented at a 2001 conference at the Deutsches Museum, Munich, organised under the auspices of the SIC. And in 2004, SIC President Jim Bennett edited five papers from the 2004 Athens and 2005 Newport News symposia on *Instrument Case Studies, 1850–1950*.

Scattered within the pages of a comprehensive journal devoted to the history of science, these three collections probably did little to enhance the scholarly

18 See the chapter by Sara Schechner in this volume.

19 SIC, *Minutes of the 19th Plenary Session of the SIC, Oxford, 7 September 2000*, p. 2; SIC, *Minutes of the 20th Plenary Session of the SIC held at the Royal Academy of Sciences, Stockholm, on 19 October 2001*, p. 1. Founded in 1986, *Nuncius* by 2000 had become the leading Italian journal for history of science and Galilean studies. The SIC sections were comprised of papers arranged sequentially in the journal, but not as separate issues.

profile of the SIC among wider audiences.[20] At the 2004 AGM in Dresden, the *Nuncius* arrangement came under scrutiny. A new editor, seeking to slim down and focus *Nuncius*, had offered the SIC its own separate issue but in a different series (*Bibliotheca di Nuncius*). Participants in Dresden called for the creation of a more formalised SIC editorial board; some wondered how many libraries or individuals were subscribing to the Italian journal.[21] At the 2005 AGM in Beijing, the SIC officers reported that they had accepted the offer to publish in the *Bibliotheca di Nuncius* and had created the SIC's first editorial board.[22] But over the next several years, negotiations with *Nuncius's* publisher hit various snags and finally in 2008 the Editorial Board concluded an agreement with Brill in Leiden to publish one or more hard-cover volumes per year in a new series entitled Scientific Instruments and Collections (SICo).[23] Giorgio Strano of the Instituto e Museo di Storia della Scienza became General Editor of SICo. The first volume in the series, *European Collections of Scientific Instruments, 1550–1750*, appeared in 2009 and published papers presented at the 2004 Dresden Symposium. SICo vol. 2, *From Earth-Bound to Satellite: Telescopes, Skills and Networks*, was issued in 2012 with papers from the 2007 Cambridge, 2008 Lisbon and 2009 Budapest Symposia. SICo 5, *How Scientific Instruments Have Changed Hands*, was published in 2017 and presented additional papers from 2009 Budapest. (Volumes 3 and 4 drew from meetings in the SIC's fourth decade and are not here examined.)

Two additional volumes were assembled from papers given at conferences sponsored by the SIC in its third decade. In 2007, an international workshop on Historic Scientific Instrument Collections in Universities (SICU2) was held at the University of Mississippi. Steven Turner edited eleven of these papers in a special issue of *Rittenhouse: Journal of the Scientific Instrument Enterprise*. An unrelated conference, also dealing with instruments in university contexts, was held in 2009 at the University of Regensburg. Fourteen of its papers,

20 In 2001–2004, *Nuncius* was published as two massive issues totalling 8–900 pages per year; the three SIC sections are comprised of 90, 220 and 50 pages.

21 SIC, *Minutes of the XXIII International Scientific Instrument Symposium, Dresden, 6–11 September 2004*, pp. 3–4.

22 SIC, *Minutes of the XXIV International Scientific Instrument Symposium, Beijing, 27–28 July 2005*, p. 2. The editors included Giorgio Strano and Maria Miniati, Instituto e Museo di Storia della Scienza; Alison Morrison-Low, National Museums Scotland; and Stephen Johnston, Museum of the History of Science, Oxford.

23 SIC, *Minutes of the XXVII International Scientific Instrument Symposium, Lisbon, 15–21 September 2008*, p. 2.

edited by Peter Heering and Roland Wittje, appeared in *Learning by Doing: Experiments and Instruments in the History of Science Teaching*.[24]

In sum, from the SIC-related meetings held between 2000 and 2009, a total of ninety-two papers by eighty authors were published in eight volumes or journal sections. No other Commission of the DHST matched this record of scholarly publication during that decade. No other Commission has sustained a publication series such as SICo. With its efficient organisational structure, well-attended annual symposia, robust digital communication, and well-edited publication series, the SIC by the end of its third decade had become perhaps the leading international hub for scholarship on, and curation of, historic scientific instruments.

5 Wider Impacts

Assessing the impact of the SIC beyond its own community of symposium participants or Rete users would require a more comprehensive survey than can be attempted in this short chapter. But we can offer a few samples of the broader responses to SIC scholarship.

For example, the three SICo volumes devoted to papers delivered in the 2000s were reviewed in leading history of science journals. The reviewers, from both within and without the SIC, consistently praised the SIC scholarship for showing the heterogeneity of contexts in which instruments have acquired historical meaning. And they underscored the innovative use of newly available digital tools, chiefly databases of ephemeral, printed materials such as trade catalogues and newspapers, for instrument studies. But none of the reviewers commented specifically on the role of the SIC in stimulating novel or particular approaches to the study of the material culture of science.[25]

A digital database of publications in the history of science, compiled by the discipline's flagship journal *Isis*, offers another means to explore the impact of the SIC. Although biased toward English-language publications, the *Isis Cumulative Bibliography* does include many non-English journals and monographs and generates, if not an absolute, at least a relative image of trends over

24 *Rittenhouse* 22 (2008): 1–224; Peter Heering and Roland Wittje, eds. *Learning by Doing: Experiments and Instruments in the History of Science Teaching* (Stuttgart: Franz Steiner Verlag, 2011).

25 SICo1 was reviewed in *Annals of Science* 70 (2013): 423–425; SICo2 in *Annals of Science* 71 (2014): 579–581, *Isis* 104 (2013): 387, and *Renaissance Quarterly* 65 (2013): 1232–1233; SICo5 in *Annals of Science* 75 (2018): 60–61, Isis *110* (2019): 410–411, and *British Journal for the History of Science* 50 (2017): 149–150.

time.[26] An N-gram of records tagged with the subject "instruments" (excluding musical and medical), as a percentage of all records in the database, shows a sharp increase around the year 2000 (*Fig. 3.3*). From 1970–2000, only 0.3 percent of all history of science publications considered instruments; from 2000–2020, that percentage hovered around 1.0 percent. More research would be required to explain this increase; but surely the expansion of the SIC, both in numbers of participants and in geographical reach, was involved.[27]

If we ask the Isis CB database for the most prolific authors of works on "instruments" we find many long-time SIC participants on the list (*Table 3.2*). Judging by such rough, albeit quantitative measures, we must conclude that the SIC community since 2000 has featured prominently in the scholarly

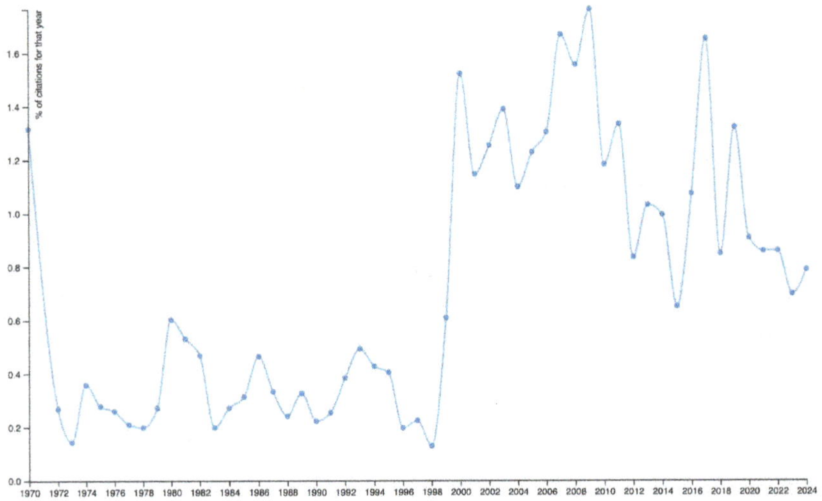

FIGURE 3.3 Publications on "instruments" (excluding medical and musical) as a percentage of all records in the *Isis Cumulative Bibliography* database, 1970–2024
CHART: BY AUTHOR

26 The ontology of such databases is the "published work", which might be a book review, a journal article, or a multi-volume monograph; hence they cannot measure scholarly impact or the quantity of scholarly production. But they can point to change over time in what we might call "publishing activity." This comment is also relevant to *Table 3.2* below.

27 We might guess that instrument studies after 2000 were also stimulated by the rise of the internet, digital catalogues of museum collections, digital scans of large trade catalogue collections, and the "turn", more generally in the history of science, to "practice" and the "material culture of science."

TABLE 3.2 Isis CB database, rank-ordered most prolific authors of publications on "instruments," 1970–2024
Frequent SIC participants in bold and the number of titles in parentheses

Brenni, Paolo (53)	**Bedini, Silvio A. (21)**
King, David A. (47)	**Mills, Allan A. (20)**
Warner, Deborah J. (41)	Greenslade, Thomas (19)
Turner, Gerard (41)	**Zuidervaart, Huibert (16)**
Turner, Anthony J. (40)	Schaffer, Simon (16)
Morrison-Low, Alison (38)	**Clercq, Peter de (16)**
Bennett, Jim (38)	**Miniati, Mara (15)**
Orchiston, Wayne (22)	Van Helden, Albert (15)
McConnell, Anita (22)	**Hackmann, Willem D. (15)**

investigation of historic scientific instruments, especially in English-language publications.[28]

6 Conclusions

As might be expected, the SIC during its third decade experienced both continuity and change. Attendance at its annual symposia remained roughly constant from 2000 to 2009, averaging around 80–90 per year, with most participants coming from Europe or North America. Most of the presentations focused on individual instruments or collections, with hands-on investigation central to much of the research. Visits to collections and handling of instruments in stores featured prominently in all the symposia. And the same SIC Board served for most of the decade. For many participants, the SIC of the 2000s carried on traditions established in the first two decades of the organisation.

28 For a recent study of articles on "objects" (a much more capacious category than "instruments") in two flagship English-language journals for the history of science and technology (*Isis*, *Technology and Culture*) from 2007–2017, see S.J.M.M. Alberti *et al.*, "The Immaterial Turn? How Historians of Science and Technology use Material Culture," in *Understanding Use: Objects in Museums of Science and Technology*, ed. by Tim Boon *et al.* (Washington, DC: Smithsonian Scholarly Press, 2024), 210–226. Finding only 11 articles (out of 343) dealing with "museum objects", these authors conclude that historians of science and technology remain committed primarily to paper rather than 3-D sources, a conclusion somewhat orthogonal to that implicit in my *Fig. 3.3*.

Yet significant change also occurred during the 2000s. For many SIC scholars, the objects of investigation became increasingly mobile, as instruments were recognised as more than devices with discrete, technical functions, i.e., as courtly gifts, mechanisms for colonial projects, global commodities, or sites for Creole combinations of local and distant cultural practices. Internal SIC communications shifted from paper newsletters to digital listservs and websites. The SIC created an Editorial Board for its scholarly publications, with an edited Brill series replacing the earlier occasional proceedings. And in the discipline of the history of science more generally, the number of publications on "instruments" jumped dramatically. Indeed, we might conclude that in the 2000s scholarly activity on scientific instruments quickened as well as broadened.

And these trends surely were nourished by the culture of the SIC as the Commission welcomed students and younger scholars, curators from smaller as well as national museums, dealers and collectors. The Commission championed a climate of informality, in which members generously shared knowledge of collections, photographs, and sources, and provoked increasingly vigorous discussion of research methods and historiography. Developing this culture as an intellectual and social community, the SIC emerged from its third decade as a mature professional organisation and a professional identity for persons from heterogeneous geographical places, professional backgrounds, institutions, and interests … and for historians of science who happen to inherit collections of historic scientific instruments.

Bibliography

Alberti, S.J.M.M. *et al.* "The Immaterial Turn? How Historians of Science and Technology Use Material Culture." Pp. 210–226 in *Understanding Use: Objects in Museums of Science and Technology*, edited by Tim Boon *et al.* Washington, D.C.: Smithsonian Scholarly Press, 2024.

Baird, Davis. *Thing Knowledge: A Philosophy of Scientific Instruments*. Berkeley: University of California Press, 2004.

Becker, Barbara J. [Review of SIC02]. *Annals of Science* 71 (2014): 579–580.

Boner, Patrick J. [Review of SIC02]. *Renaissance Quarterly* 65 (2013): 1232–1233.

Bourguet, Marie-Noëlle, Christian Licoppe, and H. Otto Sibum, eds. *Instruments, Travel and Science: Itineraries of Precision from the Seventeenth to the Twentieth Century*. London: Routledge, 2002.

Carroll-Burke, Patrick. "Tools, Instruments and Engines." *Social Studies of Science* 21 (2001): 593–625.

Chadarevian, Soraya de, and Nick Hopwood, eds. *Models: The Third Dimension of Science*. Stanford: Stanford University Press, 2004.

Daston, Lorraine, ed. *Biographies of Scientific Objects*. Chicago: University of Chicago Press, 2000.

Dunn, Richard. [Review of SIC05]. *British Journal for the History of Science* 50 (2017): 149–150.

Gooday, Graeme. *The Morals of Measurement: Accuracy, Irony and Trust in Late Victorian Electrical Practice*. Cambridge: Cambridge University Press, 2004.

Granato, Marcus, and Marta Lourenço, eds. *Scientific Instruments in the History of Science: Studies in the Transfer, Use and Presentation*. Rio de Janeiro: Museum de Astronomie e Ciências Afine, 2014.

Heering, Peter and Roland Wittje, eds. *Learning by Doing: Experiments and Instruments in the History of Science Teaching*. Stuttgart: Steiner, 2011.

Hoffmann, Christoph. *Unter Beobachtung; Naturforschung in der Zeit der Sinnesapparate*. Göttingen: Wallstein, 2006.

Holmes, Frederic L., and Trevor H. Levere, eds. *Instruments and Experimentation in the History of Chemistry*. Cambridge, MA: MIT Press, 2000.

Jardine, Boris. [Review of SIC05]. *Annals of Science* 75 (2018): 60–61.

Joerges, Bernward, and Terry Shinn, eds. *Instrumentation between Science, State and Industry*. Dordrecht: Kluwer, 2001.

Kremer, Richard L. [Review of SIC01]. *Annals of Science* 70 (2013): 423–425.

Oudshoorn, Nelly, and Trevor Pinch, eds. *How Users Matter: The Co-Construction of Users and Technologies*. Cambridge, MA: MIT Press, 2003.

Radder, Hans, ed. *The Philosophy of Scientific Experimentation*. Pittsburgh: University of Pittsburgh Press, 2003.

Schmidgen, Henning. *Die Helmholtz-Kurven: Auf der Spur der verlorenen Zeit*. Berlin: Merve-Verlag, 2009.

Scientific Instrument Commission. *Minutes of the 19th Plenary Session of the SIC, Oxford, 7 September 2000*.

Scientific Instrument Commission. *Minutes of the 20th Plenary Session of the SIC held at the Royal Academy of Sciences, Stockholm, on 19 October 2001*.

Scientific Instrument Commission. *Minutes of the XXIII International Scientific Instrument Symposium, Dresden, 6–11 September 2004*.

Scientific Instrument Commission. *Minutes of the XXIV International Scientific Instrument Symposium, Beijing, 27–28 July 2005*.

Scientific Instrument Commission. *Minutes of the XXV International Scientific Instrument Symposium, 10–14 September 2006*.

Scientific Instrument Commission. *Minutes of the XXVII International Scientific Instrument Symposium, Lisbon, 15–21 September 2008*.

Scientific Instrument Commission. *Newsletter*, nos. 1–22 (1980–2001).

Scientific Instrument Commission. [Programme 1977] *International Symposium on Scientific Instruments, Greenwich 1977, Summaries of Papers.*

Scientific Instrument Commission. [Programme 1982] *2nd Greenwich Scientific Instruments Symposium, 1982, Summaries of Papers.*

Scientific Instrument Commission. [Programme 2007] *General Schedule of the XXVI SIC Symposium 2007.*

Scientific Instrument Commission. [Programme 2008] *Program of the XXVII Symposium of the SIC 2008.*

Strano, Giorgio, Stephen Johnston, Mara Miniati, and A.D. Morrison-Low, eds. *European Collections of Scientific Instruments, 1550–1750*. Scientific Instruments and Collections, 1. Leiden: Brill, 2009.

Turner, Steven, ed. "Special issue, SICU2." *Rittenhouse* 22 (2008): 1–224.

Warner, Deborah. [Review of SIC02]. *Isis* 104 (2013): 387.

Weiss, Martin P.M. [Review of SIC05]. *Isis* 110 (2019): 410–411.

Wyka, Ewa, Maciej Kluza and Anna Karolina Zawada, eds. *Proceedings of the XXV Scientific Instrument Symposium: East and West, the Common European Heritage.* Cracow: Jagiellonian University Museum, 2006.

CHAPTER 4

One Object, Many Worlds: The Scientific Instrument Commission in the 2010s

David Pantalony

In February 2013, I tweeted this image (*Fig. 4.1*) to Chris Hadfield, the Canadian astronaut who was commander of the International Space Station at that time.[1] Hadfield had made the news for his continuous flow of images and videos from space, many of them including dramatic images of Earth. I posted this image to suggest that my view of Earth, a 1960s Rand McNally Globe in our (now former) museum storage rooms, might be as beautiful and sublime as his from a celestial perch. This particular globe has multiple histories to tell, from its many siblings found at institutions across North America to its iconic status during the Cold War and Space Age, its production (much of it by hand using 1960s techniques and materials), its use of Soviet satellite data to complete some of the mapping, and its initial exhibition history dating back to the opening of our museum in 1967.[2] The Rand McNally Globe is a human-made Earth as complex and vast in narrative possibilities as the one seen from space by Hadfield. This pluralistic view of Earth connects to another quote posted on Twitter at the same time. In a March 2013 lecture "Understanding (through) Things" at the University of Cambridge, historian Simon Schaffer commented on the British Museum's project, "The History of the World Through 100 Objects," asserting that he would rather talk about the history of 100 worlds in one object.[3]

1 David Pantalony (@SciTechCurator), "@Cmdr_Hadfield Similar to Your View? Draped Earth from a Balcony in Storage @SciTechMuseum. 1967 RandMcNally Globe," Twitter, 22 February 2013, http://T.Co/2lJMV5fNB4.

2 Simon Whitehouse, "The Rand-McNally Geo-Physical Globes: How the Earth Was Constructed to Display Post-War Sciences" (MA thesis, Florida State University, 2015).

3 Simon Schaffer, "Understanding (through) Things," lecture at "The Location of Knowledge: A Mellon CDI Conference," Centre for Research in the Arts, Social Sciences and Humanities (CRASSH), University of Cambridge, 8 March 2013; Mirjam Brusius, "100 Histories of the World in One Object: Itineraries, Islam and Indices," *Material Religion* 15, no. 2 (2019): 243–245; Rebekah Higgitt (@beckyfh), "Schaffer Says Things Become Objects When They Object & That There Are 100 Worlds in One Object. #locknowl Ht @SpoonsOnTrays @atinybitwaring," Twitter, 8 March 2013, https://twitter.com/beckyfh/status/309963980976316416.

 | DOI:10.1163/9789004737587_006

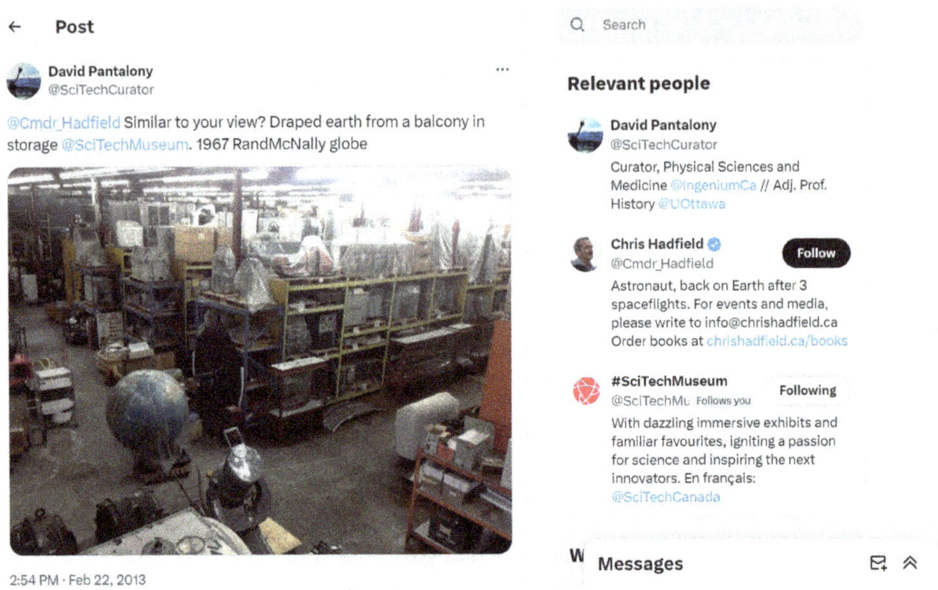

FIGURE 4.1 Tweet of a Rand McNally globe in storage at the Canada Science and Technology Museum, from the curator David Pantalony to astronaut Chris Hadfield, then in orbit on the International Space Station
SCREENSHOT OF TWITTER FEED, 22 FEBRUARY 2013

Schaffer's remark nicely captures the emerging practices and conceptions in our field during the decade of the 2010s. First, I am explicitly mentioning Twitter (now X), the social media platform that had a large impact on our community in the 2010s; for a time, it exemplified a new openness to serendipitous, interdisciplinary access to our collections and research. The SIC opened Twitter and Facebook accounts in 2019 (combined they had over 1800 followers); but during the 2010s it was the individual community members who made active use of social media for sharing collection discoveries, mystery instruments, conference proceedings and for connecting with other disciplines. Paolo Brenni's Facebook account, for example, provides a rich trove of history of the SIC at this time.[4]

At a deeper thematic level, going over past SIC meetings, publications, exhibits, collecting practices, and conference excursions I kept seeing this idea of "One Object, Many Worlds" being expressed in multiple forms and practices. In the 2010s, SIC members increasingly approached hands-on close

4 Paolo Brenni's Facebook pages, Facebook, accessed 26 June 2024, https://www.facebook.com/paolo.brenni.

examination of objects with new kinds of questions. The growing digitisation of archives and collections sparked an increase in creative mixing of new collection-based methods, disciplines, and sources. Paradoxically, even with this emphasis on the digital realm, the objects themselves became the centre of histories with rich connections to people, associated materiality, values, biographical time, and a variety of spaces and geographies. In short, we started to see scientific instruments as vessels in a sea of relations; and we imagined the "vessels" themselves coming to life as complex objects with seemingly infinite research possibilities. This chapter on the history of the SIC during the 2010s explores the main elements behind this shift, the implications for issues like "what is a scientific instrument" or "making instruments count," and what this tells us about science, instruments, and society.[5]

1 Seeking New Sources, Themes, and Contexts for Instruments

The "Many Worlds" approach derives from a constant search for new contexts and themes. During the 2010s, organisers and hosts of the annual SIC symposia prodded the community to seek new contextual and methodological currents. The SIC Florence symposium in 2010 focused on historicising the display of scientific instruments, revealing additional functions for instruments beyond the simply technical. "We encourage you to submit papers about how instruments have been presented in scientific collections, museums and permanent and temporary exhibitions in past and more recent time."[6] At Kassel 2011, there was exploration of instruments in relation to corresponding sources, images, and text, as well as a special focus on the history of cabinets of physics as a source in their own right.[7] In Rio 2012, the theme addressed instruments as knowledge-bearing objects in and out of different contexts – "use, trade, transfer of

5 Deborah Jean Warner, "What Is a Scientific Instrument, When Did It Become One, and Why?" *British Journal for the History of Science* 23, no. 1 (1990): 83–93; R.G.W. Anderson, J.A. Bennett, and W.F. Ryan, eds., *Making Instruments Count: Essays on Historical Scientific Instruments Presented to Gerard L'Estrange Turner* (Aldershot, Hampshire: Variorum, 1993); Liba Taub, "What Is a Scientific Instrument, Now?," *Journal of the History of Collections* 31, no. 3 (2019): 453–467.

6 "XXIX Scientific Instrument Symposium," SIC, accessed 11 October 2024, https://scientific-instrument-commission.org/event/xxix-scientific-instrument-symposium/.

7 "XXX Scientific Instrument Symposium," SIC, accessed 11 October 2024, https://scientific-instrument-commission.org/event/xxx-scientific-instrument-symposium/.

instruments between Europe and the Americas."[8] The International Congress at Manchester 2013 had an overall theme of "Knowledge at Work," ideal for considering scientific instruments within the context of a historic industrial city.[9] SIC Tartu 2014 had a theme that best encapsulated many themes in the 2010s with their call for "new views on old instruments," i.e., new analytic methods, ways of public engagement, political aims, challenges posed by post-war instruments, emerging networks, and previously unknown sources.[10] In Turin 2015, the SIC community explored "Instruments in Conflict," whereby participants considered conflict beyond the obvious notion of war: "We invite submissions of papers that identify and discuss approaches to the study of scientific instruments and their collections which focus on conflicts around the creation, use, interpretation, preservation and study of instruments," wrote the organisers. "These might include conflicting communities of scholars, users and makers, scientific instruments for and in armed conflict, conflicts of space and resources in the treatment of twentieth-century instruments, and the possible cross-fire from attempting interdisciplinary perspectives."[11]

The SIC community launched an ambitious travel schedule in the second half of the 2010s, a schedule that itself revealed the seeking of new contexts for collection-based research. The organisers for SIC 2016 in Istanbul used the theme of "Instruments Between East and West" to encourage dialogue not only between Western and Eastern histories of scientific instruments but also on other "East-West relationships."[12] The 2016 conference turned out to be a watershed moment for the SIC, leading to an increase in active participation from scholars from the Middle East. Back to Rio for 2017, SIC participants explored the situatedness of knowledge between the local and global.[13] In Leiden 2018, the organisers focused on a theme of instruments and the "Empire of Man

8 "XXXI Scientific Instrument Symposium," SIC, accessed 11 October 2024, https://scientific-instrument-commission.org/event/xxxi-scientific-instrument-symposium/.

9 "XXXII Scientific Instrument Symposium," SIC, accessed 11 October 2024, https://scientific-instrument-commission.org/event/xxxii-scientific-instrument-symposium/.

10 "XXXIII Scientific Instrument Symposium," SIC accessed 11 October 2024, https://scientific-instrument-commission.org/event/xxxiii-scientific-instrument-symposium/.

11 "XXXIV Scientific Instrument Symposium," SIC, accessed 11 October 2024, https://scientific-instrument-commission.org/event/xxxiv-scientific-instrument-symposium/.

12 "XXXV Scientific Instrument Symposium.," SIC, accessed 11 October 2024, https://scientific-instrument-commission.org/event/xxxv-scientific-instrument-symposium/. SIC XXV Symposium in Kracow (2006) also used the theme "East and West" but with the subtitle "The Common European Heritage" and focus on Europe.

13 "XXXVI SIC Symposium," SIC, accessed 11 October 2024, https://scientific-instrument-commission.org/event/xxxvi-sic-symposium/. The 2017 International Congress of the History and Philosophy of Science and Technology was held in Rio de Janeiro.

over Things" (inspired by Francis Bacon) and asked participants to consider "How have instruments of the 'arts and sciences' been deployed in efforts to 'command nature' for social, economic, political or personal purposes?"[14] In Havana 2019, participants explored the theme of "Instruments at Crossroads," and the ways in which instrument creation, adaptation, and use reflects multiple conditions in their surroundings, from climatic to political.[15]

Instrument scholars sought and discovered new kinds of sources, which in turn provoked new questions and contexts. During the 2010s, this was illustrated by the SIC's Brill series, Scientific Instruments and Collections, edited by Giorgio Strano with an editorial board of Mara Miniati, Sara J. Schechner, Stephen Johnston, and Alison Morrison-Low. In the second volume from 2012, historians delved into a creative variety of new sources related to the earliest telescope making, from shopping lists to clients to skills to networks.[16] In the next volume, historians and curators used cabinets of experimental philosophy as a collective material resource for understanding the culture of physics in eighteenth-century Europe.[17] In a fourth volume, instrument historians probed public display contexts to reinterpret the function of instruments in history.[18] Authors in this volume used historic museum images, institutional records, and local instrument histories to investigate these display narratives.

In the preface of the fifth volume published in 2016, *How Scientific Instruments Have Changed Hands*, the editors explicitly pointed to a trend of researchers, fuelled by the Internet, seeking alternate sources for instrument research.[19] I quote their remarks in full as they illustrate specific researchers creatively accessing new online tools:

> These papers show that new sources provide interesting insights into their subjects. The rise of the accessibility of online material has enabled modern researchers to find more detail and often new case studies to

14 "XXXVII Scientific Instrument Symposium," SIC, accessed 11 October 2024, https://scientific-instrument-commission.org/event/xxxvii-scientific-instrument-symposium/.

15 "XXXVIII Scientific Instrument Symposium," SIC, accessed 11 October 2024, https://scientific-instrument-commission.org/event/xxxviii-scientific-instrument-symposium/.

16 A.D. Morrison-Low *et al.*, *From Earth-Bound to Satellite: Telescopes, Skills and Networks* (Leiden: Brill, 2012).

17 Jim Bennett, Sofia Talas, and Giorgio Strano, eds., *Cabinets of Experimental Philosophy in Eighteenth-Century Europe* (Leiden: Brill, 2013).

18 Silke Ackermann, Richard L. Kremer, and Mara Miniati, eds., *Scientific Instruments on Display* (Leiden: Brill, 2014).

19 A.D. Morrison-Low, Sara J. Schechner, and Paolo Brenni, eds., *How Scientific Instruments Have Changed Hands* (Leiden: Brill, 2016).

> uncover the past. As well as using conventional archives in the Guildhall Library, [Alexi] Baker has used online Old Bailey court records and contemporary newspapers; [Joshua] Nall and [Liba] Taub have consulted a large archive of trade literature, now made available online. [Alison] Morrison-Low accessed newspaper archives consulted through the National Library of Scotland's website; [Paolo] Brenni looked at online sources for trade catalogues, making the point that these are indispensable tools for research into the economics of the trade. [Gloria] Clifton has been compiling a database, building on and supplementing her 1995 publication; [Sara] Schechner has been able to access instrument collections and archaeological sites for comparative material, as well as making extensive use of digitised collections of early maps in order to determine the locations of forts and trading posts, and [Richard] Kremer also used online early American newspaper archives. All these recently-available literary and visual sources often supply information that the instruments themselves cannot: yet, scrutinising the instruments – as Schechner in particular shows here – enables the scholar to pull together intriguing and useful pictures of the past. All the papers presented here are – of course – based on the primary source of the artefacts themselves. Although this volume illuminates only some aspects of the scientific instrument trade, we hope that nevertheless they reveal the complexities and the multifaceted nature of this intriguing subject.[20]

Original themes inspired the integration of old sources in new ways. Long neglected for numerous historiographic reasons, astrology emerged as a theme in the 2010s for the SIC community to interrogate more fully in its own right. In order to explore the history of astrology and instruments, the contributors to the sixth volume brought together printed and manuscript texts, iconography, and the instruments themselves to uncover a "range of astrological contexts" within this history.[21]

The 2016 SIC conference in Istanbul inspired the exploration of the broad theme for the seventh volume, *Scientific Instruments between East and West* (2019).[22] For the SIC community, this volume introduced novel sources and contexts for familiar instruments, as well as instruments previously

20 Morrison-Low, Schechner, and Brenni, "Preliminary Material," *How Scientific Instruments Have Changed Hands*, viii.

21 Richard Dunn, Silke Ackermann, and Giorgio Strano, eds., *Heaven and Earth United: Instruments in Astrological Contexts* (Leiden: Brill, 2018).

22 Neil Brown, Silke Ackermann, and Feza Günergun, eds., *Scientific Instruments between East and West*, (Leiden: Brill, 2019)

outside the attention of Western scholars. Meltem Kocaman, for example, used local sources related to retail of instruments to look at scientific culture in nineteenth-century Constantinople.[23] Panagiotis Lazos studied the educational context of French scientific instruments at a Greek School in Istanbul. These teaching instruments are found in collections around the world, but Lazos was investigating their local context of use through education.[24] Janet Laidla studied seismographs made in Tartu, and their network of practice at various sites around the world. The same instrument found use in multiple geopolitical contexts.[25]

The eighth volume published in 2021 derived from the continued innovative work during the 2010s in contemporary science classroom and museum spaces, which served as resources for understanding historical context in science education. From drama to YouTube videos to self-directed learning with objects, researchers and students used hands-on experiences to ask new questions, and above all, to see the humanity of scientific instruments in didactic settings.[26]

2 Place, Materials, and Biography as Context

The movement of instruments attracted new attention from instrument scholars in the 2010s, as demonstrated by several examples. In order to understand the history of an instrument from the perspective of making, practice, and surroundings, Edward J. Gillin took an original, still functional Fox dipping needle on an expedition to southern African countries to take magnetic measurements.[27] By doing this, he hoped to replicate not only practice but also the nineteenth-century "magnetic crusades" in a larger sense. Just off the coast of Namibia, aboard a ship, Gillin addressed his YouTube audience with descriptions of how the dip circle worked and the practical challenges as well

23 Meltem Kocaman, "Scientific Instrument Retailers in Istanbul in the Nineteenth Century, and Verdoux's Optical Shop," in *Scientific Instruments between East and West*, ed. Neil Brown, Silke Ackermann, and Feza Günergun (Leiden: Brill, 2019), 240–256.

24 Panagiotis Lazos, George N. Vlahakis, and Constantine Skordoulis, "Instruments and Laboratories in the Schools of the Greek Community of Istanbul, 1850–1960," in *Scientific Instruments between East and West*, 168–186.

25 Janet Laidla, "Hugo Masing's Golitsyn-Vilip Seismographs: From Tartu to Five Continents," in *Scientific Instruments between East and West*, 154–167.

26 Elizabeth Cavicchi and Peter Heering, eds., *Historical Scientific Instruments in Contemporary Education* (Leiden: Brill, 2021).

27 Edward J. Gillin, "The Instruments of Expeditionary Science and the Reworking of Nineteenth-Century Magnetic Experiment," *Notes and Records: The Royal Society Journal of the History of Science* 76, no. 3 (2022): 565–592.

as broader context of its use in the empire-building voyages of the nineteenth century.[28]

In 2017, Giorgio Strano, curator at the Museo Galileo in Florence, brought a Divini telescope (*Fig. 4.2*) to Ottawa for display at the newly opened Canada Science and Technology Museum. The telescope has an incredibly complex history of materials, manufacturing, style, modifications, and attributions. Dating from circa 1665, it is attributed to Eustachio Divini (1610–1685) but was associated with other makers such as Evangelista Torricelli (1608–1647) and Jacopo Mariani (seventeenth century) during other periods in its life.[29]

FIGURE 4.2 Telescope attributed to Eustachio Divini, Rome, circa 1665, on loan from the Museo Galileo, Florence to the Canada Science and Technology Museum
PHOTO: MICHEL LABRECQUE, 2017

28 Edward J. Gillin, "Weighing the Earth's Magnetic Field with Fox's Dip Circle," posted 30 January 2020, by James Clark Ross, YouTube, 4:20, https://www.youtube.com/watch?v=Xt9dyEEy9oc.

29 Museo Galileo, "Cannocchiale Terrestre," inv. no. 2555, accessed 4 April 2024, https://opac.museogalileo.it/imss/resource?uri=16353&v=l&dcnr=3. "The diaphragm securing the objective has the inscription 'Hippolytvs [?] fecit ANN 1666.' The missing surname cannot be that of Francini, since he died in 1653. In a Medici inventory of the late seventeenth century, the instrument is described as a telescope by Evangelista Torricelli, complete with objective and three double eyepiece lenses, made, like the tube, by Eustachio Divini. It appears, then, that this instrument was built by Divini around the 1660s and that the original objective was replaced with a Torricelli lens. The latter was replaced, in the second half of the eighteenth century, with an objective made by Jacopo Mariani."

The telescope also has a long biography of inventory and display that speaks to its function and status beyond any specific scientific uses. In the late seventeenth century, we find it mentioned in the Medici inventory of instruments.[30] Exhibition and storage records are quite sparse, but we know it was on display in the former Istituto e Museo di Storia della Scienza in Florence until 2008. It was then moved to storage because, not being one of the major telescope icons, there was not enough room for it in the telescope displays of the newly organised Museo Galileo. From March 2010 to February 2011 it was on loan to the exhibition *Science Storms* at the Museum of Science and Industry in Chicago. And just prior to its visit to Canada, the telescope was loaned to the Mori Art Museum in Tokyo from July 2016 to January 2017 for an exhibition entitled *The Universe and Art*. In short, between storage and public display, the telescope has moved through many contexts of use and place. In Canada, its presence inspired the local exhibition team to explore non-European contexts such as Indigenous star knowledge in Canada. In the gallery near the Divini case there was a wall displaying this knowledge, entitled *One Sky, Many Astronomies*, showing Cree, Objibwe, and Dakota representations of constellations.

Antique instrument dealers, long an important part of the SIC community, contributed to the growth of interest in instrument biography. In 2017, Charles Miller Antiques in London advertised a Ramsden sextant (*c.*1772) signed by the Pacific explorer George Vancouver (1757–1798). It sold for £34,720.[31] At the same time, the sale information revealed large gaps in the provenance of the instrument; Miller could only report that it had been "acquired unidentified [in] 2005 from the estate of a retired Royal Navy officer in Devon."[32] At the same time, another navigation object of nearly unknown provenance entered the market. Long-time SIC contributors and instrument dealers David and Yola Coffeen advertised a Marshall Islands stick chart in their 2012 *Tesseract* catalogue.[33] This chart, dated *c.*1950 but lacking specific information on maker(s) or use, reveals the complex wayfinding traditions of Marshall Islanders related to wave patterns on oceans surrounding the islands.[34]

---

30 Albert Van Helden, *Istituto e Museo di Storia della Scienza: Catalogue of Early Telescopes* (Florence: Giunti Editore, 1999), 34–36.

31 Charles Miller Ltd., "Maritime and Scientific Models, Instruments and Art ('Fearless'), 7 November 2017," George Vancouver's Sextant, lot 42, https://www.charlesmillerltd.com/auction/lot/42-george-vancouvers-sextantbrbra-14in-radius/?lot=3254&sd=1.

32 Charles Miller Ltd., "Maritime."

33 *Tesseract*, catalogue no. 94 (Hastings-on-Hudson, 2012).

34 Joseph H. Genz, "Resolving Ambivalence in Marshallese Navigation: Relearning, Reinterpreting, and Reviving the 'Stick Chart' Wave Models," *Structure and Dynamics* 9, no. 1 (2016): 8–40; Joseph H. Genz, *Breaking the Shell: Voyaging from Nuclear Refugees to*

With such increased attentiveness to place, materials, use, and biography as context for instruments, shifting and multiple notions of function came into play. In Samuel Gessner's 2014 paper on the gnomonic inventions of V. Estancel, he provided a method for analysing many different dimensions of use for mathematical instruments – operational, empirical, representational, didactic, experimental, symbolic, and economic.[35] On a broader level, this is a rich historiographic model for all scientific instruments, reflecting the emerging multiplicity of meanings and stories that can be found in single objects.

Provenance research may lead to questions about authenticity. In a paper at the SIC conference in Turin, Boris Jardine, Joshua Nall, and James Hyslop explored the value of fakes as historical resources.[36] Similarly, in the Scientific Instrument Society's 2019 Gerard Turner Memorial Lecture, Nicolàs de Hilster called for studying the history of "fakes, replicas, and reconstructions."[37] Much can be learned about public views of science, instruments, materiality, and society from this spectrum of authenticity.

Finally, curators have been questioning long-iconic scientific instruments in new ways. Nall used provenance and careful object examination to question aspects of the history of the Ramsden dividing engine.[38] Tiemen Cocquyt deployed a combination of historical and forensic material approaches, including neutron tomography, to solve long-standing questions surrounding Leeuwenhoek's earliest microscopes. In a rare moment of international attention for the history of scientific instruments, these findings were reported by major scientific outlets and newspapers such as *Le Monde*.[39]

People of the Sea in the Marshall Islands (Honolulu: University of Hawai'i Press, 2018); Ingenium Collection – Objects – 2018.0113.001, https://collection.ingeniumcanada.org/en/id/2018.0113.001/.

35 Samuel Gessner, "The Use of Useless Instruments: The Gnomonic Inventions by V. Estancel (SJ) in Transit through the Portuguese Empire (1650–1680)," in *Scientific Instruments in History of Science: Studies in Transfer, Use and Preservation*, ed. M. Granato and M.C. Lourenço (Rio de Janeiro: Museu de Astronomia e Ciências Afins, 2014), 53–68.

36 Boris Jardine, Joshua Nall, and James Hyslop, "More than Mensing? Revisiting the Question of Fake Scientific Instruments," *Bulletin of the Scientific Instrument Society*, no. 132 (2017): 22–29.

37 Nicolàs de Hilster, "Fakes, Replicas and Reconstructions: Their Place in Scientific Research of Early Modern Instruments," *Bulletin of the Scientific Instrument Society*, no. 144 (2020): 2–11.

38 Joshua Nall, "'Certainly Made by Ramsden': The Long History of the Whipple Museum's Dividing Engine," *Bulletin of the Scientific Instrument Society*, no. 137 (2018): 40–43. For another example, see Ben Russell, "A Process of Ventriloquism? James Watt's Micrometer," *Bulletin of the Scientific Instrument Society*, no. 148 (2021): 30–35.

39 Hervé Morin, "En image: le secret éventé des microscopes de van Leeuwenhoek," *Le Monde*, May 17, 2021. Tiemen Cocquyt, *et al.*, "Neutron Tomography of Van Leeuwenhoek's Microscopes," *Science Advances* 7, no. 20 (2021).

3 Access and Context

This multiplication of contexts resulted, in part, from a surging access to historic scientific instruments. In the 2010s, there were numerous digitisation projects of scientific instrument collections. New online tools for trade catalogues, city records, and patents emerged, as did descriptions of previously hidden university collections. Thousands of posts about instruments flooded social media. Access and circulation of instruments increased through the loans of instruments to more traditional exhibitions. And new media spread information about instruments through open journals (e.g., *Science Museum Journal*) and YouTube videos. In 2009, for example, SIC President Paolo Brenni and Anna Giatti created a YouTube channel for sharing professionally-edited videos of working instrument demonstrations from the collection of the Fondazione Scienza e Tecnica in Florence. By the end of the 2010s, their channel had over 4500 subscribers and 100 videos. The video showing Hertz radiowave experiments has had over 75,000 views.[40]

Even storage space, long restricted to curators, started forming its own identity for research and access. At the ICOM training facility at the Palace Museum in Beijing, participants at the semi-annual museum training workshop had hands-on training with previously hidden objects in their storage facilities.[41] Richard L. Kremer and I were both faculty at these workshops and helped the students develop "reading artefacts" methods and skills to examine and interpret artefacts.

4 Inventories … Plus

"Inventories of instruments, both of those lost and of those surviving, are one of the corner-stones upon which any history of scientific instruments must rest."[42] In March 2010, Denis Beaudouin, Paolo Brenni, and Anthony Turner announced an ambitious undertaking to develop a dictionary of French scientific instrument makers. Their opening assertion (above) explicitly situated the project within one of the central historical missions of the SIC, viz., to build fundamental and lasting inventories. By the time they launched the project

40 Fondazione Scienza e Tecnica, "florencefst," https://www.youtube.com/@florencefst.

41 International Council of Museums, "International Training Centre for Museum Studies (ICOM-ITC)," accessed 26 June 2024, https://icom.museum/en/capacity-building/international-training-centre-for-museum-studies-icom-itc/.

42 Denis Beaudouin, Paolo Brenni, and Anthony Turner, "For a Dictionary of French Scientific Instruments-Makers," *Bulletin of the Scientific Instrument Society*, no. 104 (2010): 9–11.

online in 2018, it had the name *Dictionary of Precision Instrument Makers and Related Craftsmen*, pointing towards the broad use of precision and associated artisanal skills.[43] This dictionary was the result of the accumulation of decades of work by each scholar.

The inventories of the 2010s surely benefited from a fuller sense of the cultural possibilities hidden within instruments. The individual instruments themselves became the source of their "Many Worlds" within and were not just types to be tabulated and catalogued. For example, in 2019 Sara Schechner published a catalogue of the Adler Planetarium's 268 sundials dating from the fifteenth through the twentieth centuries. The pages overflow with context related to materials, makers, skills, religion, commerce, and geography.[44]

Likewise, in 2022 Marco Beretta and the late Paolo Brenni published a catalogue of Antoine Lavoisier's collection of apparatus now at the Conservatoire National des Arts et Métiers in Paris.[45] Begun in the 2010s, this inventory project involved far more than simply listing the instruments and apparatus; rather, it offered a comprehensive reconstruction of Lavoisier's laboratory, including the technical, social and cultural contexts of these objects and practices. This inventory will be a lasting foundation for scholarship on the Chemical Revolution. It is all the more heroic when compared to equivalent work in art history; countless catalogues have been devoted to Jacques Louis David, Lavoisier's artistic contemporary in Paris. This asymmetry of scholarly attention speaks to how important our work is at the SIC and to the need to raise the profile of the history of scientific instruments.

During the 2010s, long-time SIC contributors Marvin Bolt and Michael Korey continued and expanded *Dioptrice*, an inventory of telescopes constructed before 1775.[46] They have visited and worked with curators and collectors in Italy, the United States, the United Kingdom, China, Russia, the Netherlands, Germany, and Sweden. Similar to the examples above, they are examining the objects carefully, drawing from them new contexts and findings, and raising new questions.

---

43 Denis Beaudouin, Paolo Brenni, and Anthony Turner, *The Dictionary of Precision Instrument-Makers and Related Craftsmen (France & Switzerland, 1430–1960)*, last modified July 2018, https://bibnum.explore.psl.eu/s/dictionarypim/page/prolegomena.

44 Sara J. Schechner, *Time of Our Lives: Sundials of the Adler Planetarium* (Chicago: Adler Planetarium, 2019). This is volume 1; volume 2 cataloguing the remaining dials is in preparation.

45 Marco Beretta and Paolo Brenni, *The Arsenal of Eighteenth-Century Chemistry: The Laboratories of Antoine Laurent Lavoisier (1743–1794)* (Leiden: Brill, 2022).

46 Marvin P. Bolt and Michael Korey, "Dioptrice: Examining and Cataloguing the World's Oldest Surviving Telescopes," *Journal of Glass Studies* 61 (2019): 270–276. *Dioptrice*, accessed 20 July 2025, https://dioptrice.org/.

> What was a telescope supposed to look like? What should it be made of? How should it be decorated? Who could get access to such a device? Where was it used, and for what purposes? How many lenses should it have? How should the lenses be shaped? What kind of glass should be used? And where should that glass come from? None of these questions had a canonical, a priori answer.[47]

Along with their Indiana Jones activities to locate and access these telescopes, they have also been employing cutting-edge scientific techniques to examine the instruments, using forensic tools such as Tiemen Cocquyt's portable interferometer.

By the end of the decade, there were plans afoot to expand one of the more important long-standing inventories, Gloria Clifton's Scientific Instrument Makers, Observations and Notes (SIMON) dataset, from which a volume had been published in the 1990s.[48] The *Tools of Knowledge* (TOK) project will apply new relational and analytic database technology to the existing and growing set of data.

> This database will be rigorously remodelled in a semantic form, migrated from an aging Access format and enriched with data from multiple museum collections together with newly generated data from computational analysis of text and image sources and metallurgical analysis of materials. This 'SEMSIM' (augmented, semantic SIMON) database will map to an established cultural heritage ontology and vocabularies, offering interoperability with other resources.[49]

At the same time, the TOK team are expanding the scope of the dataset through a fuller sense of the scientific instrument eco-system by considering materials, skills, workshops, archives, supplies, and geographies. This relational perspective partly grew out of the Science Museum's *Metropolitan Science* research project that examined the dense infrastructure of instruments and science in London's urban spaces in the seventeenth and eighteenth centuries.[50]

---

47 Bolt and Korey, "Dioptrice," 270.

48 Gloria Clifton, *Dictionary of British Scientific Instrument Makers, 1550–1851* (London: National Maritime Museum, 1995).

49 *Tools of Knowledge*, accessed 24 April 2024, https://toolsofknowledge.org/.

50 Rebekah Higgitt, Jasmine Kilburn-Toppin, and Noah Moxham, "Science and the City: Spaces and Geographies of Metropolitan Science," *Science Museum Group Journal* (Spring 2021), https://journal.sciencemuseum.ac.uk/article/science-and-the-city-spaces-and-geographies-of-metropolitan-science/.

5 Exhibits and the Splendour of Scientific Instruments

The 2010s saw several notable exhibitions curated by people in the SIC community. From 2010 to 2011, the Oxford Museum for the History of Science (recently renamed the History of Science Museum) mounted an exhibition *Al-Mizan,* that used their Islamic instrument collection to expand the theme of "balance" from the familiar use as a weighing instrument to broader metaphorical notions of justice and harmony, and to consider other issues such as materials, craft skills, texts, courtly culture, inscriptions, and engraving. The instruments themselves served as the source for these revealing analyses: "The links between scientific inquiry and artistic beauty are vividly revealed through the decorative and practical work of the craftsman."[51]

During the decade, university collections continued to be among the most prominent places to see historic scientific instruments on display. In 2014, the historic Colegio de San Ildefonso in Mexico City hosted an exhibition of nineteenth-century optical instruments.[52] The instrument displays at the University Museum in Tartu served as a backdrop for the 2014 SIC symposium. Similarly, in 2016, Istanbul University hosted an exhibition of instruments and manuscripts as part of the SIC symposium there.[53]

Time is one of those abstract concepts that lends itself well to creative exploration through exhibitions. In 2013 Sara Schechner curated an exhibition at Harvard's Collection of Historical Scientific Instruments on the material culture of time. The selection of objects – from seventeenth-century compendiums to artefacts from the Hopi people – elucidated interdisciplinary themes about the creation of time, time through celestial, geological, and biological objects, time flow, atomic time, labour, popular culture, and business. An online version of the catalogue is a lasting resource from the exhibit.[54] In 2016,

---

51 History of Science Museum, Oxford, "Al-Mizan: Sciences and Arts in the Islamic World," 2010–2011. https://www.mhs.ox.ac.uk/almizan/.

52 Naix'ieli Castillo, "Inauguran Exposición de Instrumentos Ópticos del Siglo XIX," report on the exhibit, *La enseñanza de la óptica a fines del siglo XIX, instrumentos antiguos de la Escuela Nacional Preparatoria,* 22 May–30 June 2014, at the Antiguo Colegio de San Ildefonso, Universidad Nacional Autónoma de México (UNAM), 2014, https://ciencia.unam.mx/leer/351/Inauguran_exposicion_de_instrumentos_opticos_del_siglo_XIX.

53 Feza Günergun, ed., *Pursuing Knowledge: Scientific Instruments, Manuscripts and Prints from Istanbul University Collections* [*Bilginin izinde: İstanbul Üniversitesi koleksiyonlarından bilimsel aletler, yazmalar, baskılar*] (Istanbul: Istanbul Üniversitesi, 2016).

54 Sara J. Schechner, *Time and Time Again: How Science and Culture Shape the Past, Present and Future* (Cambridge, MA: Collection of Historical Scientific Instruments, Harvard University, 2014); exhibition catalogue and related materials, https://chsi.harvard.edu/exhibitions/time-time-again. Sara J. Schechner, curator, *Radio Contact: Tuning in to Politics, Technology, and Culture,* exhibition at the Collection of Historical Scientific

the Harvard Collection also developed an exhibition called *Radio Contact: Tuning in to Politics, Technology, and Culture*, which brought together historical radio equipment from multiple traditions, an immersive 1930s HAM radio shack, a working radio station, and a soundscape of old radio clips. In Rio, the Museum of Astronomy and Related Sciences (MAST), set on the original site of precision time measurement in Brazil, mounted an exhibit on time which also highlighted Indigenous perspectives. At the National Museum of African History in Washington, DC there was a groundbreaking exhibition on African star knowledge; the exhibit and catalogue revealed a rich and diverse material culture of astronomy from many contexts across Africa.[55]

In the 2010s, there were several major museum renovations. In 2016, National Museums Scotland (NMS) opened several new galleries with many displays devoted to their science and technology collections. This was an era where curators began systematically collecting and displaying the contemporary sublime.[56] For their opening, NMS displayed a 2.5-tonne copper cavity from CERN's Large Electron Positron Collider.[57] For Tacye Phillipson, senior curator of Modern Science at NMS what "this particular object illustrates above all is the sheer quality of engineering required to enable the kind of scientific experiments carried out at CERN."[58] In a similar approach, the Science Museum London mounted an exhibit, *Collider*, that took visitors deep into the material culture of the Large Hadron Collider. The exhibit was a success with visitors (over 53,000 attended) and reviews, and travelled to Manchester, Paris, Sydney, and Brisbane.[59]

Two national museums, the Rijksmuseum Boerhaave (Leiden) and the Canada Science and Technology Museum (Ottawa), were completely renovated and reopened in 2017. They both chose to base their exhibits on open-ended themes, rather than chronology and traditional subject areas. In April

---

Instruments at Harvard, 2016, accessed 11 October 2024, https://chsi.harvard.edu/radio_contact.html.

55 Christine Mullen Kreamer *et al.*, *African Cosmos: Stellar Arts* (Washington, DC: National Museum of African Art, Smithsonian Institution, 2012).

56 David Pantalony, "Field Notes: Challenges and Approaches for Collecting Recent Material Heritage of Science and Technology," *Museologia e Patrimônio* 8, no. 2 (2015): 80–103.

57 Samuel J.M.M. Alberti, *Curious Devices and Mighty Machines: Exploring Science Museums* (London: Reaktion Books, 2022).

58 "National Museums Scotland Acquires CERN Component for New Science Galleries," *Museums and Heritage Advisor*, 7 May 2015, accessed 11 October 2024, https://museumsandheritage.com/advisor/posts/national-museums-scotland-acquires-cern-component-for-new-science-galleries/.

59 Alison Boyle and Harry Cliff, "Curating the Collider: Using Place to Engage Museum Visitors with Particle Physics," *Science Museum Group Journal* (Autumn 2014), https://dx.doi.org/10.15180/140207.

2013, the Mathematisch-Physikalischer Salon, Dresden, reopened after significant refurbishment of its home in the Zwinger. The exhibits, architecture, and context display together the full splendour of historical instruments at the Saxon Court through four themes – the Cosmos of the Prince, the Universe of Globes, Instruments of the Enlightenment, and the Course of Time. The beauty and culture of scientific instruments were also prominently showcased in the exhibit *Making Marvels: Science and Splendor at the Courts of Europe* at the Metropolitan Museum in New York. The exhibit and catalogues paid close attention to material, function, and style to draw out history and culture from their little-shown collection of scientific instruments.[60]

The SIC community continues to benefit from the encyclopaedic treatment of collections and display. In their exhibit and thoroughly researched catalogue, *Metronomes and Musical Time*, Tony Bingham and Anthony Turner showed the value of the sheer diversity of expressions for one technology, one that had been previously under-valued and studied in musical, scientific, and technological collections.[61] In 2018, Anthony Turner and colleagues published a comprehensive catalogue of the mathematical instrument collections at the Bibliothèque nationale de France, the first such description of these rich holdings.[62]

6 Conclusions: What Are We Collecting?

As described above, the "One Object, Many Worlds" approach took hold during the 2010s, in research, exhibits, and conferences. This perspective will undoubtedly have implications for the kinds of collecting museum curators are doing. We are seeing less encyclopaedic collecting; in fact, for a host of reasons most curators and institutions simply can not collect in this manner anymore. We are collecting less, but with more stories and themes wrapped into the selected objects. We are also collecting the processes and practices of science and not just the final products and icons. We are collecting for provenance, for

60 Wolfram Koeppe, ed., *Making Marvels: Science and Splendor at the Courts of Europe* (New York: The Metropolitan Museum of Art, 2019).

61 Tony Bingham (metronome collector) and Anthony Turner, *Metronomes and Musical Time: Catalogue of the Tony Bingham Collection of Metronomes in the Exhibition AUF TAKT! Held in the Museum für Musik, Basel, 20 January to 20 August 2017* (Antrim, Northern Ireland: W & G Baird, 2017).

62 Anthony J. Turner, *Mathematical Instruments in the Collections of the Bibliothèque Nationale de France* (Paris: BNF Éditions, 2018).

the recognition that ownership, use, and geography are essential to drawing out the culture from the instruments.

When Richard Kremer visited the opening of the new Canada Science and Technology Museum in Ottawa, he noted that it nowhere mentioned the word "science" in its galleries. As a curator, I was worried about that – we did not want to minimise the science within scientific instruments. However, our conception of science has been shifting over the past several decades, and instrument studies have played a significant role in shaping these changes. As we have seen in the SIC community during the 2010s, instrument scholarship and curation have been pointing in many directions, starting from the objects themselves and their specific biographies (materials, skills, supplies, commerce, labour, transportation, geography, marketing, aesthetics, modifications) and expanding to new contexts of science and society, of making and knowing, of representing and concealing. We are now seeing aspects of scientific instruments that show the valuable human stories and many worlds within these objects.

Bibliography

Ackermann, Silke, Richard L. Kremer, and Mara Miniati. *Scientific Instruments on Display*. Leiden: Brill, 2014.

Alberti, Samuel J.M.M. *Curious Devices and Mighty Machines: Exploring Science Museums*. London: Reaktion Books, 2022.

Anderson, R.G.W., J.A. Bennett, and W.F. Ryan, eds. *Making Instruments Count: Essays on Historical Scientific Instruments Presented to Gerard L'Estrange Turner*. Aldershot, Hampshire: Variorum, 1993.

Beaudouin, Denis, Paolo Brenni, and Anthony Turner. "For a Dictionary of French Scientific Instruments-Makers." *Bulletin of the Scientific Instrument Society*, no. 104 (2010): 9–11.

Beaudouin, Denis, Paolo Brenni, and Anthony Turner. *The Dictionary of Precision Instrument-Makers and Related Craftsmen (France & Switzerland, 1430–1960)*, last modified July 2018. https://bibnum.explore.psl.eu/s/dictionarypim/page/prolegomena.

Bennett, Jim, Sofia Talas, and Giorgio Strano, eds. *Cabinets of Experimental Philosophy in Eighteenth-Century Europe*. Leiden: Brill, 2013.

Beretta, Marco, and Paolo Brenni. *The Arsenal of Eighteenth-Century Chemistry: The Laboratories of Antoine Laurent Lavoisier (1743–1794)*. Leiden: Brill, 2022.

Bingham, Tony, and Anthony Turner. *Metronomes and Musical Time: Catalogue of the Tony Bingham Collection of Metronomes in the Exhibition AUF TAKT! Held in the*

Museum für Musik, Basel, 20 January to 20 August 2017. Antrim, Northern Ireland: W & G Baird, 2017.

Bolt, Marvin P., and Michael Korey. "Dioptrice: Examining and Cataloguing the World's Oldest Surviving Telescopes." *Journal of Glass Studies* 61 (2019): 270–276.

Boyle, Alison, and Harry Cliff. "Curating the Collider: Using Place to Engage Museum Visitors with Particle Physics." *Science Museum Group Journal* (Autumn 2014), https://dx.doi.org/10.15180/140207.

Brown, Neil, Silke Ackermann, and Feza Günergun, eds. *Scientific Instruments between East and West*. Scientific Instruments and Collections, 7. Leiden: Brill, 2020.

Brusius, Mirjam. "100 Histories of the World in One Object: Itineraries, Islam and Indices." *Material Religion* 15, no. 2 (March 15, 2019): 243–245.

Castillo, Naix'ieli. "Inauguran Exposición de Instrumentos Ópticos del Siglo XIX." Report on the exhibit, *La enseñanza de la óptica a fines del siglo XIX, instrumentos antiguos de la Escuela Nacional Preparatoria*, 22 May–30 June 2014, at the Antiguo Colegio de San Ildefonso, Universidad Nacional Autónoma de México (UNAM). 2014. Ciencia UNAM, https://ciencia.unam.mx/leer/351/Inauguran_exposicion_de_instrumentos_opticos_del_siglo_XIX.

Cavicchi, Elizabeth and Peter Heering, eds. *Historical Scientific Instruments in Contemporary Education*. Leiden: Brill, 2021.

Cocquyt, Tiemen, *et al.* "Neutron Tomography of Van Leeuwenhoek's Microscopes." *Science Advances* 7, no. 20 (2021).

Dunn, Richard, Silke Ackermann, and Giorgio Strano, eds. *Heaven and Earth United: Instruments in Astrological Contexts*. Leiden: Brill, 2018.

Fondazione Scienza e Tecnica, Florence. "florencefst." http://www.youtube.com/@florencefst.

Genz, Joseph H. *Breaking the Shell: Voyaging from Nuclear Refugees to People of the Sea in the Marshall Islands*. Honolulu: University of Hawai'i Press, 2018.

Genz, Joseph H. "Resolving Ambivalence in Marshallese Navigation: Relearning, Reinterpreting, and Reviving the 'Stick Chart' Wave Models." *Structure and Dynamics* 9, no. 1 (2016): 8–40.

Gessner, Samuel. "The Use of Useless Instruments: The Gnomonic Inventions by V. Estancel (SJ) in Transit through the Portuguese Empire (1650–1680)." Pp. 53–68 in *Scientific Instruments in History of Science: Studies in Transfer, Use and Preservation*, edited by M. Granato and M.C. Lourenço. Rio de Janeiro: Museu de Astronomia e Ciências Afins, 2014.

Gillin, Edward J. "The Instruments of Expeditionary Science and the Reworking of Nineteenth-Century Magnetic Experiment." *Notes and Records: The Royal Society Journal of the History of Science* 76, no. 3 (2022): 565–592.

Günergun, Feza, ed. *Pursuing Knowledge: Scientific Instruments, Manuscripts And Prints from Istanbul University Collections* [*Bilginin izinde: İstanbul Üniversitesi koleksiyonlarından bilimsel aletler, yazmalar, baskılar*]. Istanbul: Istanbul Üniversitesi, 2016.

Hilster, Nicolàs de. "Fakes, Replicas and Reconstructions: Their Place in Scientific Research of Early Modern Instruments." *Bulletin of the Scientific Instrument Society*, no. 144 (2020): 2–11.

Higgitt, Rebekah, Jasmine Kilburn-Toppin, Noah Moxham, "Science and the City: Spaces and Geographies of Metropolitan Science." *Science Museum Group Journal* (Spring 2021). https://journal.sciencemuseum.ac.uk/article/science-and-the-city-spaces-and-geographies-of-metropolitan-science/.

History of Science Museum, Oxford. "Al-Mizan: Sciences and Arts in the Islamic World." 2010–2011. https://www.mhs.ox.ac.uk/almizan/.

International Council of Museums. "International Training Centre for Museum Studies (ICOM-ITC)." https://icom.museum/en/capacity-building/international-training-centre-for-museum-studies-icom-itc/.

Jardine, Boris, Joshua Nall, and James Hyslop. "More than Mensing? Revisiting the Question of Fake Scientific Instruments." *Bulletin of the Scientific Instrument Society*, no. 132 (2017): 22–29.

Kocaman, Meltem. "Scientific Instrument Retailers in Istanbul in the Nineteenth Century, and Verdoux's Optical Shop." Pp. 240–256 in *Scientific Instruments between East and West*, edited by Neil Brown, Silke Ackermann, and Feza Günergun. Leiden: Brill, 2019.

Koeppe, Wolfram, ed. *Making Marvels: Science and Splendor at the Courts of Europe*. New York: The Metropolitan Museum of Art, 2019.

Kreamer, Christine Mullen, Erin Haney, Katharine Monsted, Karel Nel, and Randall Bird. *African Cosmos: Stellar Arts*. Washington, DC: National Museum of African Art, Smithsonian Institution, 2012.

Laidla, Janet. "Hugo Masing's Golitsyn-Vilip Seismographs: From Tartu to Five Continents." Pp. 154–167 in *Scientific Instruments between East and West*, edited by Neil Brown, Silke Ackermann, and Feza Günergun. Leiden: Brill, 2019.

Lazos, Panagiotis, George N. Vlahakis, and Constantine Skordoulis. "Instruments and Laboratories in the Schools of the Greek Community of Istanbul, 1850–1960." Pp. 168–186 in *Scientific Instruments between East and West*, edited by Neil Brown, Silke Ackermann, and Feza Günergun. Leiden: Brill, 2019.

Morrison-Low, A.D., Sven Dupré, Stephen Johnston, and Giorgio Strano, eds. *From Earth-Bound to Satellite: Telescopes, Skills and Networks*. Leiden: Brill, 2012.

Morrison-Low, A.D., Sara J. Schechner, and Paolo Brenni, eds. *How Scientific Instruments Have Changed Hands*. Leiden: Brill, 2016.

Morin, Hervé. "En image: le secret éventé des microscopes de van Leeuwenhoek." *Le Monde*, May 17, 2021.

Nall, Joshua. "'Certainly Made by Ramsden': The Long History of the Whipple Museum's Dividing Engine." *Bulletin of the Scientific Instrument Society*, no. 137 (2018): 40–43.

Pantalony, David. "Field Notes: Challenges and Approaches for Collecting Recent Material Heritage of Science and Technology." *Museologia e Patrimônio* 8, no. 2 (2015): 80–103.

Russell, B., "A Process of Ventriloquism? James Watt's Micrometer," *Bulletin of the Scientific Instrument Society*, no. 148 (2021): 30–35.

Schaffer, Simon. "Understanding (through) Things." Lecture at the "The Location of Knowledge: A Mellon CDI Conference," Centre for Research in the Arts, Social Sciences and Humanities (CRASSH), University of Cambridge, 8 March 2013.

Schechner, Sara J. *Time and Time Again: How Science and Culture Shape the Past, Present and Future*. Cambridge, MA: Collection of Historical Scientific Instruments, Harvard University, 2014.

Schechner, Sara J. *Time of Our Lives: Sundials of the Adler Planetarium*. Chicago: Adler Planetarium, 2019.

Scientific Instrument Commission. Past symposia webpages. https://scientific-instrument-commission.org/events/category/sic-conference/?eventDisplay=past.

Taub, Liba. "What Is a Scientific Instrument, Now?" *Journal of the History of Collections* 31, no. 3 (2019): 453–467.

Tesseract: Early Scientific Instruments. Catalogue no. 94. Hastings-on-Hudson, 2012.

Tools of Knowledge: Modelling the Creative Communities of the Scientific Instrument Trade, 1550–1914. https://toolsofknowledge.org/.

Turner, Anthony J. *Mathematical Instruments in the Collections of the Bibliothèque Nationale de France*. Paris: BNF Éditions, 2018.

Van Helden, Albert. *Istituto e Museo di Storia della Scienza: Catalogue of Early Telescopes*. Florence: Giunti Editore, 1999.

Warner, Deborah Jean. "What Is a Scientific Instrument, When Did It Become One, and Why?" *British Journal for the History of Science* 23, no. 1 (1990): 83–93.

Whitehouse, Simon. "The Rand-McNally Geo-Physical Globes: How the Earth Was Constructed to Display Post-War Sciences." MA thesis, Florida State University, 2015.

CHAPTER 5

Transactional Tales: Historiography and Scientific Instrument Studies

Boris Jardine

In his classic 2004 paper "Knowledge in Transit," Jim Secord singled out one major trend in the historiography of science in the late twentieth-century:

> The move to study practice has, in my view, been the single most significant transformation in our field during the past twenty years. It breached old boundaries between "internal" and "external" and opened up a view of science as a process, including inquiries into experiment, fieldwork, and theory making. Most fundamentally, it broke down old distinctions between words and things, between texts, books, instruments, and images.[1]

Of note here is the even-handedness; this is not one paradigm replacing another or an entirely new set of concerns, but rather a pluralistic view of "science as a process," with a wide range of spaces, kinds of activity, products or "results" and – for us – sources. And, as indicated in that final list, scientific instruments would seem to loom large in this new programme of historical inquiry.

While Secord is squarely in the field of History and Philosophy of Science (HPS) his more sociologically-minded colleagues in Science and Technology Studies (STS) are also keen to place instruments at the heart of their researches. For example, Mario Biagioli's "Introduction" to the 1999 Routledge *Science Studies Reader* claims that

> As a result of a widespread shift in focus from scientific theory to scientific practices, the study of instruments, experiments, and replication has become one of the most practiced areas of science studies.[2]

1 James A. Secord, "Knowledge in Transit," *Isis* 95 (2004): 654–672, see 658.
2 Mario Biagioli, ed., *The Science Studies Reader* (New York and Abingdon: Routledge, 1999), "Introduction," xv.

 | DOI:10.1163/9789004737587_007

Welcome as these sentiments are, we might pause over Biagioli's confidence and Secord's chronology. Of the thirty-six essays in the *Reader*, only one closely concerns instrumentation: Simon Schaffer's "Late Victorian Metrology and Its Instrumentation: A Manufactory of Ohms." The main journal dedicated to science studies is *Social Studies of Science*, which as far as I can tell has published only three articles on instrumentation: Steven Shapin's "Pump and Circumstance" – a great paper to be sure but one that indicates in its subtitle that it will be dealing with "literary" rather than material technology; a paper on "Databases as Scientific Instruments," which regardless of its merits does not justify the use of "instrument" over "tool" or "technology;" and Patrick Carroll-Burke's "Tools, Instruments and Engines," a very important paper for instrument historians but one that is avowedly "conceptual and methodological."[3]

Meanwhile, Secord's claim that the "turn to practice" is a phenomenon of "the past twenty years" takes us back to the middle of the 1980s, when a group of provocative and heavily sociological books were published, notably Andrew Pickering's *Constructing Quarks* (1984), Steven Shapin's and Simon Schaffer's *Leviathan and the Air-Pump* (1985), and Trevor Pinch's *Confronting Nature* (1986). To augment this male-dominated list we might usefully add, *inter alia*, Sharon Traweek's *Beamtimes and Lifetimes* (1988). Secord himself acknowledges the decisive intervention of feminist science studies, notably Donna Haraway's paper on "Situated Knowledge."[4]

Yet this cannot be the whole story. First, we can look further back, and point out that while the contributions of feminists and sociologists were certainly new in the 1980s, the internal/external divide itself has a vastly complex history, with many early works in the history of science at least attempting to avoid the either/or position and paying close attention to scientific practice,

3 Steven Shapin, "Pump and Circumstance: Robert Boyle's Literary Technology," *Social Studies of Science* 14 (1984): 481–520; Christine Hine, "Databases as Scientific Instruments and Their Role in the Ordering of Scientific Work," *Social Studies of Science* 36 (2006): 269–298; Patrick Carroll-Burke, "Tools, Instruments and Engines: Getting a Handle on the Specificity of Engine Science," *Social Studies of Science* 31 (2001): 593–625.

4 Donna Haraway, "Situated Knowledge: The Science Question in Feminism as a Site of Discourse on the Privilege of Partial Perspective," *Feminist Studies* 14 (1988): 575–599; Andrew Pickering, *Constructing Quarks: A Sociological History of Particle Physics* (Edinburgh: Edinburgh University Press, 1984); Steven Shapin and Simon Schaffer, *Leviathan and the Air-Pump: Hobbes, Boyle, and the Experimental Life* (Princeton: Princeton University Press, 1985); Trevor Pinch, *Confronting Nature: The Sociology of Solar-Neutrino Detection* (Dordrecht: D. Reidel Pub. Co., 1986); Sharon Traweek, *Beamtimes and Lifetimes: The World of High Energy Physicists* (Cambridge, MA: Harvard University Press, 1988).

especially instruments.[5] Moreover, "distinctions between words and things" have been broken down and set back up again many times through the twentieth century. It is no doubt true that book history and iconographic analysis are relatively new. But the point remains: the turn to practice was not *simply* a new phenomenon in the 1980s.

We can also point in the other direction, and ask why, if this renewed attention to practice dates from the 1980s, we in fact see an *increasing* divide between scientific instrument studies and the history of science as studied at the major university departments? Naturally, this question contains a lot of assumptions – that there is such a thing as "instrument studies" and that there is a distinct "mainstream" university history of science; and that the gulf is growing – but I believe these can be quite easily defended, for example with reference to the introductions to the many journal special issues that have been dedicated to scientific instruments since the supposed beginning of the turn to practice forty years ago. The first of these special issues was the 1994 edition of *Osiris*, followed quickly by the *British Journal for the History of Science* (1995), and *Journal of the History of Collections* (1995). The first of the *Nuncius* issues with papers from a Scientific Instrument Commission symposium appeared in 2004. Then were special issues of *History of Science* (2006), *Studies in History and Philosophy of Science* (2009), and *Isis* (Focus Section, 2011), and most recently another *JHC* issue in 2019. The introductions to these special issues tend to make a similar claim: the "old" historiography had no time for instruments, but new developments have shown how integral they have been to developments in the sciences; it is now time for those with first-hand experience of working with instruments as sources to unite with historians of science working in other areas, to create a new practice-oriented and instrument-oriented historiography.

No doubt these are worthy aims – but if the case needs to be made anew each time, then perhaps we need a new perspective on the matter. Another way to frame this is to turn to those early pathbreaking sociological studies by Pickering, Shapin and Schaffer, Pinch, and Traweek. We might think that they

5 Because "internalist" and "externalist" have generally been used pejoratively, we shouldn't be surprised to find that many on either side did not understand themselves as such. Indeed, because they lacked a rigorous sociology of knowledge, the most doctrinaire mid-century Marxist theorists of science still retained an "idealist" conception of science, and were therefore all too happy to understand science as a set of entirely internally consistent solutions to problems set by society. This is, of course, a simplification, but see Gary Werskey, *The Visible College: The Collective Biography of British Scientific Socialists of the 1930s* (London: Allen Lane, 1978), ch. 6.

deal, at least in part, with instrumentation; yet their concerns are very different from those of instrument historians. Where the latter might look for surviving artefacts or examine the ways in which instruments were made and by whom, the former tend to work with published descriptions of instruments and apply critique or sociological discussion at a level above that of actual physical practice, typically focusing on the argumentative strategies of actors rather than the technical substrate itself. This disjunction continues with very little exception through journals, conferences, edited volumes and monographs between the 1980s and the present day. Each reader of this chapter will be able to identify exceptions – but I stand by the broad claim and would simply point to the major papers and books in HPS and STS as evidence. Even works ostensibly concerned with instruments deal with them in a way quite alien to the approach of curators and instrument historians.[6]

So we have a solution and its problem: Secord and Biagioli suggest that, following the turn to practice, instruments have taken historiographic centre-stage, but instrument historians will, I suspect, see things differently. This is akin to the frequent claims made for the importance of history by scientists and science writers – who nevertheless ignore the work of actually existing historians of science.[7]

---

6 The exception that comes to my mind is not a paper or a book, but a person. Jim Bennett may be said to have dedicated his career to working across the divide I identify here. He made his own statements of the problem – notably in his 2002 presidential address to the British Society for the History of Science. He worked with the medium of the curator (the exhibition) to reach a vast range of audiences, and he made many contributions that cannot easily be placed either into the "instrument studies" or the "mainstream HPS" categories. The BSHS address was published as "Knowing and Doing in the Sixteenth Century: What Were Instruments for?," *British Journal for the History of Science* 36 (2003): 129–150. Jim's contributions to major edited volumes stand out for their "conciliatory" role, e.g., "The Challenge of Practical Mathematics," in *Science, Culture and Popular Belief in Renaissance Europe*, ed. S. Pumfrey, P. Rossi and M. Slawinksi (Manchester: Manchester University Press, 1991), 176–190; "Wind-Gun, Air-Gun or Pop-Gun: The Fortunes of a Philosophical Instrument," in *The Mindful Hand: Inquiry and Invention From the Late Renaissance to Early Industrialisation*, ed. L. Roberts, S. Schaffer and P. Dear (Amsterdam: Koninklijke Nederlandse Akademie van Wetenschappen, 2007), 221–246; "Cosmography and the Meaning of Sundials," in *Nature Engaged: Science in Practice from the Renaissance to the Present*, ed. Mario Biagioli and Jessica Riskin (New York: Palgrave Macmillan, 2012), 249–262.

7 See the brilliant essay review by David Philip Miller, titled (in full) "The 'Sobel Effect': The Amazing Tale of How Multitudes of Popular Writers Pinched All the Best Stories in the History of Science and Became Rich and Famous while Historians Languished in Accustomed Poverty and Obscurity, and How This Transformed the World. A Reflection on a Publishing Phenomenon," *Metascience* 11 (2002): 185–200.

Instead of restating the case for instrument studies, then, I want to do something different here – something more provocative. I want to suggest that instrument studies is *not* in fact well accounted for in the historiography of science as it is generally understood in the academy, as it is represented in august journals, and as it is so ably summarised in position papers like Secord's "Knowledge in Transit." In order to make this case I begin with the SIC itself – with its founding aims and what I take to be the key feature of the vast proportion of presentations at its meetings and (therefore) publications. The case can be stated very simply: instrument studies is founded upon the close study of surviving artefacts. Where there are exceptions – most instruments, after all, do not survive – the methods involved are, in my view, derived from expertise in dealing with artefacts.

A caveat should be presented immediately: this is not a "gatekeeping" paper. In emphasising the separate identity of instrument studies – and suggesting that this identity has its origins in curatorial training – I am not claiming that anyone is barred from entry or will not have the necessary skills. In fact, instrument studies is (in my experience) a welcoming area of study, and I hope that this will continue. Anyone interested enough to concern themselves with these obscure devices has surely already proved their worth. This is the sense I get from others, and I hope also to convey it in what follows.

My proposal for building on the curatorial (even antiquarian) foundations of instrument studies is to explore more deeply the history and meaning of the sub-discipline itself. Then, in a more constructive mode, and drawing on recent work presented at SIC symposia, I suggest that we borrow not from HPS or STS but rather from art history in developing a curatorially-minded historiographic programme, in which the "transactional" nature of instruments comes to the fore. I focus on the work of Michael Baxandall, although there are other relevant models from art history, cultural and literary studies, and allied fields.

1 Tunnels and Funnels for Scientific Instrument Studies

To understand how deeply rooted is the concern for surviving artefacts amongst the community of instrument historians, we need only quote the founding aims of the Scientific Instrument Commission itself:

1. To stimulate the production of inventories of historic scientific instruments.
2. To stimulate the publication of catalogues of such instruments by individuals who have no institutional support.

3. To facilitate the exchange of information between research workers in the field of historic scientific instruments.[8]

These are worthy aims, but they can hardly be said to lay out a programme of historical research. If anything, they point to a separation of the history of instruments from other academic disciplines. For one thing, "research workers in the field of historic scientific instruments" have tended to come largely from the professional ranks of the museums housing instruments – a virtue of the instrument studies, to be sure, but also a means of estrangement from the concerns of academic history of science. There is also an implicit "foundationalism" here: the SIC's main concern was to lay the groundwork for historical research by producing inventories, publishing catalogues, and sharing information.

If we turn to the papers presented at SIC meetings in its first two decades, we can see some distinct trends emerge, generally commensurate with these founding aims.[9] First and most obviously, most papers emerge from close engagement with artefacts in museums and, more rarely, private collections. Second, instrument typologies (e.g., astrolabes, spectrometers) tend to guide research, either through the case-study method or through a survey of a specific kind of instrument or an episode in which competing designs were in play. Third, there is an emphasis on the craft of instrument making, either as part of an organised trade or within specific experimental traditions. Fourth, there are case studies of individuals and places. Of these, the emphasis on craft is (to my mind) by far the most substantial contribution that instrument studies has made; it is faithful to the emphasis on surviving artefacts, and distinct from the concerns of historians and philosophers of science. In what follows I attempt to show the hidden or implicit theoretical claims made by instrument historians and to bring these together into a full account of the significance of craft for those in other disciplines, as well as for researchers in the field of historic scientific instruments.

As Anthony Turner and others have pointed out, the first true histories of instrumentation date from the first half of the nineteenth century.[10] A decisive contribution was then made with the construction of the great private

8 Quoted in Gerard L'E. Turner, "A Decade in the Study of Scientific Instruments," in *An Apparatus of Instruments: The Role of the Scientific Instrument Commission*, ed. R.G.W. Anderson and G.L'E. Turner (London: The Scientific Instrument Commission of the International Union of the History and Philosophy of Science, 1993), 5–14, see 5.

9 These comments are based on the series of SIC *Newsletters* (1980–1996).

10 Anthony Turner, "Interpreting the History of Scientific Instruments," in *Making Instruments Count: Essays on Historical Scientific Instruments Presented to Gerard L'Estrange Turner*, ed. R.G.W. Anderson, J.A. Bennett, and W.F. Ryan (Aldershot: Variorum, 1993), 17–26.

and national collections of early instruments.[11] Mature reflection upon these collections in the early twentieth century led to a series of monographs that constitute the origins of "instrument studies." Classics in the field include Gould on marine chronometers, Gunther on astrolabes, and Clay and Court on microscopes.[12] Later examples include Zinner's monumental survey of astronomical instruments, King's history of the telescope, and Middleton's of the barometer.[13] Turner calls these evolutionary narratives focused on a specific instrument type "tunnel" histories, "in which, following an occult logic all of its own, an instrument evolves towards its modern form."[14] These histories, of necessity, were relatively light on social, economic, or even sometimes scientific context – yet as Turner points out they are crucial to the health of the discipline, perhaps akin to scholarly editions of important texts in intellectual history.

When the question of the scientific context arose, it did so in what I would like to call a "funnel" approach, in which instruments were straightforward responses to the demands of scientists – as if all the pressures of scientific innovation were pressed through a funnel and the appropriate instrument popped out the other end. The key works in this tradition were Maurice Daumas's *Scientific Instruments of the Seventeenth and Eighteenth Centuries and their Makers* and E.G.R. Taylor's two-volume *Mathematical Practitioners*.[15] In very

11 For the longer history, see the articles in the special issue of the *Journal of the History of Collections* on historical collections of scientific instruments, in particular A.J. Turner, "From Mathematical Practice to the History of Science: The Pattern of Collecting Scientific Instruments," *Journal of the History of Collections* 7 (1995): 135–150. For the major British collections see R.G.W. Anderson, "Connoisseurship, Pedagogy or Antiquarianism? What Were Instruments Doing in the Nineteenth-Century National Collections in Great Britain?," *Journal of the History of Collections* 7 (1995): 211–225. For the Oxford and Cambridge collections, see Jim Bennett, "Museums and the Establishment of the History of Science at Oxford and Cambridge," *British Journal for the History of Science* 30 (1997): 29–46.

12 R.T. Gould, *The Marine Chronometer: Its History and Development* (London: J.D. Potter, 1923); R.T. Gunther, *Astrolabes of the World*, 2 vols. (Oxford: Oxford University Press, 1932); R.S. Clay and T.H. Court, *The History of the Microscope, Compiled from Original Instruments and Documents, Up to the Introduction of the Achromatic Microscope* (London: Charles Griffin and Company, 1932).

13 Ernst Zinner, *Deutsche und niederländische astronomische Instrumente des 11.–18. Jahrhunderts* (Munich: Beck, 1956); Henry C. King, *The History of the Telescope* (London: Charles Griffin and Company, 1955); W.E. Knowles Middleton, *The History of the Barometer* (Baltimore: The Johns Hopkins Press, 1964).

14 Turner, "Interpreting the History of Scientific Instruments," 19.

15 Maurice Daumas, *Les Instruments scientifiques aux xviie et xviiie siècles* (Paris, 1953); Maurice Daumas, *Scientific Instruments of the Seventeenth and Eighteenth Centuries and their Makers*, trans. M. Holbrook (London, 1972); E.G.R. Taylor, *The Mathematical Practitioners*

different ways these books initiated entire research programmes: Daumas initiated the study of workshop culture, so that while broadly positivistic with regard to the problems to be solved, he was incredibly innovative in his attention both to the actual products of particular workshops, the codification of craft know-how, and the passing on of skills from one generation to the next.[16] This idea, of a true craft lineage, has been hugely influential, though as Jim Bennett has pointed out, Daumas's work had no real predecessors and has not been surpassed.[17]

Daumas's idea of a separate history of instrument making has its parallel in Eva Taylor's practice-oriented account of the mathematical communities of the early-modern period.[18] Taylor was a geographer who had taken an interest in the antecedents of modern surveying techniques, and this had led her to study "practical mathematics" in the early-modern period. While Daumas explored the store rooms of the CNAM, Taylor scoured the literary products of early-modern science, finding that the select few well-known works on experimentation floated on a vast ocean of books about land measurement, gunnery, navigation, architecture, and allied fields. Taylor's story had little to do with the "Scientific Revolution," and surprisingly little to do even with the history of mathematics as popularly understood – but it had everything to do with the development of an instrumental culture, especially in London in the period

---

of Tudor and Stuart England (Cambridge, 1954); E.G.R. Taylor, *The Mathematical Practitioners of Hanoverian England, 1714–1840* (Cambridge, 1966). Another important early work is Alfred Rohde, *Die Geschichte der wissenschaftlichen Instrument vom Beginn der Renaissance bis zum Ausgang des 18. Jahrhunderts* (Leipzig: Klinkhardt & Biermann, 1923).

16 Following Daumas's work, there has been an enormous amount of research into workshop culture. Amongst the many classic studies are Silvio Bedini, *Early American Scientific Instruments and Their Makers* (Washington, DC: Museum of History and Technology Smithsonian Institution, 1964); Silvio Bedini, *Giuseppe Campani, "Inventor Romae": An Uncommon Genius*, ed. by Cristiano Zanetti (Leiden: Brill, 2021); Gerard L'E. Turner, *Elizabethan Instrument Makers: The Origins of the London Trade in Precision Instrument Making* (Oxford: Oxford University Press, 2000); A.D. Morrison-Low, *Making Scientific Instruments in the Industrial Revolution* (Aldershot: Ashgate, 2007). Three invaluable resources are Gloria Clifton, *Directory of British Scientific Instrument Makers, 1550–1851* (London: Philip Wilson, 1994); *The Webster Signature Database*, available online at http://historydb.adlerplanetarium.org/signatures; Denis Beaudouin, Paolo Brenni, and Anthony Turner, *The Dictionary of Precision Instrument-Makers and Related Craftsmen (France & Switzerland, 1430–1960)*, last modified July 2018, https://bibnum.explore.psl.eu/s/dictionarypim/page/prolegomena.

17 Jim Bennett, "What is the History of Scientific Instruments? Reflections on Maurice Daumas, *Les Instruments scientifiques aux xviie et xviiie siècles*," *Sphaera Occasional Paper*, no. 3 (Oxford: Museum of the History of Science, 1997).

18 Taylor, *Mathematical Practitioners* (1954, 1966).

1550–1700.[19] There have been some critiques of her community of "mathematical practitioners" – but these seem frankly insignificant against the monumental achievement of successfully showing that if we pay attention to the widest possible range of sources we can bring coherent communities of practice back to life.[20] Those who find the idea of the "mathematical practitioner" too homogeneous and Anglo-centric should instead counter by defining and contrasting different kinds of communities in other places and other times. Nevertheless, her view of instrument making specifically is of a "handmaiden" activity, guided from on high by scholars and experimenters. As with Daumas, instruments emerge from the funnel of knowledge.

Naturally there are also early histories of instruments that have come from within historical and sociological accounts of science, and these are distinct from both the "tunnel" and "funnel" approaches suggested above. By far the most significant early contribution in this vein is Edgar Zilsel's "The Sociological Roots of Science," which gave a decisive role to artisans in the origin of science itself, said to have occurred in the years around 1600. The "potential energy" of Zilsel's pithy and suggestive thesis is far from being exhausted, yet there is sense in which this most radical contribution to the debate over early modern science appears, in some ways, conservative when set against the achievements of the early historians of instruments themselves. Zilsel and others were so concerned to identify the origins of "Science" and to determine the causes of the "Scientific Revolution" that they made an absolute exception of the period

19 Like Daumas's work, Taylor's *Mathematical Practitioners* has been hugely influential. See, for example, Bennett, "The Challenge of Practical Mathematics;" Lesley B. Cormack, "Mathematics for Sale: Mathematical Practitioners, Instrument Makers, and Communities of Scholars in Sixteenth-Century London," in *Mathematical Practitioners and the Transformation of Natural Knowledge in Early Modern Europe*, ed. by Lesley B. Cormack, Steven A. Walton, and John A. Schuster (Cham: Springer, 2017), 69–85; Stephen Johnston, "Making Mathematical Practice: Gentlemen, Practitioners and Artisans in Elizabethan England" (unpublished PhD thesis, University of Cambridge, 1994); Philip Beeley, "Practical Mathematicians and Mathematical Practice in Later Seventeenth-Century London," *British Journal for the History of Science* 52 (2019): 225–248; Stephen Johnston, "The Identity of the Mathematical Practitioner in Sixteenth-Century England," in *Der "mathematicus": Zur Entwicklung und Bedeutung einer neuen Berufsgruppe in der Zeit Gerhard Mercators*, ed. Irmgarde Hantsche (Bochum: Brockmeyer, 1996), 93–120; Hester Higton, "Does Using an Instrument Make You Mathematical? Mathematical Practitioners of the 17th Century," *Endeavour* 25 (2001): 18–22; Katherine Hill, "'Juglers or schollers?' Negotiating the Role of a Mathematical Practitioner," *British Journal for the History of Science* 31 (1998): 253–274; Stephen Johnston, "Mathematical Practitioners and Instruments in Elizabethan England," *Annals of Science* 48 (1991): 319–344.

20 The most substantial critique of Taylor's work is Eric Ash, *Power, Knowledge, and Expertise in Elizabethan England* (Baltimore: Johns Hopkins University Press, 2004), ch. 4.

with which they dealt, and looked only for actors and events that could tell a celebratory story about the triumph of European reason. Historians such as Daumas, dealing first and foremost with instrumentation – or, like Taylor, coming from an unrelated field of study – have been far more agnostic about matters of chronology, European exceptionalism, or even the sheer scholarly importance of the instruments they have looked at. This means their work has aged particularly well as historians have (as noted above) turned from theory to practice, from the academy to the field and urban setting, when they have told longer stories of the transmission of techniques, and when they have looked beyond the geography of the old idea of a singular Scientific Revolution.

This last point, about the *scale* of instrument studies, is particularly important. The collections on which most of this work is based were amassed in the age of Empire – but the great prominence of Islamic artefacts (especially astrolabes) immediately leads us outside of the modern colonial context. The emphasis on craft lineage and the autonomy of making has resulted in incredibly long durée histories, notably in the work of Derek J. De Solla Price.[21] So the canonical works of instrument scholarship have long been set against the trends of mainstream history and philosophy of science: they celebrated artisans, explored "low" practice against "high" theory and, significantly, created a distinctive two-part story of making and using that is in no way beholden to the dominant narratives of the Scientific Revolution, the development of theory, experimental method, or even "scientific knowledge" itself.

2 The What and When of Scientific Instruments

This necessarily simplified picture should nevertheless make clear that we are dealing with something quite well defined and unusual in instrument studies. Yet while the curatorial community of historians and a thorn-like relation to history of science are important, we have yet to learn anything about instruments themselves. Here we can turn to a more recent "foundational" study, namely Deborah Warner's essay "What is a Scientific Instrument, When Did It Become One, and Why?"[22] Warner directly confronts the question of anachronism: instruments could hardly be "scientific" before the nineteenth century,

21 See, in particular, D.J. Price, "Philosophical Mechanism and Mechanical Philosophy: Some Notes toward a Philosophy of Scientific Instruments," *Annali dell'Istituto e Museo di Storia della Scienza di Firenze* 5 (1980): 75–85.

22 Deborah Warner, "What Is a Scientific Instrument, When Did It Become One, and Why?" *British Journal for the History of Science* 23 (1990): 83–93.

owing to the lack of a term or concept that maps on to our concept of "science" (and we do no better with "Wissenschaftliche Instrumente" or "instrumens a l'usage des sciences," also dating to the early-nineteenth century).[23] In fact the formal definition of a "scientific instrument" – specifically contrasted with "apparatus" – is ascribed to James Clerk Maxwell, writing in the 1870s. Maxwell's definition was strictly to do with experiment, and was *functional*: "A piece of apparatus constructed specially for the performance of experiments is called an Instrument."[24] But Maxwell's definition also contains the germ of its own destruction – at least from the point of view of curators of history of science museums. If we are to collect "instruments" and we seek "apparatus constructed specially for the performance of experiments," then we are in fact thrown back into the more general world of tools and supporting objects – back, that is, to "apparatus," because we will naturally become interested in all of the *other* specially made tools and vehicles and buildings and materials concocted for the prosecution of science.

What, we might ask, is Maxwell's own electromagnetic induction model, made for him by Elliott Brothers in 1876 in order to demonstrate the phenomena of induced currents? (*Fig. 5.1*) The baggy term "apparatus" does us no favours here, and even though the model produced no measurements and is not used in a process of discovery or even investigation, it is clearly a legitimate object of analysis for instrument historians.[25]

The point is more than semantic, as Maxwell's immediate successors realised. Emerging from the "sealing-wax-and-string" tradition of Maxwell's Cavendish Laboratory, subsequent experimentalists increasingly acknowledged that what they were involved in was more like technological system building than homely instrumental practice.[26] This development has had

23 Warner, "What Is a Scientific Instrument," 87.

24 Quoted in Warner, "What Is a Scientific Instrument," 88.

25 Although as far as I know the only treatment of it is the non-instrumental paper by Otto Mayr, "Maxwell and the Origins of Cybernetics," *Isis* 62 (1971): 424–444. The primary sources are Maxwell's letter to Lewis Campbell, Christmas 1876, in *The Life of James Clerk Maxwell*, ed. Lewis Campbell and William Garnett (London: Macmillan and Co., 1882), 396–397, as well as the description on pp. 551–554, and J.J. Thomson's footnote to the 3rd edition of Maxwell's *Treatise on Electricity and Magnetism*, 2 vols. (Oxford: Clarendon Press, 1892), 2: 228.

26 For the development of the Cavendish tradition see Jeff Hughes, "Plasticine and Valves: Industry, Instrumentation and the Emergence of Nuclear Physics," in *The Invisible Industrialist: Manufactures and the Construction of Scientific Knowledge*, ed. Jean-Paul Gaudillière and Ilana Löwy (Basingstoke: Macmillan, 1998), 58–101.

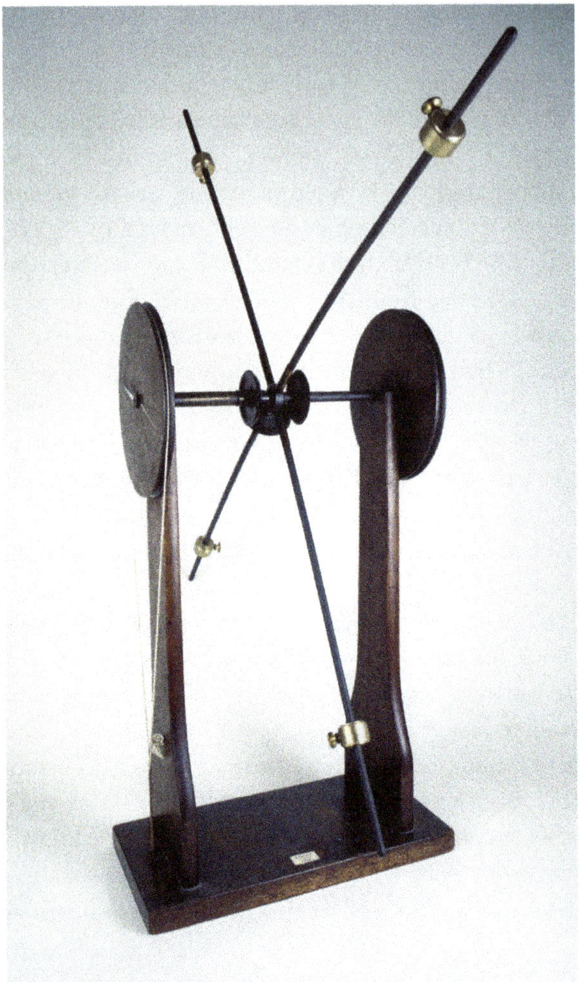

FIGURE 5.1
Maxwell's electromagnetic induction demonstration apparatus, made by Elliott Brothers, London, 1876
WHIPPLE MUSEUM OF THE HISTORY OF SCIENCE, WH.2455

important historiographic consequences for instrument historians.[27] The tools given to us by Gunther, Zinner, Daumas, Taylor, Price, and others, honed on the study of craft communities, are inadequate to the task of describing many of the phenomena of modern techno-science: mass produced instrumentation; large-scale "engine" or "tron" science; the elaborate "hacking" and "cannibalisation" of instruments within the laboratory; survey and consortium science, and so on.

We have now come full circle, to see how instrument studies developed and why the ontological question leads to difficulties as we progress into the

27 See Peter Galison and Bruce Hevly, eds., *Big Science: The Growth of Large-Scale Research* (Stanford: Stanford University Press, 1992).

twentieth century. We might pause to wonder why the same difficulties have not been encountered if we travel backwards in time. The methodological legacy of Daumas and Taylor has been surprisingly resilient against the charges of anachronism and demarcation that have bedevilled the history and philosophy of science. This, in my view, is a consequence first of our reliance on surviving instruments, and second on the emphasis on the craft of instrument making. The collections that have come down to us seem like a kind of historiographical bedrock on which the foundations of instrument studies can be built, and the makers in whom we are interested can be traced back – and when we reach the beginning of the lineage we look further back, at texts and earlier survivals, situating these in new contexts. To raise questions of anachronism would be to undermine those foundations, and this is more or less unthinkable.

Is this a helpful way to proceed? Yes and no. Avoiding the question of anachronism is a virtue: the ways around anachronism in the history and philosophy of science are now well mapped, but they are tortuous.[28] In my view, the philosophical and sociological questions I have just raised are far more productive than questions of demarcation and terminology. We do well to rely on collections. The problem, however, as J.V. Field pointed out in what may be considered a counterpart to Warner's essay, is that if we rely on surviving artefacts we are more likely to *ignore* large areas of the history of science than we are to be overly inclusive in our definitions.[29] Architecture and scale are the big losers here: especially in the history of astronomy, size is a virtue for reasons of accuracy – but large instruments rarely survive. We have very little between the portable astrolabe and the surviving buildings that functioned as instruments, and this tends to lead to a neglect of the latter.[30] Could instrument studies become more capacious in the kinds of objects it considers? Richard Sorrenson's "The Ship as a Scientific Instrument" could be the foundation work of a new strand of research here.[31] I return to this crucial question in my conclusion below. Just as important, the reliance on surviving artefacts and craft relations leads to obvious lacunae where textual traditions, lost instruments, and earlier and non-western instrument making are concerned.

28 See, for example, Nick Tosh, "Anachronism and Retrospective Explanation: In Defence of a Present-Centred History of Science," *Studies in History and Philosophy of Science*, Part A 34 (2003): 647–659.

29 J.V. Field, "What is Scientific About a Scientific Instrument?" *Nuncius* 3 (1988): 3–26.

30 This point is extensively discussed in Liba Taub, "What is a Scientific Instrument, Now?" *Journal of the History of Collections* 31 (2019): 453–467.

31 Richard Sorrenson, "The Ship as a Scientific Instrument in the Eighteenth Century," *Osiris* 11 (1996): 221–236.

3 Making History

So far I have presented an outline of the history of instrument studies, and raised some of the methodological questions that seem to me to be implicit in its focus on surviving artefacts. The history of collections themselves shapes the discipline and its concerns; it provides certain limits on what we might consider to *be* a scientific instrument, and it has, for better or worse, given us an apparently endless source of case studies that appear to need little or no theoretical justification or historiographic reflection. I have suggested that the philosophy of science and the history and sociology of technology can usefully help us to situate our researches. Now it is time to turn to the constructive aspect of my analysis.

Again we can begin with the SIC's founding aims: to compile inventories, produce catalogues, and share knowledge. At the outset I said that these do not amount to a programme of historical research. But that is not to say that they do not have significant historiographic consequences. We have seen that Daumas proposed instrument making itself as the core of instrument studies. As his contemporary Derek Price pointed out (in his famous exposé of fake antique scientific instruments) clarification of the surviving artefactual record must form the basis of a history of instrument making.[32] Inventories, catalogues, and the sharing of information are, therefore, absolutely central to instrument studies.[33] But catalogues do far more than this. They provide a means of collective analysis of places, instrument types, and makers. Again, Price's paper on fakes makes this point very eloquently. Price was discussing a group of instruments ascribed to the maker Johannes Bos, in which at least some had to be

32 D.J. Price, "Fake Antique Scientific Instruments," in *Actes du VIIIe Congrès International d'Histoire des Sciences: Florence – Milan 3–9 Septembre 1956* (Vinci: Gruppo Italiano di Storia delle Scienze, 1958), 380–394. See also, *inter alia*, G.L'E. Turner and A.D. Morrison-Low, "Zinner's Ghosts and a Curious Date: 1576," *Bulletin of the Scientific Instrument Society*, no. 50 (1996): 6–10. For this point applied to E.G.R. Taylor's somewhat imprecise work, see M.A. Crawford, "Makers and Dates," *Bulletin of the Scientific Instrument Society*, no. 13 (1987): 2–8. An important "meta-catalogue" is M. Holbrook, R.G.W. Anderson, and D.J. Bryden, *Science Preserved: A Directory of Scientific Instruments in Collections in the United Kingdom and Eire* (London: HMSO, 1992).

33 The introductions to catalogues naturally make this point – but see also Jim Bennett, "Mathematical Instruments in Florence: A Cataloguing Project and Its Relevance for Historians of the Sixteenth Century," in *Mathematical Instruments between Material Artifacts and Ideal Machines: Their Scientific and Social Role before 1950*, ed. S. Gessner, U. Hashagen, J. Peiffer, and D. Tournès, Oberwolfach Research Institute for Mathematics Reports, 58 (Oberwolfach, Germany: Oberwolfach Research Institute for Mathematics, 2017), 7–10.

fakes. As Price put it, Bos "becomes incomprehensible as an historic person unless the fakes are identified and discounted from the record."[34] We can turn this around and say that instrument makers become comprehensible precisely by consideration of instruments that we have in our collections. And this is taken even further now that we have such extraordinarily rich data on instrument makers themselves.[35]

But how does this actually work? What kinds of evidence do instruments provide? As Gerard Turner has pointed out, Penelope Gouk's *Ivory Sundials of Nuremberg* gives an exemplary statement of how to proceed in the analysis of instruments as they are found in collections.[36] These are the seven questions Gouk attempts to answer in her book:

1. What are they and how do they work?
2. Why were they made of ivory and where did it come from?
3. Why were they made at Nuremberg?
4. Who made them?
5. How were they made?
6. What was the technical and decorative context?
7. What was the market for them?

This is strongly reminiscent of Michael Baxandall's account of the general applicability of an art historical method in his *Patterns of Intention*.[37] This is no coincidence. Aside from the fact that Baxandall's father was a curator at the Science Museum, there is a general confluence of interests in art history and the history of instruments, owing to an interest in the history of making.[38] Like instrument studies, art history has its origins in curatorial expertise and

34 Price, "Fake Antique Scientific Instruments," 393.

35 Notably Clifton, *Directory of British Scientific Instrument Makers;* and Beaudouin, Brenni, and Turner, *Dictionary of Precision Instrument-Makers and Related Craftsmen (France & Switzerland)*. See also Bedini, *Early American Scientific Instruments and their Makers*; D.J. Bryden, *Scottish Scientific Instrument-Makers* (Edinburgh: Royal Scottish Museum, 1972); J.E. Burnett and A.D. Morrison-Low, *"Vulgar and Mechanick": The Scientific Instrument Trade in Ireland 1650–1921* (Edinburgh: National Museums Scotland, 1989); T.N. Clarke, A.D. Morrison-Low, and A.D.C. Simpson, *Brass and Glass: Scientific Instrument Making Workshops in Scotland as Illustrated by Instruments from the Frank Collection at the Royal Museum of Scotland* (Edinburgh: National Museums Scotland, 1989); and Morrison-Low, *Making Scientific Instruments in the Industrial Revolution.*

36 Penelope Gouk, *The Ivory Sundials of Nuremberg, 1500–1700* (Cambridge: Whipple Museum of the History of Science, 1988).

37 Michael Baxandall, *Patterns of Intention: On the Historical Explanation of Pictures* (New Haven and London: Yale University Press, 1985).

38 See "Obituary Notices: Baxandall, David," *Monthly Notices of the Royal Astronomical Society* 99 (1939): 299.

connoisseurship.[39] In each case this has resulted in close attention to the ways in which artefacts were constructed, and an interest in the forms of expertise and professional identity of craftspeople. But while instrument studies has had an uneasy relationship with what we might broadly call the "reception" side of instrumentation – i.e. the history of science itself – art history has no such problems, as exemplified in the work of Baxandall. *Patterns of Intention* in fact moves directly from a consideration of the artefact as it comes down to us to an account of the social context within which the creation of art works can be explained. Importantly, he points out that no amount of minute focus on the actions of the maker will be sufficient to *explain* the resulting artwork, which will always require some understanding of the intellectual, stylistic, and even emotional environment in which it was meant to be experienced.[40]

What are the lessons to be learnt from Baxandall's approach? First, he is insistent that even though the minutiae of artistic creation are not sufficient for the explanation of a painting, they are necessary; each gesture and motif is embedded in a social whole, and the closer one looks the more one finds. Here there are strong analogies with the kind of close attention instrument historians have paid to particular instruments. At the SIC meeting that prompted this volume, for example, Stephen Johnston showed how we can begin to reconstruct the uses to which astrolabes were put if we look closely at the reduced form of the instrument bearing only the Regiomontanus house divisions. Here we might fruitfully invoke Baxandall's famous notion of the "period eye."[41] Paintings, for Baxandall, can only be understood if we already know something of the way in which people approached them – specifically what kinds of visual interpretations they were primed to favour. In fact his canonical statement of this concerns mathematical instruments. In *Painting and Experience in Fifteenth-Century Italy* he argues that the intense interest in practical forms of measurement – with instruments in the German lands, rules of thumb in the Italian city-states – explains the highly formalised architectural construction

39 Another important area of study considering similar questions is material culture studies; see Jules David Prown, "Mind in Matter: An Introduction to Material Culture Theory and Method," *Winterthur Portfolio* 17 (1982): 1–19; Ivan Gaskell and Sarah Anne Carter, eds., *The Oxford Handbook of History and Material Culture* (Oxford: Oxford University Press, 2020).

40 A similar (and similarly useful) argument is made by D.F. McKenzie in his *Bibliography and the Sociology of Texts* (Cambridge: Cambridge University Press, 2004).

41 See Michael Baxandall, *Painting and Experience in Fifteenth Century Italy* (Oxford: Oxford University Press, 1972); Michael Baxandall, *The Limewood Sculptors of Renaissance Germany* (New Haven: Yale University Press, 1980). Stephen Johnston, "Astrolabes and Astrology: Instrumental Transformations in 16th century Europe," paper given at the Scientific Instrument Symposium XLI, Athens, September 2022.

of many Renaissance paintings, and gives us a means of understanding how they would have been seen.

In certain cases the art historical approach has dovetailed more fully with the history of scientific instruments. Take, for instance, the bizarre "square astrolabe" of Johann Stabius (*Fig. 5.2*). This astonishing astronomical diagram, known as the "Astrolabium imperatorium," was printed at Nuremberg in 1515, and at first glance defies interpretation. The underlying grid is at least comprehensible: it represents lines of right ascension, celestial longitude, and (diagonally) zodiacal divisions. But what can be the meaning of the wavy lines that dominate the image? These, as Richard Kremer explains in his useful account of the diagram, represent astrological houses, and their winding paths are a function of the square format of the diagram. For Kremer, the instrument is a "geometrical tool," or even a combination of geometrical tools – supra textual devices that could yield results and further understanding of the celestial sphere (though not necessarily provide results of sufficient accuracy for astrological practice). Kremer's technical analysis is complemented by

FIGURE 5.2 Johann Stabius, *Astrolabium imperatorium* (Nuremberg: Johann Stuchs, 1512)
BAYERISCHE STAATSBIBLIOTHEK, EINBL. VIII,12

Alexander Marr's equally fine analysis, in which the print is discussed as an example of an epistemic image, gaining its value to early viewers and users through its successful integration of parts, uniting "clarity (perspicuitas) … ornateness (ornatus); and … appropriateness (decorum)."[42] Most strikingly, Marr situates the print in the context of what Baxandall called the characteristic "Nuremberg line" – the calligraphic *non*-geometrical line so conspicuous in the golden locks of hair in Albrecht Dürer's famous self-portrait in the guise of Christ.[43]

Another entirely different route to this kind of integrated understanding of the craft of the instrument maker is the "scientific" or "forensic" analysis of instruments themselves. Here the studies by Tiemen Cocquyt, Michael Korey and Marvin Bolt on early optical instruments have been exceptionally revealing.[44] They have shown, for example, that the crucial innovation of early telescope makers was to find the "sweet spot" of the lens and create an aperture around that spot, dramatically improving optical performance. Cocquyt and his collaborators have also made exciting discoveries about Leeuwenhoek's famous microscopes, showing that his methods were not so formally inventive as had previous been supposed – though his skill in creating lenses to others' designs was still of course superlative.[45]

What emerges from these selected cases is that the figure of the instrument maker is of profound importance for instrument studies. This sounds obvious but it is not. The instrument maker is an unusual kind of craftsperson, concerned as he or she is not only with creating an instrument that measures, demonstrates, or creates some phenomenon, but also with the relationship between that phenomenon and the nature of the device being created. This can be illustrated by comparing instruments with other material objects of interest to historians of science and technology; a building must perform certain functions, and these are related to the architect's and builder's knowledge of material and structure – but the building doesn't record or register or demonstrate or create phenomena in any complex sense beyond its aesthetics and functionality. If, say, the building is the Monument in London, which was

42 Richard L. Kremer, "Playing with Geometrical Tools: Johannes Stabius's *Astrolabium imperatorium* (1515) and its Successors," *Centaurus* 58 (2016): 104–134; Alexander Marr, "Ingenuity in Nuremberg: Dürer and Stabius's Instrument Prints," *The Art Bulletin* 100 (2018): 48–79.

43 Marr, "Ingenuity," 70.

44 Marvin Bolt, Tiemen Cocquyt, and Michael Korey, "Johannes Hudde and his Flameworked Microscope Lenses," *Journal of Glass Studies* 60 (2018): 207–222; Marvin Bolt and Michael Korey, "Dioptrice," *Journal of Glass Studies* 61 (2019): 270–276.

45 Tiemen Cocquyt, *et al.*, "Neutron Tomography of Van Leeuwenhoek's Microscopes," *Science Advances* 7, no. 20 (2021).

supposed to function as a transit instrument, then the architect and builder face a new problem – the problem of precision measurement. All the issues of calibration and working knowledge rush back in. This is missed – and missed badly – when theoretical statements about "materiality" and "practice" conflate instruments with other kinds of objects.

Here we have a more concrete version of an argument made as long ago as the seventeenth century by Robert Hooke – that instruments are both tools and objects of inquiry.[46] The instrument maker crafts rules and dividers, but the rule and divider are emblematic of craft itself – they are the foundation of geometric practice, which underlies the creation of ... rules and dividers (*Fig. 5.3*).

But the question is not just about individuals at particular times; rather we are interested in *kinds* of craft skill, about *patterns* of making, *communities* of expertise. Another of Baxandall's terms comes in useful here. In *Patterns of Intention,* he talked of "troc," which can be roughly translating as "the barter" – a looser term for a market transaction:

> It is no more than a *form* of relation in which two classes of people, both within the same culture, are free to make choices in the course of an exchange, any choice affecting the universe of the exchange and so the other participants.[47]

When we think of instruments, I suggest that we should change the order of Gouk's list, beginning not with function and working our way towards the market, but instead beginning with this dualistic understanding of the object and its social, political, and intellectual context; it must be something made as part of a negotiation, between (usually) a maker and a user. Here Baxandall's remarkable analysis of the construction of the Firth of Forth Bridge comes to mind, in which he gives a list of some twenty-four "cause-suggesting features of the narrative" of the bridge.[48] What, we might ask, are the *possible* "cause-suggesting features" of the story of the Stabius astrolabe, or the simplified astrolabes? Baxandall's list is intentionally Borgesian: "1. East-coast estuaries of Scotland. ... 4. Railways independent companies. ... 7. Inchgarvie Islet in the Forth. ... 10. The Tay Bridge disaster. ... 16. Cantilever model in Orient. ... 22. William Arrol's executive virtuosity." The point is to widen the scope of the

46 For Hooke's concept of an instrument, see Jim Bennett, "Instruments and Ingenuity," in *Robert Hooke: Tercentennial Studies*, ed. Michael Cooper and Michael Hunter (Aldershot: Routledge, 2006), 65–76.

47 Baxandall, *Patterns of Intention*, 48.

48 Baxandall, *Patterns of Intention*, 25.

FIGURE 5.3 Elias Allen (d. 1653), etching by Wenceslaus Hollar after Hendrik van der Borcht II, 1666. Allen, the leading London craftsman in the first half of the seventeenth century, is shown here holding dividers typical of his style METROPOLITAN MUSEUM OF ART, 49.95.2674

"curatorial eye." The cataloguing impulse and forensic attention to materials, marks, and points of construction are not necessarily distinct from this kind of catholic causal inquiry.

Returning to "troc" – the negotiation we are trying to reconstruct – sometimes we will be in the world of a relatively straightforward commercial transaction, with a manufacturing firm, various markets, and consumers. Modern mass-produced microscopes might be a good example. Here we can zoom out to look at sales figures, prices, and other data, or we can zoom in to look at individual objects and their journeys.[49] For earlier instruments the utility of "troc" will perhaps be greater, because the agency of individual firms and craftspeople will be tied in complex ways to various other interested parties such as science popularisers, retailers, patrons, and purchasers, not to mention individual users and their needs. "Troc" functions too in the use and reuse of instruments over time – their creation, maintenance, "cannibalisation" and ultimate destruction or survival. Each of these moments involves a form of bartering – what should be kept, who should do the mending, and to what end, and so on.

Personally I prefer to think of instrument studies as requiring some blend of Baxandall's slightly awkward Francophone "troc" and Jim Secord's perhaps overly epistemological "knowledge in transit" – the word that comes to mind is *transactional* – but I am not bold enough to propose and name a new paradigm for instrument studies. The idea of "transaction" nicely captures the fact that when we are talking about instruments we are always of necessity talking about forms of reified labour – who was it that had the motivation, the time, the financial incentive to create this object and for whom? Unlike the purely aesthetic object, as I said above, we do not need to worry so much about the affective or the qualitative features of the artefact. Unlike the scientific text or image, we are not necessarily dealing with something that is well understood in terms of its propositional content. And, again, we are dealing with something that embodies not just forms of knowledge but also social and economic

49 For the wider picture I am thinking of an analysis such as Paolo Brenni's "Some Considerations about the Prices of Physics Instruments in the Nineteenth Century," in *How Scientific Instruments Have Changed Hands*, ed. A.D. Morrison-Low, Sara J. Schechner, and Paolo Brenni (Leiden: Brill, 2016), 57–87. Studies of individual instruments are numerous, but the example that comes to mind is Pedro Ruiz-Castell, "An Instrument from Communist Europe in Franco's Spain," in *East and West: The Common European Heritage, Book of Abstracts of the xxv Scientific Instrument Symposium*, ed. Ewa Wyka et al. (Cracow: Jagiellonian University Museum, 2006), 59–64.

relations. These insights come directly from engagement with the world of craft and use and provenance that is specific to the curatorial mind.

4 The View from the Museum

Now we can return to the SIC's founding aims once more, and ask whether it has proved possible to extract a historiographic programme from the study of surviving artefacts? Ultimately, of course, it will be possible to tell all kinds of stories about instruments without ever setting eyes on them. But as I said at the outset, it is also possible to approach instruments and indeed other topics *as if* one is dealing with surviving artefacts. What the curatorial experience gives us is a tool-kit, of the sort enumerated by Gouk above.

In general my approach here has been to argue that instrument studies has a distinct identity, and works from the premise that surviving artefacts matter. I have explored some of the consequences of this and tried to show ways in which a more satisfying theory of scientific instruments could be constructed around our case studies. Another far more radical approach is to investigate historiography itself, from the point of view of those who inventory, catalogue, and share information about instruments. What have been their historical concerns? Why do they do what they do?

A simple answer is that SIC members generally are involved in a project of salvaging and organising the material remnants of science, just as they find them. This is something like the project that Derek Price had in mind when he argued forcefully that inventories, catalogues, and information sharing were necessary for any kind of materially-oriented history of science. Price even took this further, proposing a form of "scientometrics" that would prove certain facts about the history of science – an approach that has rarely been taken up since.[50] These positivistic aims aside, might it be possible to argue that cataloguing as an activity in its own right constitutes historical work? Certainly catalogues are shorn of any coherent historical narrative in the traditional sense. They tend to be typological – but within typology they also tend to be

50 See, for example, D.J. Price, "International Checklist of Astrolabes," *Archives internationales d'histoire des sciences* 32 (1956): 243–263 and 33 (1956): 363–381. A more recent attempt in this direction is Gerard Turner's "The Microscope as a Technical Frontier in Science," in *Historical Aspects of the Microscope*, ed. S. Bradbury and G.L'E. Turner (Cambridge: Heffers, 1967), 175–199. A more similar though more historiographically subtle approach is taken in Allan Chapman, "A Study of the Accuracy of Scale Graduations on a Group of European Astrolabes," *Annals of Science* 40 (1983): 473–488.

chronological. It is possible to "read" catalogues, especially if one is interested in the development of a particular instrument type. This curious activity has a dual effect; one learns a kind of framework technological history at the same time as one gains tremendous insight into the specifics of individual objects. This of course tends towards the "tunnel" approach mentioned above, so it does need to be treated with caution. Many members of the SIC, and hopefully some readers from further afield, will know this experience. It is history from the point of view of objects themselves – or perhaps better, from the point of view of the museum, because what one is really also learning about is the institutional culture of collecting.

If there is an embedded history in the catalogue, then not only do we need to reassess the kinds of work that go into cataloguing (drawing on Penelope Gouk's list of questions to ask of an object, above), but we also need to consider other kinds of invisible labour in museum work – by which I mean labour that is visible within the museum but then is occluded upon public presentation of research work. Curators, educators, accessibility experts, conservators, and exhibition creators are all involved in making meaning out of historical collections, yet they may not think of themselves as historians, owing in part to Big Categories like "historiography" itself.

In this sense my proposal is that we embrace the accusation of "antiquarianism" that is sometimes made about instrument studies. As others have pointed out, not only is the antiquarian tradition an important and neglected aspect of historiography, but it can also advance a surprisingly radical programme of inquiry.[51] In this vein, and returning to the demarcation question that I left hanging above, I conclude with a second proposal – that historians of instruments embrace a far wider range of objects than have previously been considered legitimate.

The "transactional" approach that I have been advocating brings the demarcation question into closer focus. "What is a scientific instrument?" can be changed (echoing Gwyn Williams's famous book *When Was Wales?*) to: "When is a scientific instrument?"[52] What are the historical circumstances in which some device or artefact comes to be seen as "scientific"? This is not the same as asking when "science" begins (and ends) and then placing instruments into the story. Rather it is a plea for understanding instrumentation as an emergent property of certain kinds of inquiry. Some extreme examples might help to

51 See, for example, John Berger, "Walter Benjamin, Antiquarian and Revolutionary," in *Landscapes: John Berger on Art*, ed. T. Overton (London: Verso Books, 2016), 54–59.

52 Gwyn A. Williams, *When Was Wales?* (London: British Broadcasting Corporation, 1979).

clarify this point. In his 1633 *Circles of Proportion*, William Oughtred explains how to turn the fabric of a city into a mathematical instrument:

> The declination of an vpright wall, or Window being given, it may be found at what houre the Sunne vpon any day in the yeare will come to that Wall, or Window, and when it will goe from it. As in the former example, There is about London a Northerne wall declining Westward 35 deg. I would know at what time of the day the Sunne will begin to shine vpon it on the 24th day of March.[53]

This is not simply a trick that can be played by the cosmographically literate – rather, as, Simon Werrett points out, it is an example of "making use" of everyday items (in this case buildings) that was characteristic of the "domestic oeconomy" of early modern experimental investigations.[54]

The ultimate extension of this – in terms of scale, complexity, societal impact, and conceptual ambition – might be Roger Revelle's claim that in the Anthropocene era the Earth itself has been instrumentalised:

> In one respect, however, human beings during the next few decades may, almost in spite of themselves, be doing something that will have a major effect on the climate of the earth. I refer to the combustion of coal, oil, and natural gas by our worldwide industrial civilization, which adds carbon dioxide to the atmosphere. … From the standpoint of meteorologists and oceanographers we are carrying out a tremendous geophysical experiment of a kind that could not have happened in the past or be reproduced in the future.[55]

SIC members are not necessarily expected to apply their cataloguing zeal to the Earth's altered atmosphere. But the rhetoric of this statement is instructive. What, we might ask, are the limits of "instrumentalisation?" Elsewhere,

53 William Oughtred, *The Circles of Proportion and the Horizontal Instrument* (London, 1633), 137.

54 Personal communication; see also Simon Werrett, *Thrifty Science: Making the Most of Materials in the History of Experiment* (Chicago: University of Chicago Press, 2019).

55 *Second Supplemental Appropriation Bill, 1956: Hearings before Subcommittees of the Committee on Appropriations, House of Representatives*, 84th Congress, 2nd session (1956), 467 (testimony of Roger Revelle, oceanographer). The claim was repeated a year later in Revelle's famous paper, co-authored with Hans Suess, "Carbon Dioxide Exchange Between Atmosphere and Ocean and the Question of an Increase of Atmospheric CO, during the Past Decade," *Tellus* 9 (1957): 18–27.

animals and humans have become the registering apparatus of experiments, and, more benignly, all sorts of natural and artificial objects have been incorporated into instrumental set-ups. Bringing the rigour and attention to materiality to bear on these kinds of real and rhetorical instruments might be a fruitful extension of this idea of the transaction.

An attentive reader might at this point suggest that transactions are not so different from transitions, and that Jim Secord's essay on the latter suggests many of the same themes that I have been emphasising. No doubt this is true. It is also true that much current literature in STS, *pace* Biagioli, deals with the ways in which socio-material negotiations happen, especially across borders and between social groups with different agendas.[56] In this broader sense, instrument studies – or at least my take on it – only swims with the tide, though this doesn't mean that all swimmers use the same stroke.

Bibliography

Anderson, R.G.W. "Connoisseurship, Pedagogy or Antiquarianism? What Were Instruments Doing in the Nineteenth-Century National Collections in Great Britain?" *Journal of the History of Collections* 7 (1995): 211–225.

Anderson, R.G.W., and G.L'E. Turner, eds. *An Apparatus of Instruments: The Role of the Scientific Instrument Commission*. London: The Scientific Instrument Commission of the International Union of the History and Philosophy of Science, 1993.

Anderson, R.G.W., J.A. Bennett, and W.F. Ryan, eds. *Making Instruments Count: Essays on Historical Scientific Instruments Presented to Gerard L'Estrange Turner*. Aldershot: Variorum, 1993.

Ash, Eric. *Power, Knowledge, and Expertise in Elizabethan England*. Baltimore: Johns Hopkins University Press, 2004.

Baxandall, Michael. *Painting and Experience in Fifteenth Century Italy*. Oxford: Oxford University Press, 1972.

Baxandall, Michael. *Patterns of Intention: On the Historical Explanation of Pictures*. New Haven: Yale University Press, 1985.

Baxandall, Michael. *The Limewood Sculptors of Renaissance Germany*. New Haven: Yale University Press, 1980.

Beaudouin, Denis, Paolo Brenni, and Anthony Turner. *The Dictionary of Precision Instrument-Makers and Related Craftsmen (France & Switzerland, 1430–1960)*.

56 For an outstanding example, see Annemarie Mol, *The Body Multiple: Ontology in Medical Practice* (Durham, NC: Duke University Press, 2002).

Last modified July 2018. https://bibnum.explore.psl.eu/s/dictionarypim/page/prolegomena.

Bedini, Silvio. *Early American Scientific Instruments and Their Makers*. Washington, DC: Museum of History and Technology, Smithsonian Institution, 1964.

Bedini, Silvio. *Giuseppe Campani, "Inventor Romae": An Uncommon Genius*. Edited by Cristiano Zanetti. Leiden: Brill, 2021.

Beeley, Philip. "Practical Mathematicians and Mathematical Practice in Later Seventeenth-Century London." *British Journal for the History of Science* 52 (2019): 225–248.

Bennett, Jim. "Cosmography and the Meaning of Sundials." Pp. 249–262 in *Nature Engaged: Science in Practice from the Renaissance to the Present*, edited by Mario Biagioli and Jessica Riskin. New York: Palgrave Macmillan, 2012.

Bennett, Jim. "Instruments and Ingenuity." Pp. 65–76 in *Robert Hooke: Tercentennial Studies*, edited by Michael Cooper and Michael Hunter. Aldershot: Ashgate, 2006.

Bennett, Jim. "Knowing and Doing in the Sixteenth Century: What Were Instruments for?" *British Journal for the History of Science* 36 (2003): 129–150.

Bennett, Jim. "Mathematical Instruments in Florence: A Cataloguing Project and Its Relevance for Historians of the Sixteenth Century." Pp. 7–10 in *Mathematical Instruments between Material Artifacts and Ideal Machines: Their Scientific and Social Role before 1950*, edited by Samuel Gessner, Urs Hashagen, Jeanne Peiffer, and Dominique Tournès. Oberwolfach Research Institute for Mathematics Reports, 58. Oberwolfach: Oberwolfach Research Institute for Mathematics, 2017.

Bennett, Jim. "Museums and the Establishment of the History of Science at Oxford and Cambridge." *British Journal for the History of Science* 30 (1997): 29–46.

Bennett, Jim. "The Challenge of Practical Mathematics." Pp. 176–190 in *Science, Culture and Popular Belief in Renaissance Europe*, edited by S. Pumfrey, P. Rossi and M. Slawinksi. Manchester: Manchester University Press, 1991.

Bennett, Jim. "What is the History of Scientific Instruments? Reflections on Maurice Daumas, *Les Instruments scientifiques aux xviie et xviiie siècles*." *Sphaera Occasional Paper*, no. 3. Oxford: Museum of the History of Science, 1997.

Bennett, Jim. "Wind-Gun, Air-Gun or Pop-Gun: The Fortunes of a Philosophical Instrument." Pp. 221–246 in *The Mindful Hand: Inquiry and Invention from the Late Renaissance to Early Industrialisation*, edited by Lissa Roberts, Simon Schaffer and Peter Dear. Amsterdam: Koninkliijke Nederlandse Akademie van Wetenschappen, 2007.

Berger, John. "Walter Benjamin, Antiquarian and Revolutionary." Pp. 54–59 in *Landscapes: John Berger on Art*, edited by T. Overton. London: Verso Books, 2016.

Biagioli, Mario, ed. *The Science Studies Reader*. New York and Abingdon: Routledge, 1999.

Bolt, Marvin, Tiemen Cocquyt, and Michael Korey. "Johannes Hudde and his Flame-worked Microscope Lenses." *Journal of Glass Studies* 60 (2018): 207–222.

Bolt, Marvin, and Michael Korey. "Dioptrice." *Journal of Glass Studies* 61 (2019): 270–276.

Brenni, Paolo. "Some Considerations about the Prices of Physics Instruments in the Nineteenth Century." Pp. 57–87 in *How Scientific Instruments Have Changed Hands*, edited by A.D. Morrison-Low, Sara J. Schechner, and Paolo Brenni. Leiden: Brill, 2016.

Bryden, D.J. *Scottish Scientific Instrument-Makers, 1600–1900*. Edinburgh: Royal Scottish Museum, 1972.

Burnett, J.E., and A.D. Morrison-Low. *"Vulgar and Mechanick": The Scientific Instrument Trade in Ireland 1650–1921*. Edinburgh: National Museums Scotland, 1989.

Campbell, Lewis, and William Garnett. *The Life of James Clerk Maxwell*. London: Macmillan, 1882.

Carroll-Burke, Patrick. "Tools, Instruments and Engines: Getting a Handle on the Specificity of Engine Science." *Social Studies of Science* 31 (2001): 593–625.

Chapman, Allan. "A Study of the Accuracy of Scale Graduations on a Group of European Astrolabes." *Annals of Science* 40 (1983): 473–488.

Clarke, T.N., A.D. Morrison-Low, and A.D.C. Simpson. *Brass and Glass: Scientific Instrument Making Workshops in Scotland as Illustrated by Instruments from the Arthur Frank Collection at the Royal Museum of Scotland*. Edinburgh: National Museums Scotland, 1989.

Clay, Reginald S., and Thomas H. Court. *The History of the Microscope, Compiled from Original Instruments and Documents, up to the Introduction of the Achromatic Microscope*. London: Griffin, 1932.

Clifton, Gloria. *Directory of British Scientific Instrument Makers, 1550–1851*. London: Philip Wilson, 1994.

Cocquyt, Tiemen *et al.* "Neutron Tomography of Van Leeuwenhoek's Microscopes." *Science Advances* 7, no. 20 (2021).

Cormack, Lesley B. "Mathematics for Sale: Mathematical Practitioners, Instrument Makers, and Communities of Scholars in Sixteenth-Century London." Pp. 69–85 in *Mathematical Practitioners and the Transformation of Natural Knowledge in Early Modern Europe*, edited by Lesley B. Cormack, Steven A. Walton, and John A. Schuster. Cham: Springer, 2017.

Crawforth, Michael A. "Makers and Dates." *Bulletin of the Scientific Instrument Society*, no. 13 (1987): 2–8.

Daumas, Maurice. *Les Instruments scientifiques aux XVIIe et XVIIIe siècles*. Paris: Presses universitaires de France, 1953.

Daumas, Maurice. *Scientific Instruments of the Seventeenth and Eighteenth Centuries and their Makers*. Translated by Mary Holbrook. London: Batsford, 1972.

Field, J.V. "What is Scientific About a Scientific Instrument?" *Nuncius*, 3 (1988): 3–26.

Galison, Peter, and Bruce Hevly, eds. *Big Science: The Growth of Large-Scale Research.* Stanford: Stanford University Press, 1992.

Gaskell, Ivan, and Sarah Anne Carter, eds. *The Oxford Handbook of History and Material Culture.* Oxford: Oxford University Press, 2020.

Gouk, Penelope. *The Ivory Sundials of Nuremberg, 1500–1700.* Cambridge: Whipple Museum of the History of Science, 1988.

Gould, R.T. *The Marine Chronometer: Its History and Development.* London: J.D. Potter, 1923.

Gunther, R.T. *Astrolabes of the World.* 2 vols. Oxford: Oxford University Press, 1932.

Haraway, Donna. "Situated Knowledge: The Science Question in Feminism as a Site of Discourse on the Privilege of Partial Perspective." *Feminist Studies* 14 (1988): 575–599.

Higton, Hester. "Does Using an Instrument Make You Mathematical? Mathematical Practitioners of the 17th Century." *Endeavour*, 25 (2001): 18–22.

Hill, Katherine. "'Juglers or Schollers?' Negotiating the Role of a Mathematical Practitioner." *British Journal for the History of Science*, 31 (1998): 253–274.

Hine, Christine. "Databases as Scientific Instruments and Their Role in the Ordering of Scientific Work." *Social Studies of Science*, 36 (2006): 269–298.

Holbrook, Mary, with additions and revisions by R.G.W. Anderson and D.J. Bryden. *Science Preserved: A Directory of Scientific Instruments in Collections in the United Kingdom and Eire.* London: HMSO, 1992.

Hughes, Jeff. "Plasticine and Valves: Industry, Instrumentation and the Emergence of Nuclear Physics." Pp. 58–101 in *The Invisible Industrialist: Manufactures and the Construction of Scientific Knowledge*, edited by Jean-Paul Gaudillière and Ilana Löwy. Basingstoke: Macmillan, 1998.

Johnston, Stephen. "Astrolabes and Astrology: Instrumental Transformations in 16th century Europe," paper given at the Scientific Instrument Symposium XLI, Athens, September 2022.

Johnston, Stephen. "Making Mathematical Practice: Gentlemen, Practitioners and Artisans in Elizabethan England." Unpublished PhD thesis, University of Cambridge, 1994.

Johnston, Stephen. "Mathematical Practitioners and Instruments in Elizabethan England." *Annals of Science* 48 (1991): 319–344.

Johnston, Stephen. "The Identity of the Mathematical Practitioner in Sixteenth-Century England." Pp. 93–120 in *Der "mathematicus": Zur Entwicklung und Bedeutung einer neuen Berufsgruppe in der Zeit Gerhard Mercators*, edited by Irmgarde Hantsche. Bochum: Brockmeyer, 1996.

King, Henry C. *The History of the Telescope.* London: Griffin, 1955.

Kremer, Richard L. "Playing with Geometrical Tools: Johannes Stabius's *Astrolabium imperatorium* (1515) and its Successors." *Centaurus* 58 (2016): 104–134.

McKenzie, D.F. *Bibliography and the Sociology of Texts*. Cambridge: Cambridge University Press, 2004.

Marr, Alexander. "Ingenuity in Nuremberg: Dürer and Stabius's Instrument Prints." *The Art Bulletin* 100 (2018): 48–79.

Maxwell, James Clerk, and J.J. Thomson. *A Treatise on Electricity and Magnetism*, 3rd ed. 2 vols. Oxford: Clarendon Press, 1892.

Mayr, Otto. "Maxwell and the Origins of Cybernetics." *Isis* 62 (1971): 424–444.

Middleton, W.E. Knowles. *The History of the Barometer*. Baltimore: Johns Hopkins University Press, 1964.

Miller, David Philip. "The 'Sobel Effect': The Amazing Tale of How Multitudes of Popular Writers Pinched All the Best Stories in the History of Science and Became Rich and Famous while Historians Languished in Accustomed Poverty and Obscurity, and How This Transformed the World. A Reflection on a Publishing Phenomenon." *Metascience* 11 (2002): 185–200.

Mol, Annemarie. *The Body Multiple: Ontology in Medical Practice*. Durham NC: Duke University Press, 2002.

Morrison-Low, A.D. *Making Scientific Instruments in the Industrial Revolution*. Aldershot: Ashgate, 2007.

"Obituary Notices: Baxandall, David." *Monthly Notices of the Royal Astronomical Society*, 99 (1939): 299.

Oughtred, William. *The Circles of Proportion and the Horizontal Instrument*. London: Augustine Matthewes, 1633.

Pickering, Andrew. *Constructing Quarks: A Sociological History of Particle Physics*. Edinburgh: Edinburgh University Press, 1984.

Pinch, Trevor. *Confronting Nature: The Sociology of Solar-Neutrino Detection*. Dordrecht: D. Reidel Pub. Co., 1986.

Price, Derek J. de Solla. "An International Checklist of Astrolabes." *Archives internationales d'histoire des sciences* 32 (1956): 243–263 and 33 (1956): 363–381.

Price, Derek J. de Solla. "Fake Antique Scientific Instruments." Pp. 380–394 in *Actes du VIIIe Congrès International d'Histoire des Sciences: Florence – Milan 3–9 Septembre 1956*. Vinci: Gruppo Italiano di Storia delle Scienze, 1958.

Price, Derek J. de Solla. "Philosophical Mechanism and Mechanical Philosophy: Some Notes toward a Philosophy of Scientific Instruments." *Annali dell'Istituto e Museo di Storia della Scienza di Firenze* 5 (1980): 75–85.

Prown, Jules David. "Mind in Matter: An Introduction to Material Culture Theory and Method." *Winterthur Portfolio* 17 (1982): 1–19.

Revelle, Roger, and Hans Suess. "Carbon Dioxide Exchange Between Atmosphere and Ocean and the Question of an Increase of Atmospheric CO_2 during the Past Decades." *Tellus* 9 (1957): 18–27.

Rohde, Alfred. *Die Geschichte der wissenschaftlichen Instrumente vom Beginn der Renaissance bis zum Ausgang des 18. Jahrhunderts*. Leipzig: Klinkhardt & Biermann, 1923.

Ruiz-Castell, Pedro. "An Instrument from Communist Europe in Franco's Spain." Pp. 59–64 in *East and West: The Common European Heritage, Book of Abstracts of the XXV Scientific Instrument Symposium*, edited by Ewa Wyka *et al.* Cracow: Jagiellonian University Museum, 2006.

Secord, James A. "Knowledge in Transit." *Isis* 95 (2004): 654–672.

Shapin, Steven. "Pump and Circumstance: Robert Boyle's Literary Technology." *Social Studies of Science* 14 (1984): 481–520.

Shapin, Steven, and Simon Schaffer. *Leviathan and the Air-Pump: Hobbes, Boyle, and the Experimental Life*. Princeton: Princeton University Press, 1985.

Sorrenson, Richard. "The Ship as a Scientific Instrument in the Eighteenth Century." *Osiris* 11 (1996): 221–236.

Taylor, E.G.R. *The Mathematical Practitioners of Hanoverian England, 1714–1840*. Cambridge: Cambridge University Press, 1966.

Taylor, E.G.R. *The Mathematical Practitioners of Tudor and Stuart England*. Cambridge: Cambridge University Press, 1954.

Taub, Liba. "What Is a Scientific Instrument, Now?" *Journal of the History of Collections* 31 (2019): 453–467.

Tosh, Nick. "Anachronism and Retrospective Explanation: In Defence of a Present-Centred History of Science." *Studies in History and Philosophy of Science*, Part A, 34 (2003): 647–659.

Traweek, Sharon. *Beamtimes and Lifetimes: The World of High Energy Physicists*. Cambridge, MA: Harvard University Press, 1988.

Turner, Anthony. "From Mathematical Practice to the History of Science: The Pattern of Collecting Scientific Instruments." *Journal of the History of Collections* 7 (1995): 135–150.

Turner, Anthony. "Interpreting the History of Scientific Instruments." Pp. 17–26 in *Making Instruments Count: Essays on Historical Scientific Instruments Presented to Gerard L'Estrange Turner*, edited by R.G.W. Anderson, J.A. Bennett, and W.F. Ryan. Aldershot: Variorum, 1993.

Turner, Gerard L'E. "A Decade in the Study of Scientific Instruments." Pp. 5–14 in *An Apparatus of Instruments: The Role of the Scientific Instrument Commission*, edited by R.G.W. Anderson and G.L'E. Turner. London: The Scientific Instrument Commission of the International Union of the History and Philosophy of Science, 1993.

Turner, Gerard L'E. *Elizabethan Instrument Makers: The Origins of the London Trade in Precision Instrument Making*. Oxford: Oxford University Press, 2000.

Turner, Gerard L'E. "The Microscope as a Technical Frontier in Science." Pp. 175–199 in *Historical Aspects of the Microscope*, edited by S. Bradbury and G.L'E. Turner. Cambridge: Heffers, 1967.

Turner, Gerard L'E., and A.D. Morrison-Low. "Zinner's Ghosts and a Curious Date: 1576." *Bulletin of the Scientific Instrument Society*, no 50 (1996): 6–10.

Warner, Deborah. "What Is a Scientific Instrument, When Did It Become One, and Why?" *British Journal for the History of Science* 23 (1990): 83–93.

The Webster Signature Database. http://historydb.adlerplanetarium.org/signatures.

Werrett, Simon. *Thrifty Science: Making the Most of Materials in the History of Experiment*. Chicago: University of Chicago Press, 2019.

Werskey, Gary. *The Visible College: The Collective Biography of British Scientific Socialists of the 1930s*. London: Allen Lane, 1978.

Williams, Gwyn A. *When Was Wales?* London: British Broadcasting Corporation, 1979.

Zinner, Ernst. *Deutsche und niederländische astronomische Instrumente des 11.–18. Jahrhunderts*. Munich: Beck, 1956.

CHAPTER 6

Expansion, Globalisation, and National Trends

Sofia Talas

The Commission for Scientific Instruments, also called the Scientific Instrument Commission (SIC), was reconstituted in 1977 by ten individuals attending the XV International Congress of the History of Science in Edinburgh, which was organised by the International Union of the History and Philosophy of Science (IUHPS). One of its aims was to "facilitate the exchange of information between research workers in the field of historic scientific instruments."[1] This led to the development of an active international network of people interested in or working on historical scientific instruments.

In this short essay, I will outline how the Commission expanded in terms of institutions and countries involved. Quite naturally, the growing exchange of knowledge and experiences contributed to internationally shared practices in scientific instruments studies, preservation, and display. However, as the paper will show, some specific national trends also emerged in relation, for instance, to research and teaching practices, inventories, and the legal protection of scientific heritage.

Let me underline that this essay is a very preliminary analysis of the topic, based on the SIC official documents, on a general survey of historical scientific instrument-related literature, and on personal impressions based on the talks presented at the SIC meetings I attended. It means to be a starting point which could, hopefully, stimulate further in-depth studies.

1 From the UK-Driven Beginnings to the SIC International Expansion … and Its Limits

The ten members who revivified the Commission in 1977 were representatives not only of the Western countries – the United Kingdom (UK), United States (USA), the Netherlands, the Federal Republic of Germany (FRG), Denmark, and France – but also of the Eastern ones, including the German Democratic Republic (GDR in English, DDR in German), Poland, and the Soviet Union

1 SIC, *Newsletter*, no. 1 (May 1980), in Gerard Turner Papers, Whipple Museum, Cambridge. See also Robert Anderson's chapter on the 1980s in this volume.

 | DOI:10.1163/9789004737587_008

(USSR).[2] And there was quite a wide participation at the symposium organised on the "Role of Scientific Instruments in the Development of Science" within the XVI International Congress of the International Union of the History and Philosophy of Science, held in Bucharest in 1981.[3] Nineteen papers were presented, with speakers from Romania, the UK, France, the USSR, USA, FRG, Hungary, the GDR, Czechoslovakia and the Netherlands, thus showing the early diffused interest in scientific instrument studies. Interestingly, of the nineteen papers presented, ten talks were held in English, seven in French, and two in German.[4]

However, in spite of this geographical breadth at the Bucharest Congress, the first years of the Commission for Scientific Instruments were undoubtedly UK-driven.[5] The first presidents (Derek Howse, from 1977 to 1981, and Robert G.W. Anderson from 1981 to 1998) and the secretary, Gerard L'E. Turner, were all based in the UK. Furthermore, the first official booklet published by the Commission in 1993 had only reports and papers by authors based in the UK – i.e., Anderson and Turner, respectively SIC chairman and secretary, Jim Bennett, who worked at the time at the Whipple Museum in Cambridge, and Robert R. Fox, who was in Oxford. The booklet's title was *An Apparatus of Instruments: The Role of the Scientific Instrument Commission*.[6] It was distributed both to the delegates at the XIXth International Congress of the History of Science, held in 1993 in Zaragoza, and to those attending the International Conference on Technological Change, held in Oxford that same year. The leading role played by UK scholars in the Commission should not be that surprising; it derived not only from the personal energy of the first presidents and secretary

2 SIC, *Newsletter*, no. 1 (May 1980). These organisers were listed as Mr H.D. Howse (UK)-Chairman, Dr S. Bedini (USA), Dr A.F. Gogelein (Netherlands), Dr Grötsch (DDR), Dr J. Jaciuk (Poland), Dr L.E. Maistrov (USSR), Dr K. Møller Pedersen (Denmark), Professor E. Poulle (France), Professor L. Von Mackensen (FRG), and Mr G.L'E. Turner (UK)-Secretary. The first SIC *Newsletter* explicitly uses the word "membership" in reference to these ten organisers, and subsequent *Newsletters* identify the participants of SIC meetings as "members." From its beginnings to today, the SIC has been a free-membership organisation, which one could join simply by participating in events and meetings.

3 SIC, *Newsletter*, no. 3 (December 1981), in Gerard Turner Papers, Whipple Museum, Cambridge.

4 During the Cold War years, the International Congress actually had four official languages: Russian, French, German, and English.

5 While the administration was UK-driven, the programme for creating national inventories was international since the first founding of the Commission. See chapters by Alison Morrison-Low and Robert Anderson in this volume for efforts to create national inventories of instruments in the 1950s, which were focused on mainland Europe and not just the UK.

6 R.G.W. Anderson and G.L'E. Turner, eds., *An Apparatus of Instruments: The Role of the Scientific Instrument Commission* (London: Scientific Instrument Commission, 1993).

but also from the gap existing at the time between the UK and the rest of the world in terms of scientific instrument-related institutions, museums, collections, exhibitions, meetings, conferences, and studies. As for UK institutions, the early SIC newsletters mention, for instance, the Magic Lantern Society of Great Britain, formed in 1976,[7] and the Royal Microscopical Society, founded in 1839.[8] In the list of new publications about scientific instruments provided in the second SIC newsletter in 1981, seven of the nineteen books and papers were published in the UK. Furthermore, the UK national inventory that Mary Holbrook had been setting up since 1968 described objects from more than 200 collections, including only those "reasonably accessible to the public," besides a summary of the well-known larger collections at the Science Museum, Wellcome Institute, National Maritime Museum, British Museum, Oxford Museum of the History of Science, Whipple Museum in Cambridge, and National Museums Scotland.[9] In other countries there were only a few museums and collections acknowledged and known at the time.

The situation was to change in a very few years thanks to the activities promoted by the SIC board and members. Starting in 1980, the SIC published a *Newsletter*, which was meant to "provide a forum for the exchange of information, and to make scholars in the field aware of others in different parts of the world."[10] While the first *Newsletter* was sent to thirty-three institutions in eighteen countries in 1980, the third *Newsletter*, distributed in 1981, apparently reached sixty-five institutions in twenty-three countries. From 1981, the Commission also inaugurated a series of international symposia and, to foster the diffusion of knowledge in the field, it started publishing and distributing an annual list of publications on historical scientific instruments. Furthermore, the Commission encouraged the publication of inventories and catalogues, underlining their value as a basic means to make historical scientific instruments emerge, make them known to other scholars, and allow their preservation and study.

These activities produced amazing results in a short time, showing that they not only met the needs of people already working in the field but also contributed to the emergence of what Robert Anderson called, in 1993, a "small but enthusiastic core of historians of instruments who are mutually encouraged

7 SIC, *Newsletter*, no. 1 (May 1980). See also Richard Kremer's chapter on the 2000s in this volume.
8 SIC, *Newsletter*, no. 10 (June 1989).
9 SIC, *Newsletter*, no. 1 (May 1980).
10 SIC, *Newsletter*, no. 1 (May 1980): 1.

by learning of each other's research and publications."[11] After the first two symposia, held, respectively, in Bucharest in 1981, as part of the XVI International Congress of the History of Science, and in Greenwich in 1982,[12] all subsequent annual SIC Symposia were hosted each year by institutions of different countries, showing a growing interest for the field in various places. Furthermore, the number of participants quickly increased and remained stable at about a hundred per meeting, with a lower participation when the symposia were held outside of Europe or were part of one of the congresses of the IUHPS. The distribution of the newsletters, the minutes of the SIC meetings (later called "plenary sessions" or "annual general meetings"), the programmes of speakers at the various symposia, as well as the SIC bibliographies,[13] show that not only the number of people involved in SIC activities increased but also the kind of institutions changed as well. Although only large national institutions were known and active at the very beginnings of the SIC, smaller museums and collections, such as university collections, started participating as well, thus enlarging and diversifying the SIC network. The SIC symposia, in particular, boosted the participation of new people from the hosting countries. At the 1986 SIC Symposium hosted in Florence, for example, speakers from several Italian university collections participated for the first time, whereas before then, only scholars from the Istituto e Museo di Storia della Scienza (now Museo Galileo) had been active participants. More recently, the SIC symposium held in Turkey in 2016 led to several Turkish scholars participating in subsequent SIC programmes. More and more collections have thus joined the network in each passing year. As Gerard Turner wrote in 1993, the "first Commission Newsletter of 1980 listed the active museums, and these could be counted on the fingers. Now an abacus would be needed."[14] Of course, to fully analyse the actual expansion and impact of the SIC's activities would require a thorough country-by-country study of the development of historical scientific instrument studies, preservation, use, and promotion over the past forty years.

Let me add that, although the SIC certainly was a strong driving force to stimulate new historical scientific instrument-related researches and activities, it simply acted as a "detector" in some cases, shedding light on what was

11 Robert G.W. Anderson, "Introduction," 1–3 in *An Apparatus of Instruments*.

12 This was actually the second meeting on the topic held at the National Maritime Museum, in Greenwich, after a first one organised in 1977.

13 In 1993, Gerard Turner emphasised that the first annual Bibliography, in 1983, included 61 entries, while the one published in December 1992 "added 94 new publications, to give a remarkable total of 725." See Gerard L'E. Turner, "A Decade in the Study of Scientific Instruments," 5–14 in *An Apparatus of Instruments*, see 6.

14 Turner, "Decade in the Study of Scientific Instruments," 7.

already going on and making already existing activities known to the rest of the community.[15] But this in turn contributed to boost the field. The creation of the Scientific Instrument Society in 1983, "independent of the Scientific Instrument Commission, but naturally overlapping on the people involved,"[16] certainly also contributed to further enhance the sharing of knowledge and practices in the field. So did the new Rete mailing list, launched by the Oxford Museum of the History of Science a few years later, in 1997.[17]

Thus far I have only discussed the SIC's expansion in terms of new countries involved, but it is worth underlining that some countries are still missing in SIC Symposia and other activities, even though they host scientific instrument-related activities. I am thinking, for instance, of Serbia where the Museum of Pharmacy of the University of Belgrade displays a collection of historical instruments with related documents and books.[18] Furthermore, some countries once connected to the SIC have nearly disappeared or become less represented. Let us recall, for instance, that the SIC was in touch with various institutions in India, in Aligarh, Bombay, Calcutta, and New Delhi in the early 1980s,[19] but only a few scholars from India were involved in the SIC symposia or in other activities in the following years, as far as I know. Furthermore, at the first Symposium, held in Bucharest in 1981, five papers out of fifteen were presented by Romanian scholars, which was, of course, due to the fact that Romania was the host nation. But Romanian scholars and scientific instrument museums and collections nearly disappeared from SIC programmes from that moment onwards. The USSR's participation is also interesting in this sense, as it was quite strong in the early years, then disappeared for a while, with a renewed participation by scholars of Russia and other Eastern countries in the 1990s after the fall of the Berlin Wall and dissolution of the Soviet Union.[20]

15 I am thinking for instance of the case of Cuba, where cutting-edge research on the nuclear scientific heritage of the country started being carried out before any contact was set up with the SIC community. These activities emerged thanks to the SIC Symposium organised in Havana in 2019.

16 Turner, "Decade in the Study of Scientific Instruments."

17 "Mailing List (Rete)," History of Science Museum, University of Oxford, https://www.hsm.ox.ac.uk/mailing-list. See SIC, *Newsletter*, no. 17 (January 1996): 4; and Giorgio Strano, "In Memory of James (Jim) Arthur Bennett (1947–2023)," *Nuncius* 39, no. 2 (2024): 529–536.

18 "Museum for the History of Pharmacy in Belgrade," Univerzitet u Beogradu Farmaceutski fakultet, last updated 2013, http://www.ph.bg.ac.rs/en/about-us/museum/1775/museum-for-the-history-of-pharmacy-in-belgrade/.

19 The first two SIC *Newsletters* were sent there. See SIC, *Newsletter*, no. 1 (May 1980) and SIC, *Newsletter*, no. 2 (July 1981).

20 See the chapter on the 1990s by Sara Schechner in this volume.

Indeed, Russia hosted the SIC Symposium in 1999, but Russian colleagues did not regularly attend subsequent SIC symposia.[21] These patterns can certainly be related to the political changes of the last forty years, but it is also important to consider how our field and the dynamics within the SIC have often depended on single individuals, their personal passion, and the opportunities they were offered to work, travel, and publish.[22] Today, historical instrument studies and collections have become more relevant, more people are involved and have permanent positions in the field, but the community still remains quite a small one, probably still quite fragile and dependent on single individuals. This fact underscores the importance of the SIC as a network to nourish relations among scattered and often isolated instrument scholars.

2 Globalisation versus National Trends

Quite naturally, setting up a commission for people dealing with historical scientific instruments led to the spreading of shared knowledge and practices in the way scientific instruments were studied, preserved, conserved, displayed, and more generally, "used." Changes in the larger field of history of science, as well as in the museum world (with new definitions of museums),[23] have also had a major impact on historical instrument-related activities. For instance, the sessions on gender or decolonisation organised for recent SIC symposia have reflected new trends in history of science and museum studies. Many practices and approaches are nowadays shared and globalised.

---

21 Papers from Russian colleagues presented in symposia organised in Eastern European countries: Symposium XIV, Prague, 1995: 7 papers; Symposium XXV, Crakow, 2006: 4 papers; Symposium XL, Prague, 2021: 0 papers.

22 Let us mention for instance Hemming Andersen, who seems to have carried out lots of activities by himself in Denmark: he introduced Hauch's Cabinet of Physics, in Sorø, to the SIC community at the 1983 Munich Symposium; he organised the Scientific Instrument Symposium XVII in Sorø on the topic in 1998 and renovated the display of Hauch's instruments. He was thus assigned the task of conducting a Danish national inventory, which was published in 1995: Hemming Andersen, *Historic Scientific Instruments in Denmark*, Matematisk-fysiske Meddelelser, 44, no. 2 (Copenhagen: The Royal Danish Academy of Sciences and Letters, 1995).

23 See the 2007 and 2022 ICOM definitions of museums at this website: "Museum Definition," ICOM, https://icom.nl/en/about-icom/museum-definition. For previous definitions, see ICOM Czech Republic, "224 Years of Defining the Museum," ICOM, https://icom.museum/wp-content/uploads/2020/12/2020_ICOM-Czech-Republic_224-years-of-defining-the-museum.pdf.

Nonetheless, there have been certain trends that have appeared to be tied to particular countries. A thorough analysis should be conducted by country, but let me share a few personal thoughts and examples, based on SIC documents, papers, and talks that I have attended at the SIC symposia.

First, I would like to offer a few words about national inventories of scientific instruments. As already mentioned, the first SIC board regarded inventories as crucial; one of the aims of the Commission in the early 1980s was "to stimulate the production of inventories of historic scientific instruments."[24] In the first SIC *Newsletter* (published in 1980), a specific section was devoted to national inventories, specifying that in 1946, "the concept of a world inventory of scientific instruments was proposed by the then Director of the Palais de la Découverte in Paris" and that in 1958, "the Union Internationale d'Histoire des Sciences decided to patronise the project, and constituted a Commission for the purpose."[25] By 1980, national inventories of scientific instruments kept in Belgium, France, Italy, Poland, and Russia had already been published, while the British inventory was in preparation. According to the minutes of the 1982 plenary session of the SIC, held in Greenwich, it "was agreed that the production of national inventories should be pursued actively" and it "was suggested that an approach should be made to the National Committees, proposing that firm suggestions about the production of these inventories should be made."[26] Furthermore, the "possibility of the translation of the USSR inventory into English was discussed, and it was agreed that this should be investigated."[27] However, according to Gerard Turner, these first national inventories "scarcely scratch[ed] the surface," as they were "highly selective," and it seems that no general criteria of selection were adopted.[28] Turner complained, for instance, that the French inventory had excluded "astrolabes, globes, sundials, and other gnomonic instruments," while "Pasteur's apparatus was included."[29] I have examined the Italian national inventory supported by the Union Internationale d'Histoire et Philosophie des Sciences and published by the

24 SIC, *Newsletter*, no. 1 (May 1980): 1, and SIC, *Newsletter*, no. 3 (December 1981): 2.

25 SIC, *Newsletter*, no. 1 (May 1980): 4. For more on national inventories, please see Alison Morrison-Low's chapter on cataloguing, as well as the chapters on the 1980s and 1990s by Robert Anderson and Sara Schechner in this book.

26 SIC, *Newsletter*, no. 4 (December 1982): 3. Here National Committees refer to groups of scholars from individual countries who represent their nations within the governing structure of the IUHPS. They act as a bridge between domestic programmes and policies and those of the international organisation.

27 SIC, *Newsletter*, no. 4 (December 1982): 3.

28 Turner, "Decade in the Study of Scientific Instruments," 8.

29 Turner, "Decade in the Study of Scientific Instruments," 8.

Museo Nazionale di Storia della Scienza e della Tecnica Leonardo da Vinci in 1963.[30] It describes about 400 mathematical, geographical, surveying, astronomical, navigation, and gnomonic instruments, but only 200 physics devices are included. Furthermore, as Gerard Turner underlined in 1993, "We would all like to have inventories to help us in our studies, but even with computers many difficulties still have to be overcome," such as "the problems of the thesauri and dictionaries."[31]

In spite of these difficulties, the SIC went on supporting the production of national inventories for years, and until the end of the 1990s, a section of the SIC plenary sessions was always dedicated to discussing the state-of-the-art of national inventories, reporting, for example, the publication of the British National Inventory in 1992,[32] and that of the Danish inventory in 1995, supported by the Royal Danish Academy. Projects for new inventories in Belgium and in Italy were also started in the 1990s,[33] but with no actual results, and nothing happened in other countries, as far as I know.

What about the situation today? A few countries do have ongoing projects aiming at creating a centralised database of their cultural heritage, and these databases include historical scientific instruments. Such huge works, far from being completed, are currently being carried out in France and Italy by the Ministries of Culture. In France, there is the *Inventaire général du patrimoine culturel* (general inventory of cultural heritage), which identifies, studies, and promotes awareness of French cultural, historical, and scientific heritage.[34] In Italy, thanks to the Istituto Centrale per il Catalogo e la Documentazione (ICCD – Central Institute for Catalogue and Documentation), the Ministry of Culture supports and supervises the collection of data about Italian heritage, on the basis of a specific law established in 2004.[35] In the UK, the Science

30 Union Internationale d'Histoire et Philosophie des Sciences, Commission pour l'Inventaire mondial des Appareils scientifiques d'intéret historique, *Inventaire des instruments conservés en Italie*, 5 vols. (Milan: Museo Nazionale della Scienza e della Tecnica Leonardo da Vinci, 1963).

31 Turner, "Decade in the Study of Scientific Instruments," 9.

32 The national inventory for Great Britain and Ireland was published with the title *Science Preserved: A Directory of Scientific Instruments in Collections in the United Kingdom and Eire*; see SIC, *Newsletter*, no. 14 (January 1993).

33 Reported in various SIC newsletters.

34 Created in 1964 under the leadership of André Malraux, then Minister of Cultural Affairs, it represents a vast scientific undertaking to learn more about France's heritage (see https://www.culture.gouv.fr/Thematiques/inventaire-general-du-patrimoine).

35 In recent years, the ICCD has promoted and coordinated the setting up of a variety of national cataloguing standards, including specific cataloguing standards for historical scientific instruments, and a dedicated platform, SIGECWeb, hosts the cataloguing

Museum is currently running, with AHRC (Arts and Humanities Research Council) funding, *The Congruence Engine: Digital Tools for New Collections Based Industrial Histories,* which covers scientific instruments.[36] Begun in November 2021, it is one of the five "Discovery Projects" funded by the AHRC under the *Towards a National Collection* funding stream. It would seem that such activities are not widespread at an international level, but similar projects may exist in some other countries. A further in-depth study would be useful to provide a complete framework of the field.

Likewise, the legal protection of historical scientific instruments seems to be quite different from one country to another. In Brazil, for instance, the preservation of scientific-technical heritage is anchored to the federal constitution and, according to Italian law, scientific instruments more than fifty years old are not fully acknowledged as "cultural heritage," but have to be treated as part of Italian national heritage.[37] This means that any loan and all conservation activities have to be authorised by the Italian Ministry of Cultural Heritage. In other countries, such as in the United States, no similar law exists at all. Here again, it would be interesting to launch a general international survey on the topic to get a more precise country-by-country overview of the situation.

The institutions that keep historical scientific instruments also vary according to the country they are in. In most cases, not only national or regional museums but also universities hold collections of scientific instruments, which formerly constituted their cabinets of physics or were recently collected after they had become obsolete for ongoing research and teaching activities. In some countries, secondary schools have also kept their original (now historical) teaching apparatus. There are significant differences in how these school collections have been preserved and studied from one place to another. In France and Spain, for instance, there are dedicated associations working on most of the existing school collections in the country. The French ASEISTE regularly organises conferences, workshops, and exhibitions, and publishes detailed

forms elaborated by various institutions throughout Italy (http://www.iccd.beniculturali.it/it/per-catalogare and http://www.iccd.beniculturali.it/it/istituto/chi-siamo).

36 See "The Congruence Engine," Science Museum Group, https://www.sciencemuseumgroup.org.uk/projects/the-congruence-engine; and the special edition of *Science Museum Group Journal,* special edition no. 20 (Autumn 2023): https://journal.sciencemuseum.ac.uk/issue/autumn-2023/.

37 See for instance: "Beni storico-artistici ed etnoantropologici," http://sabap_ssnu.beniculturali.it/it/571/beni-storico-artistici-ed-etnoantropologici or Decreto Legislativo 22 gennaio 2004, n. 42, http://www.bncrm.beniculturali.it/getFile.php?id=466.

catalogues of the scientific instruments kept in French schools.[38] Similar activities are carried out by the Spanish ANDPIH, which takes care not only of scientific instruments, but also of all heritage materials kept in schools.[39] In other countries, such as Greece, Italy, Cuba or Mexico, some of the school scientific instrument collections have been studied by school teachers and students or by university scholars. In many cases, this heritage is still partly unknown, so that it has been often dismissed or destroyed by school directors or teachers not interested in it. I cannot provide here a complete country by country picture of the ongoing activities carried out on school collections of scientific apparatus, but it would be extremely interesting to have researches carried out on this point as well.

It also seems to me that the practice of using historical scientific instruments in university teaching has not been a general international trend, but rather a country-specific activity, at least until a few years ago. It was pioneered in the US and UK, but this practice is now spreading internationally.[40] The 2023 UMAC Award presented by the International Council of Museums Committee for University Museums and Collections was given to an object-based teaching project carried out by the Chau Chak Wing Museum at the University of Sydney,[41] and the Erasmus+ Programme of the European Union recently funded "Teaching with Objects," which was carried out by several European institutions in 2022–2024.[42]

38 *Association de Sauvegarde et d'Etude des Instruments Scientifiques et Techniques de l'Enseignement*, i.e. Association for the preservation and study of teaching scientific and technical instruments. For a complete overview of the activities of the ASEISTE, see their website: http://www.aseiste.org/.

39 *Asociación Nacional para la Defensa del Patrimonio de los Institutos Históricos*, i.e. National Association for the Preservation of the Heritage of Historical institutes. See https://www.asociacioninstitutoshistoricos.org/.

40 See for instance the work of Sara Schechner at the Adler Planetarium and University of Chicago in the 1980s and at Harvard University in 2000–2024. The History of Science Society also has recognised innovative object-based teaching with historical instruments by awarding its Joseph H. Hazen Education Prize in 1999 and 2008 to curators Liba Taub (Whipple Museum, Cambridge) and Sara Schechner (Harvard Collection of Historical Scientific Instruments), respectively.

41 "News Archive: Category: UMAC Award," International Council of Museums (ICOM) Committee for University Museums and Collections (UMAC), accessed 8 January 2025, http://umac.icom.museum/category/umac-award/.

42 See "Teaching with Objects," accessed 8 January 2025, https://www.teachingwithobjects.org/. The Erasmus+ Programme of the European Union supports activities set out in the European Education Area, Digital Education Action Plan and the European Skills Agenda. Originally established by the European Union in 1987, it looked to promote

The replication method used to teach and carry out research on historical scientific instruments and the history of science could also be regarded as a national trend. Although there was some work done in France along these lines in the 1990s, in particular by Christine Blondel,[43] the primary centre has since the 1980s been based in Germany, starting at the University of Oldenburg.[44] And even though its scholars have helped to spread the replication method beyond Germany – for example to Geneva[45] – devotees to the practice remain strongly situated in Germany, in particular now Peter Heering and his group at the University of Flensburg.[46] But perhaps the replication method should be regarded as a university-related trend – Oldenburg and now Flensburg universities – rather than a national German specificity.

Finally, it would be interesting to check whether the training of museum staff who should deal with historical scientific instruments is different from one nation to the other. It is probably more developed in countries with a long-standing tradition in the field, like the UK or the Netherlands, but a serious study should be based on a detailed and worldwide analysis of a large number of universities' curricula.

As for the kind of instruments preserved in the various nations, they reflect the specific history of each country, region, or city, which is quite obvious. There are, for instance, courtly Renaissance instruments in Florence, at the Museo Galileo, as the Medici patronised science for nearly three centuries, or in Dresden, at the Mathematisch-Physikalische Salon, originated in the sixteenth-century Kunstkammer of the Electors of Saxony. In a similar way, nations with a long-standing tradition in navigation, like the Netherlands, Portugal, or the UK, quite naturally host important collections of navigational instruments.

closer cooperation between universities and higher education institutions across Europe. See https://erasmus-plus.ec.europa.eu/.

43 See for instance the classic work: C. Blondel and M. Dörries, eds., *Restaging Coulomb. Usages, controverse et réplications autour de la balance de torsion*, Biblioteca di Nuncius, Studi e Testi, 15 (Florence: Leo S. Olschki, 1994).

44 See for instance the collection of papers by authors affiliated with the Research Group on Higher Education and History of Science in Oldenburg: P. Heering, F. Rieß, and C. Sichau, *Im Labor der Physikgeschichte: Zur Untersuchung historischer Experimentalpraxis* (Oldenburg: BIS-Verlag, 2000).

45 See Jan Lacki, ed., "Editorial," in "The Replication Method in History of Science," special issue, *Archives des Sciences* 58 (2005): 93–95. This introductory paper also provides several bibliographical references.

46 See the activities carried out by Peter Heering and his group, presented at various SIC symposia, and their papers.

However, some national peculiarities seem to emerge, not directly related to the early life of instruments, but to their later life. Several university museums were for instance set up as the result of donations from wealthy private collectors in the twentieth century, and this trend seems to be quite specifically related to the Anglo-Saxon and northern European world. I am thinking for example of the Oxford History of Science Museum, which opened to the public in 1925 thanks to the collection offered by Lewis Evans in 1922 to the university,[47] or the Whipple Museum, which takes its name from Robert Stewart Whipple who offered his collection of one thousand scientific instruments to Cambridge University in 1944.[48] Furthermore, the amazing collection of 700 sundials held by the Harvard Collection of Historical Scientific Instruments was donated by the first curator, David P. Wheatland.[49] But Wheatland's collection was a consolidation of prior collections. For example, Joseph Drecker of Aachen, Germany, had built a collection of 243 sundials and 350 related books, which were purchased in 1928 by a Dutch historian Frederik Casparus Wieder in Noordwijk. With the assistance of the German historian of astronomy Ernst Zinner, who drew up an inventory for Wieder's son, and American sundial scholar Robert Newton Mayall who acted as a go-between, the Drecker collection was sold to Wheatland in 1959 and later donated to Harvard.[50] I do not know of such huge and rich donations to university museums in countries of the Mediterranean area.

47 The collection contained sixty-three astrolabes, besides hundreds of other devices. The Museum was at first officially named the Lewis Evans Collection. See "Lewis Evans, founder of the History of Science Museum," History of Science Museum, University of Oxford, accessed 10 January 2025, https://www.hsm.ox.ac.uk/lewis-evans-biographical-account.

48 Whipple's instruments ranged from medieval instruments for astronomical observations to early twentieth-century industrial technology. "Foundations of the Museum," Whipple Museum of the History of Science, accessed 10 January 2025, https://www.whipplemuseum.cam.ac.uk/about-us/foundations-museum.

49 David P. Wheatland of Topsfield, Massachusetts donated his personal collection of sundials in the 1980s to the Collection of Historical Scientific Instruments, which he had unofficially founded in the late 1930s and was officially recognised by the university in 1948. He had earlier donated thousands of objects from his personal collections of navigation, surveying, horology, natural philosophy, physics, telegraphy, telephony, and radio apparatus and books. Sara Schechner, David P. Wheatland Curator of the Collection of Historical Scientific Instruments, Harvard University (2000–2024), now Curator Emerita, personal correspondence with author, January 2025; and Sara J. Schechner, "David P. Wheatland (1898–1993): Scholar, Author, Avid Collector, *Sine qua non* for the Collection of Historical Scientific Instruments," at "CHSI History," Collection of Historical Scientific Instruments, Harvard University, accessed 2 February 2025, https://chsi.harvard.edu/chsi-history.

50 Unpublished research by Sara Schechner.

3 A Few Concluding Lines

I would just like to underline, once again, that this paper should be regarded less as a scholarly contribution and more as a personal set of impressions and ideas, based not only on papers I have read and conferences I have attended, but also on discussions and informal exchanges that are typical of the SIC coffee breaks, lunches, or dinners. As such, this chapter might stimulate new ideas and reflections, and I hope it will provoke further researches on international and national trends in historical scientific instruments studies and museums.

Acknowledgments

For their patience and support in the preparation of this paper, I wish to warmly thank Rich Kremer and Alison Morrison-Low. I am particularly indebted to Sara Schechner, who kindly and continuously encouraged me to write the paper, offered me her assistance, and did a great job to improve my first draft. Finally, I would like to remember and thank Jim Bennett, one of the initial editors of this volume.

Bibliography

Andersen, Hemming. *Historic Scientific Instruments in Denmark*. Matematisk-fysiske Meddelelser 44, no. 2. Copenhagen: The Royal Danish Academy of Sciences and Letters, 1995.

Anderson, R.G.W., and G.L'E. Turner, eds. *An Apparatus of Instruments: The Role of the Scientific Instrument Commission*. London: Scientific Instrument Commission, 1993.

Blondel, C., and M. Dörries, eds. *Restaging Coulomb. Usages, controverse et réplications autour de la balance de torsion*. Biblioteca di Nuncius, Studi e Testi, 15. Florence: Leo S. Olschki, 1994.

Centre national d'histoire des sciences (Belgium). *Inventaire des instruments scientifiques historiques conservés en Belgique*. vol. 1. Brussels, 1959.

Heering, P., F. Rieß, and C. Sichau. *Im Labor der Physikgeschichte: Zur Untersuchung historischer Experimentalpraxis*. Oldenburg: BIS-Verlag, 2000.

Holbrook, M., with R. G. W. Anderson and D. J. Bryden. *Science Preserved: A Directory of Scientific Instruments in Collections in the United Kingdom and Eire*. London: H.M. Stationery Office, 1992.

Lacki, Jan, ed. "The Replication Method in History of Science," special issue, *Archives des Sciences* 58 (2005).

Scientific Instrument Commission. *Newsletters*, nos. 1–18 (1980–1997). Gerard L'E. Turner Papers. Whipple Museum of the History of Science, University of Cambridge, Cambridge, UK.

Strano, G. "In Memory of James (Jim) Arthur Bennett (1947–2023)." *Nuncius* 39, no. 2 (2024): 529–536.

Union Internationale d'Histoire et Philosophie des Sciences, Commission pour l'Inventaire mondial des Appareils scientifiques d'intéret historique. *Inventaire des instruments conservés en Italie*. 5 vols. Milan: Museo Nazionale della Scienza e della Tecnica Leonardo da Vinci, 1963.

CHAPTER 7

Forty Years of Cataloguing Historic Scientific Instruments

A.D. Morrison-Low

For the forty years of the Scientific Instrument Commission's formal existence, from 1981 to 2021, there have been a number of ways to publicise what historic scientific instruments were to be found where: (1) There were efforts to map out what was held in public collections on a national scale, to produce published national inventories, gradually building up a global picture. (2) At the same time, there were ambitions to catalogue individual items in a more comprehensive manner, collections of similar instruments held in a single location, usually by the larger museums. (3) With time and radical changes in inventory methods, versions of online museum collections could be accessed by members of the public. (4) A fourth, non-institutional form of catalogue was made by dealers in historic scientific instruments, initially in printed form and then electronically, delivered straight to the likely purchaser's desk. (5) Finally, a fifth form of catalogue was that made by the collector, often as important as many of the foregoing. This paper will consider how these five forms of catalogue have developed, or not, over the past forty years.

The author of this paper – who was fortunate to hold a curatorial position at a major national museum from 1980 until 2015 – can only draw on limited international experience after a career spent almost entirely in the English-speaking sphere. A broader *resumé* or perhaps several, from other points of view, may be inspired by this piece.

1 National Inventories

The aspiration to produce national inventories, under the auspices of the International Union of the History and Philosophy of Science (today the International Union of the History and Philosophy of Science and Technology), was an attempt to foster the mapping of just where in the world a historic

 | DOI:10.1163/9789004737587_009

scientific instrument might be located.[1] By 1981 when what has come to be counted as Scientific Instrument Symposium I met in Bucharest,[2] the drive for national inventories as a global desirability had long been active, since at least 1946 just after the Second World War.[3] In fact there had been a couple of meetings before the Bucharest symposium, as Gerard Turner noted in a retrospective paper published in 1993. He states, somewhat baldly, that:

> The Scientific Instrument Commission of the International Union of the History and Philosophy of Science was established in its present form at the XV International Congress held in 1977 at Edinburgh.[4]

Turner mentions the get-together held in 1979 at the National Maritime Museum, Greenwich, and an Oxford meeting of the British Society of the History of Science later that year with the theme "Scientific Instruments and Scientific Theory." Derek Howse (1919–1998), Keeper of Navigation and Astronomy at the National Maritime Museum, was elected as Chairman (the post later became President), and Gerard Turner (1926–2012), Senior Assistant Curator at the Museum of the History of Science, Oxford (now the History of Science Museum, Oxford), became Secretary of the formalised Scientific Instrument Commission. The SIC's first *Newsletter* was published in May 1980, drawing up the three articles of its mission statement. The first of these resolutions was "to stimulate the production of inventories of historic scientific instruments;"[5] later in that same first *Newsletter*, a few paragraphs were devoted to the development and publication of national inventories, voiced first in 1946, with practical support to be provided by the International Union

1 The beginnings of this objective are described on the Scientific Instrument Commission website: "History," Scientific Instrument Commission, accessed 7 March 2023, https://scientific-instrument-commission.org/history.

2 Scientific Instrument Commission, *Newsletter*, no. 1 (May 1980), Gerard Turner Papers, Whipple Museum, Cambridge.

3 Mary Holbrook, R.G.W. Anderson, and D.J. Bryden, eds., *Science Preserved: A Directory of Scientific Instruments in Collections in the United Kingdom and Eire* (London: HMSO, 1992), 11.

4 G.L'E. Turner, "A Decade in the Study of Scientific Instruments," in *An Apparatus of Instruments: The Role of the Scientific Instrument Commission*, ed. R.G.W. Anderson and G.L'E Turner (London: Scientific Instrument Commission of the International Union of the History and Philosophy of Science, 1993), 5–14, quote 5. This paper, entirely without references, clearly draws upon the run of SIC *Newsletters*, nos. 1–18 (May 1980–April 1997).

5 Also quoted by Turner, "Decade in the Study," 5.

from 1958.[6] Belgium had produced two printed volumes in 1959 and 1960; France, one volume in 1964; Italy, five volumes in 1963; the Soviet Union, one volume in 1968; and Poland had produced a microfilm in 1970. However, in 1993 to Gerard Turner these seemed "quaint," as they were highly selective and not particularly inclusive.[7] But national inventories are an important curatorial resource if museum catalogues are not to be internal.

In 1980, the British national inventory was in the doldrums. Mary Holbrook had obtained funding to survey the scientific instrument holdings of some 200 smaller museums in Great Britain by 1968; the six largest collections (the London Science Museum, the National Maritime Museum, the British Museum, the university museums at Oxford and Cambridge and the Department of Technology at the Royal Scottish Museum) were always going to be excluded in detail, although a summary of their holdings would be listed. The typescript draft was with the Science Museum, being checked. The 1980 *Newsletter* piously hoped that the British national inventory would be published in 1981. It finally appeared in 1992 as *Science Preserved: A Directory of Scientific Instruments in Collections in the United Kingdom and Eire*, at Scientific Instrument Symposium XII held at the Royal Museum of Scotland, Edinburgh. It contains information about the collections of 228 institutions and lists some 3700 instruments. As its publication date approached, it was realised – through the work of Charles Mollan of the Royal Dublin Society and encouraged by a prestigious committee from the SIC – that there were far more instrumental riches in the island of Ireland than hitherto realised, and Mollan was able through "the advent of that miraculous aid," the computer word processor, to put together a far more detailed national inventory, including Irish-made instruments located in other countries.[8] But 1992 is way ahead in this narrative.

Two papers given at a conference at the National Railway Museum in York in April 1981 and published a year later give a pertinent starting point for a discussion of forty years of cataloguing, beginning that year. The talks were given by the new Chairman of the Scientific Instrument Commission (after Derek Howse's retirement), Robert Anderson, then of the Science Museum London, and by Gerard Turner, continuing as SIC Secretary, based at the Museum of the History of Science, Oxford. Anderson's paper, "The Museum Catalogue – a

6 Henri Michel, "Notes sur l'inventaire mondial des instruments scientifiques d'intérêt historique," *Archives internationales d'histoire des sciences* 11 (1958): 394–401.

7 Turner, "Decade in the Study," 8. The Polish National Inventory is discussed by Ewa Wyka, "The World Inventory of Historical Scientific Instruments – the Polish Contribution," *Kwartainik Historii Nauikii Techniki* 65, no. 1 (2020): 81–99; she also shows "how the concept of a global inventory of historical scientific instruments was born and developed" (p. 82), https//doi.org/10.4467/002.3589XKHNT.20.005.11621.

8 Turner, "Decade in the Study," 9.

Disappearing Art Form?" is a cry for help from a long-vanished world, not quite the Dickensian office of quill pens and dust, but with glacial movement of curators and hot-metal presses certainly before the office revolution of the 1980s with the advent of personal computers.[9] Yet catalogues still managed to be produced. From the Royal Scottish Museum, there had been Anderson's own 1978 catalogue of *The Playfair Collection* of surviving chemical apparatus there dating from before 1850, and Turner had – in collaboration with Trevor Levere – catalogued the early collections at Teylers Museum, Haarlem, published in 1973.[10] Both Anderson and Turner understood, at first-hand, the difficulties of cataloguing. Indeed, the SIC's first *Newsletter* had stated, succintly: "Catalogues are of vital importance to all those studying scientific instruments."

The terms of Anderson's 1981 York paper were:

> I shall only be dealing with catalogues which deal with collections of historical artefacts relating to science, technology and medicine. I shall deal only with physical, and not biological, science. And I shall only be talking about catalogues which list museum collections, by which I mean that I shall not be discussing catalogues of special or temporary exhibitions.[11]

My discussion shares these definitions of a museum catalogue of historic scientific instruments. Anderson then laid out the publishing history (in terms of catalogues as just defined) of the London Science Museum and its predecessor, the South Kensington Museum. He produced a graph, pointing out that "publishing activity increased as soon as the Science Museum gained its identity by having a purpose-built building provided in the 1920s" (and, additionally, after the cataclysm of the First World War). "In the twenty years, from 1920 to 1940, forty catalogues were issued, at the rate of one every six months."[12] This rate, of course, slumped during the Second World War, revived a little thereafter but collapsed again during the 1970s so that by the 1980s there was nothing happening at all. Why should this be?

---

9 R.G.W. Anderson, "The Museum Catalogue – A Disappearing Art Form?" in *Museum Publications in Science, Technology and Medicine*, ed. [R.I.C. Charlton] (Oxford and London: Ashmolean Museum and Science Museum, 1982), 24–30.

10 R.G.W. Anderson, *The Playfair Collection and the Teaching of Chemistry at the University of Edinburgh 1713–1858* (Edinburgh: Royal Scottish Museum, 1978); G.L'E. Turner and T.H. Levere, "Descriptive Catalogue of Van Marum's Scientific Instruments in Teyler's Museum," in *Martinus van Marum: Life and Work*, ed. E. Lefebvre and J.G. De Bruijn (Leiden: Noordhoff International Publishing, 1973), vol. IV, Part II, pp. 129–401.

11 Anderson, "Disappearing Art Form?" 24.

12 Anderson, "Disappearing Art Form?" 24–25, graph 41.

Anderson's argument was that, by 1981, the role of museums had changed considerably since the previous half century. The museum had come to be seen as a leisure activity, a publicly-funded entertainment, albeit a worthy one but no longer a branch of the education service, with homework (a catalogue) to take home at the end. Second, pressures of work on the curator had increased: refreshed exhibitions, more temporary displays, increased paperwork, and, as he added: "enquiries from the public have increased enormously and every owner of a bauble writes to his friendly museum curator to discover its value." Thus, "what does get neglected is research and publication".[13] Third, and this was a comment of its time, publications were often outrageously expensive; a long-due publication of the British Museum's scientific instrument catalogue was to cost £48 a copy (roughly £187 today).[14] He juxtaposed this with the National Maritime Museum's cheaply produced, loose-leaf inventory of the collections of the Department of Navigation and Astronomy as "one of the most valuable and usable publications in recent years."[15]

Turner's 1981 paper, "Do Museums of Science and Technology have a Publications Policy? A Review of the Current Situation," did what it said on the tin.[16] It was prefaced by the same tone of exasperation that had permeated Anderson's contribution. "To collect, conserve, study, and exhibit cultural artefacts is the purpose of a museum," he began, by way of explanation, quoting the International Council of Museums' (ICOM) definition. "You must study to know what to collect, and if you do not conserve then you lose your evidence and the possibility of further study at a later time."

---

13 Anderson, "Disappearing Art Form?" 29–30. Anderson also, elsewhere, pointed out that a diminution of staffing levels was also a growing problem, both qualitatively and quantitatively: Robert Anderson, "The Museum Curator as an Endangered Species," in *The Management of Change in Museums: Proceedings of a Seminar held at the National Maritime Museums, Greenwich, on 22nd November 1982*, ed. Neil Cossons (London: National Maritime Museum, 1985), 27–29. Also, a decade later: Nigel Hawkes, "Scholarship at Risk, Say Unions. Job Cuts Will Turn Science Museum 'into Theme Park,'" *The Times*, 16 October 1993. And again, into the new century: James Delingpole, "Ouch! Is This the Direction Our Museums Have to Go?" *The Times*, 18 March 2006. My thanks to Dr Anderson for these references.

14 F.A.B. Ward, *A Catalogue of European Scientific Instruments in the Department of Medieval and Later Antiquities of the British Museum* (London: British Museum, 1981).

15 [National Maritime Museum], *An Inventory of the Navigation and Astronomy Collections in the National Maritime Museum Greenwich* (London: National Maritime Museum, 1971). Unfortunately, this – and its updates – was only available inside the museum profession.

16 G.L'E. Turner, "Do Museums of Science and Technology have a Publishing Policy?" in *Museum Publications in Science, Technology and Medicine*, ed. [R.I.C. Charlton] (Oxford and London: Ashmolean Museum and Science Museum, 1982), 3–9.

> To study involves publication … There can only be exchange of information and development of ideas when results of work on collections are published. So a museum's publications policy is intimately connected with its research policy. One implies the other. The difficulty is to find out what that policy is. Each year the [UK] Museums Association sends out a questionnaire to members and museums, and there is no question on it about acquisitions or about publications – it might just as well be asking about the facilities of a hotel.[17]

However, the next part of his paper looked at the *status quo*, with a survey of museums in the United Kingdom and Europe. Lacking time to be able to cover North America, he nonetheless pointed out "that the United States national Museum, the Smithsonian, has two series of scholarly publications, the *Proceedings*, and the *Bulletin* for monographs, both series having begun in the 1870s."[18] At the Museum of the History of Science, Oxford, which opened in 1925, the one-man power-house Robert Gunther by his death in 1940 had published three increasingly large versions of a museum *Handbook*, fourteen volumes of *Early Science in Oxford*, one volume on instruments in Cambridge, as well as his *Astrolabes of the World* (1932), so important as a single instrument study that it was reprinted in 1976 and used as the basis of further astrolabe studies. More recently, there were three further subject catalogues, with four more in production.[19]

The Museum Boerhaave, Turner continued, was founded in 1928 as a state museum and had from the very beginning a series of Communications, which could be "8-page notes on someone such as Christiaan Huygens, or a 156-page book on instrument makers in Holland." Not all authors were museum employees; and because there was a "particularly rich library … for its publications this museum must be commended."[20] Moving from the Netherlands to Germany, Turner was not impressed by the bookshop or its contents at the Deutsches Museum in Munich: "If there is no encouragement for the curators to study the collection, then regression will occur." However, "in Italy, there is the outstanding museum of the history of science in Florence. This is a one-lady-band" consisting of Maria Luisa Righini Bonelli (1917–1981), who had worked there for over thirty years. She had produced an illustrated catalogue of the collections, shorter guides, articles in journals, along with editing the *Annali* (which

17 Turner, "Publishing Policy?" 3.
18 Turner, "Publishing Policy?" 3.
19 Turner, "Publishing Policy?" 4.
20 Turner, "Publishing Policy?" 4–5.

modern readers will know better under its current name, *Nuncius*), which Turner characterised as being "just about the only place to publish on antique scientific instruments."[21] Bonelli was, at the time of this presentation in early 1981, a vice-president of the International Union of the History and Philosophy of Science (IUHPS), the parent body of the SIC, but sadly she had died by the time of Turner's publication in 1982.[22]

"With larger museums, administration has a tendency to become an end in itself, and the consequent tendency to drive out scholarship must be resisted," wrote Turner. In writing his own Van Marum catalogue, he had looked at science and technology catalogues from all over the world but had found little to help him to design actual catalogue entries. He turned to other disciplines, finding inspiration specifically from a musical-instrument catalogue published by the Victoria and Albert Museum, London. "The signatures were prominent and carefully transcribed; the dimensions were given; photographs front and back [and] illustrated most items."[23] He mentions approvingly fine arts catalogues, although without mentioning that other desirable area of cataloguing – provenance.

Turner, like Anderson, wrote positively about the National Maritime Museum's loose-leaf ring-binder mode of inventory publication, and its series of bibliographies, monographs, and symposia proceedings. He spoke less favourably about the Science Museum's record of recent publications. The final part of Turner's talk was about what a museum is and what curators should be doing. Fortunately, that part of his paper has been superseded by events.

Even in 1981, no one could foresee the extraordinary office revolution of the 1980s in the form of the word processor and machines with memories being used at work. When I joined the museum world in 1980, my mother told me to let no one know that I held a typing qualification (Pitman's Elementary Typing, 40 words a minute, using all my fingers) lest I become a glorified typist; but in fact, this proved incredibly useful in an office where the typing pool for the entire institution consisted of two typists and letters (with their carbon copies) took two weeks to turn around. But I arrived at the then Royal Scottish Museum just as a series of packing cases filled with Scottish scientific instruments appeared from Jersey, purchased from the collector Arthur Frank. Although it took almost ten years (doubtless due to the excessive "administration" described by both Anderson and Turner), *Brass & Glass* finally appeared

21 Turner, "Publishing Policy?" 5.

22 Silvio A. Bedini, "Obituary: Maria Luisa Righini Bonelli 1917–1981," *Annals of Science* 39 (1982): 327–328.

23 Turner, "Publishing Policy?" 6.

in 1989 – the first publication to be produced by word processor at the National Museums Scotland.[24]

Over the next decade or so, Anderson and Turner continued to promote the idea of the necessity of cataloguing instrument collections. For instance, in 1984 Anderson published an updated version of a paper given at the 1982 SIC Symposium in Greenwich on "National Inventories of Scientific Instruments and the British Contribution."[25] Later in 1988, Turner wrote:

> To sum up so far, we have developed during the last decade an international audience and a means of communication through Symposia, the Newsletter, and the *Bulletin* of the Scientific Instrument Society. We have also defined the ends to which these are the means: namely, to discover what instruments exist, where they exist, and to list, catalogue and study them.[26]

By 1993 Gerard Turner was able to talk about additional national inventories. In the published version of the paper he had given the previous year at Scientific Instrument Symposium XII in Edinburgh, when the British inventory was ultimately published, he mentioned Hemming Andersen's work on Danish collections – finally published in 1995. National inventory work was currently being undertaken in Italy, Belgium, Hesse, and the United States.[27]

By Scientific Instrument Symposium XVI, held at Liège as part of the XXth International Congress of History of Science in 1997 and published five years later, an entire eight-paper session was devoted to national inventories.[28] It began with both Robert Anderson and Gerard Turner, after fifteen years, bowing out as office-holders of the Commission: "We have played a part in those [national inventories] of the United Kingdom, Ireland and Denmark, though of course the hard work of compilation has been conducted by the authors, not by us. The value of these and earlier inventories should be considered, as should be the Commission's *rôle* in encouraging the production of further national

24 T.N. Clarke, A.D. Morrison-Low, and A.D.C. Simpson, *Brass & Glass: Scientific Instrument Making Workshops in Scotland* (Edinburgh: National Museums Scotland, 1989).

25 R.G.W. Anderson, "National Inventories of Scientific Instruments and the British Contribution," *Bulletin of the Scientific Instrument Society*, no. 4 (1984): 9–12.

26 Gerard L'E. Turner, "Recent Advances in the History of Scientific Instruments," *Bulletin of the Scientific Instrument Society*, no. 19 (1988): 3–6, quote 3.

27 Turner, "Decade in the Study," 8–10.

28 Maurice Dorikens, ed., *Scientific Instruments and Museums* (Turnhout: Brepols, 2002).

inventories."[29] Anderson's paper suggested that the value of catalogues of different formats, as well as national inventories, and the Commission's promotion of any or all of these, needed to be debated. He mentioned monographs on single instrument types – for instance, Alan Stimson's work on the mariner's astrolabe and that of Willem Mörzer Bruyns on the cross-staff – that were perhaps also worthy of the Commission's support. And he observed that most of the "easy" national inventories had now been published; the United States or Germany, however, would provide daunting geographies for coverage.[30]

In fact, this final observation was addressed within the session. Rand B. Evans recommended "Dividing the Effort: An Inventory of Historical Scientific Instruments in America,"[31] and Paolo Brenni's paper, "The Italian Scientific Instrument Heritage: An Impossible Inventory?" offered another national case. Brenni explained that a series of circumstances peculiar to Italy had resulted in enormous amounts of redundant apparatus hidden in cupboards in public institutions of all sorts, mostly because de-accessioning these items required vast quantities of bureaucracy. "In fact for many decades," remarked Brenni, "a large part of it wasn't really preserved, it was simply abandoned and forgotten."[32] Even estimating its size – perhaps a five-digit number – was tricky. Brenni saw the answer in the internet: databases of individual collections, together with bibliographies of catalogues of other comparable collections. He gave examples of Italian collections with recent online catalogues and foresaw a future evolving out of this.

And this does indeed seem to have been the way forward. Although the large institutions with substantial scientific instrument collections – in Leiden, Florence, Chicago, Paris, Washington, DC, and London – have produced printed catalogues of their core collections, they have also helped instrument scholars by other means. *Epact*, discussed by Jim Bennett at the Liège symposium, grew out of another aspect of SIC's symposia, namely the informal networking amongst delegates. *Epact* – still available on the web – catalogues the "combined collections of mathematical instruments up to 1600" of four museums: the History of Science Museum, Oxford, the Museo di Storia della Scienza

29 R.G.W. Anderson, "Instruments, Inventories and Catalogues," in *Scientific Instruments and Museums*, ed. Maurice Dorikens (Turnhout: Brepols, 2002), 145–164, quote 145.

30 Anderson, "Instruments, Inventories and Catalogues," 151.

31 Rand B. Evans, "Dividing the Effort: An Inventory of Historical Scientific Instruments in America," in *Scientific Instruments and Museums*, ed. Maurice Dorikens (Turnhout: Brepols, 2002), 171–177.

32 Paolo Brenni, "The Italian Scientific Instrument Heritage: An Impossible Inventory?" in *Scientific Instruments and Museums*, ed. Maurice Dorikens (Turnhout: Brepols, 2002), 191–197, quote 193.

(today the Museo Galileo) in Florence, the Museum Boerhaave in Leiden, and the British Museum in London:

> It contains two levels of documentation on each instrument, drawing attention first to the basic features and points of interest, with one or two good photographs, and then offering a detailed technical description for those who want it. In the final product [promised Bennett] there will also be help and commentary, in the form of a trade geography, explanations of instrument types, a glossary of instrument terms and so on.[33]

National inventories – in the conventional, printed, and published form – are no longer discussed at the Annual General Meetings of the SIC; but the proliferation of catalogues and aspects of cataloguing certainly are. The past twenty years, from the turn of the millennium, has seen programmed publications as well as entire collections available online for the scholar as well as the curious. Despite the lack of standardisation in format, we are better off now than we were in 1981 when Robert Anderson asked: "The Museum Catalogue: A Disappearing Art Form?"

2 Catalogues of Permanent Public Collections

In 1988, Gerard Turner, Secretary of the SIC and Chairman of the Scientific Instrument Society, penned an invited editorial in the Scientific Instrument Society's *Bulletin*. Although unconnected with the SIC, the *Bulletin* was used as a channel to reach a similar, often overlapping, audience. Entitled "Collecting and Collectors," the editorial gave a rapid thumbnail sketch of collecting instruments since the Renaissance: "The fact is that we all owe a deep debt to the collecting instinct, for many of the collections amassed over the centuries by those with the resources and the opportunity are now available to everyone in the world's museums."[34] Turner forcefully concluded, "[Museums] must accept a duty to catalogue, and as far as possible to display (or at least make available for study) the objects they acquire."[35] It is no surprise then that elsewhere

33 Jim Bennett, "Cataloguing a Museum: Problems and Solutions," in *Scientific Instruments and Museums*, ed. Maurice Dorikens (Turnhout: Brepols, 2002), 165–169, quote 168. The *Epact* website is to be found at https://www.mhs.ox.ac.uk/epact/.

34 G.L'E. Turner, "Editor's Page: Collecting and Collectors," *Bulletin of the Scientific Instrument Society*, no. 19 (1988): 1–2, quote 1.

35 Turner, "Collecting and Collectors," 2.

in the same *Bulletin*, Turner effusively praised the ongoing implementation of eight desk-top computer-produced catalogues published by the Whipple Museum of the History of Science since 1982, as well as its series of special exhibitions and occasional accompanying monographs. He was particularly impressed by the Whipple's catalogue of an exhibition of ivory sundials from both the Oxford and Cambridge museum collections. Written by Penelope Gouk with essays organised by questions on whom, what, where, and how, it was "the first comprehensive and inter-disciplinary study of the Nuremberg ivory diptychs and the cultural context in which they were manufactured and marketed."[36] Turner commented: "I firmly believe that when more such studies are published, the general historian will cease to ignore the fundamental fact of the modern era: we live in a scientific age, and science develops by observation and measurement using instruments made by highly skilled craftsmen."[37] Turner's statement is hyperbolic when referring to pocket sundials, but nevertheless it is the fact that these instruments were now being catalogued to these standards that is important.

The forty years of the SIC has seen large numbers of traditionally published catalogues in book form, reviewed in the literature and read by the profession, and it would be impossible, even undesirable, to try to list them all. Most have been produced in the *lingua franca* of our day, English (for which I, for one, am truly grateful). The SIC produced an annual *Newsletter*, which between 1983 and 1995 was accompanied by a bibliography "of relevant books, pamphlets, catalogues and articles, especially those in less well known or unexpected periodicals."[38] This was succeeded the following year by a *Supplement*, and today the set joins SIC Bibliographies 14–24 (1998–2004) on the SIC website.[39] The digital revolution has made catalogue images clearer and easier

36 Penelope Gouk, *The Ivory Sundials of Nuremberg 1500–1700* (Cambridge: Whipple Museum, 1988), 7.

37 Turner, "Recent Advances;" Turner, "Decade in the Study," 10–11, quote 11. For another point of view, see the review by Richard L. Kremer, "Sundials and Related Instruments," *Isis* 81, no. 1 (1990): 104–105.

38 G.L'E. Turner and D.J. Bryden, *A Classified Bibliography on the History of Scientific Instruments* (Oxford: Scientific Instrument Commission, 1997), quote i.

39 A.V. Simcock, *A Supplement to a Classified Bibliography on the History of Scientific Instruments* (Oxford: Scientific Instrument Commission, 1998). The cessation of paper-printed bibliographies posted by the SIC was announced by Peter de Clercq, "The Scientific Instrument Commission and Its Bibliographies," *Bulletin of the Scientific Instrument Society*, no. 71 (2001): 38. For more recent developments, see: "Bibliography," Scientific Instrument Commission, accessed 20 February 2023, https://scientific-instrument-commission.org/sic-resources/bibliography where it states that "The SIC Bibliography will soon be united with IsisCB Explore database."

to understand, but the long hoped-for thesaurus of terms has not yet emerged. "There has been discussion on documentation and on the problems of nomenclature, but these are by no means solved," wrote Anderson in 1993.[40] In 1997, he added: "... it is surely desirable that this Commission should consider the purposes of inventories and catalogues, to ensure that they are produced in such a way that their value is optimised. The likely answer is that they exist to assist research; they are essentially finding tools."[41] In a pre-digital age, Anderson and Turner had essentially promoted a reasonably successful programme of locating instruments geographically so that future scholars might be able to analyse further what had been identified in this mapping exercise.

The traditional catalogue before the forty-year period of the SIC symposia generally treated a single subject – e.g., astrolabes, microscopes, or surveying instruments – across multiple museums or collections or found within a single institution. An example of a catalogue spanning multiple collections was the aforementioned Robert Gunther's 1932 *Astrolabes of the World*, which was then enlarged, or at least extended (as the format was very condensed) by Derek de Solla Price, first in 1955 and subsequently in 1973, using a computer print-out.[42] The task of cataloguing a single, long-lived instrument across time and space was becoming unmanageable, but the World Wide Web (introduced in 1991) became the answer to scholars seeking that breadth. In May 2002, David King took this approach with *The Frankfurt Catalogue of Medieval Astronomical Instruments*.[43]

More scarce instruments were perhaps an easier target. In 1972 Robert Anderson put together a temporary exhibition showing almost all of the then-known mariner's astrolabes and, unusually, produced a catalogue. In 1988 Alan Stimson of the National Maritime Museum, Greenwich published on the same topic.[44] Although not scarce items, microscopes offer a third example of

40 R.G.W. Anderson, "Introduction," in *An Apparatus of Instruments: The Role of the Scientific Instrument Commission*, ed. R.G.W. Anderson and G.L'E. Turner (London: Scientific Instrument Commission of the International Union of the History and Philosophy of Science, 1993), 1–3, quote 1.

41 Anderson, "Instruments, Inventories and Catalogues."

42 Derek J. de Solla Price, "An International Checklist of Astrolabes," *Archives International d'Histoire des Sciences* 8 (1955): 243–263 and 361–381; Sharon L. Gibbs, J.A. Henderson, and D.J. De Solla Price, *A Computerized Checklist of Astrolabes* (New Haven: Yale University, 1973).

43 David King's website, accessed 19 February 2023, http://www.davidaking.org/instrument-catalogue-TOC.htm (site discontinued).

44 R.G.W. Anderson, *The Mariner's Astrolabe* (Edinburgh: Royal Scottish Museum, 1972); Alan Stimson, *The Mariner's Astrolabe: A Survey of Known Surviving Sea Astrolabes* (Utrecht: Hes Publishers, 1988).

single-object catalogues. Working outside the museum space, collectors and dealers R.S. Clay and T.H. Court published *The History of the Microscope* (1932), which used material from museum collections as illustrations. Their narrative ended just as the history became optically interesting (with the advent of the achromatic microscope lens in the early nineteenth century); but as collectors, their focus was the development of mechanical stands.[45] By contrast, R.H. Nuttall, an academic and microscopist, in 1979 examined the crucial introductory achromatic period (with a comprehensive bibliography) when writing up Arthur Frank's private collection of fifty-eight British and European instruments, subsequently acquired by the Royal Scottish Museum.[46] In 1989 Gerard Turner produced a substantial catalogue of the microscopes belonging to the Royal Microscopical Society; its collection is now to be found (with others) at the History of Science Museum Oxford.[47] Microscope catalogues also were among the first to be put on a CD-ROM.[48]

It is more common now to find that single subject areas within a museum are catalogued by a specialist. From the 1920s to the 1940s, the specialist was the curator of the individual collection. Since 1990, some major collections have turned to outside scholars. For instance, the Museo Galileo in Florence has provided a series of such publications, in English, since the first appeared in 1991, by inviting external experts such as Gerard Turner who catalogued the microscope collection. He was followed by Paolo Brenni on mechanical instruments (1993), Willem Hackmann on pneumatic, magnetic, and electrical instruments (1995), Albert van Helden on early telescopes (1999), Elly Dekker on orbs, spheres, and globes (2004), Anthony Turner on sundials, nocturnals, and related instruments (2007), and most recently, Jim Bennett on surveying and related instruments (2022). All these catalogues have beautiful photographs,

45 R.S. Clay and T.H. Court, *The History of the Microscope* (London: C. Griffin and Co., Ltd., 1932; reprinted London: The Holland Press, 1985). For many years this was the main book on the development of the microscope available in English: Catalogues were published of the Billings collection of the Medical Museum of the Armed Forces Institute of Pathology in Washington, DC (1967; reprint, 1974); of the Whipple Museum, Cambridge (1986); of the Royal Microscopical Society's collection held in Oxford (1989); and of the Museum Boerhaave, Leiden (2004).

46 R.H. Nuttall, *Microscopes from the Frank Collection 1800–1860* (Jersey: Arthur Frank, 1979).

47 Gerard L'E. Turner, *The Great Age of the Microscope: The Collection of the Royal Microscopical Society through 150 Years* (Bristol and New York: Adam Hilger, 1989).

48 Julian Holland, "Review: M. Dingley, *A Catalogue of Portable Microscopes*, self-published CD-ROM, 2002," *Bulletin of the Scientific Instrument Society*, no. 77 (2003): 39; followed by Brian Bracegirdle, *A Catalogue of the Microscopy Collections at the Science Museum* (London: Little Imp Publications, London, CD-ROM, 2005).

both colour and monochrome, extensive descriptions, provenance, inventory details, literature, as well as an introduction, concordance, and bibliography.[49] Such catalogues did not spring into being without a history of dedicated curatorial documentation, namely, that of the Museo Galileo's first director, Maria Luisa Righini Bonelli.

Another museum that brought in outside experts to work on its collections with perhaps something of a public relations disaster was the National Maritime Museum, Greenwich. The catalogues themselves are superb, but the museum sought financial sponsorship, so that scholars could reside in Greenwich to study the collections and produce catalogues published by Oxford University Press. The sponsor was the Sackler Trust, an American charitable foundation which later became embroiled in a United States' prescription drug scandal.[50] Nevertheless, the Sackler Trust had sponsored fellowships from the mid-1990s that produced excellent high-quality catalogues of sections of the National Maritime Museum's scientific instrument collections: Elly Dekker catalogued the globes and armillary spheres (1999); Hester Higton the sundials (2002); Koenraad van Cleempoel the astrolabes (2006); W.F.J. Mörzer Bruyns the sextants and other navigation instruments (2009); and Jonathan Betts the marine chronometers (2017). "A spokesperson for the Greenwich museums told the *Observer*: 'The Sackler Trust's last donation to Royal Museum Greenwich was received in March 2018 and no additional applications are planned.'"[51]

What general trends in cataloguing historic scientific instruments have we seen over the past forty years? Catalogues have been greatly enhanced by other supportive works, for instance, geographical lists of scientific instrument makers; biographies of individual makers and firms; literature about trade cards; scans online of trade catalogues enabling the identification of individual instruments; discussions about forgeries; case-studies about individual collectors of scientific instruments. The SIC has helped all of this work by holding annual symposia and sponsoring workshops, thereby sponsoring topical sessions around themes that have helped bulk out with useful information the

49 All published by Giunti, Florence, except the final catalogue, by Sillale, Florence.

50 "The Sackler Trust is run by the Sackler family, members of which run Purdue Pharma, a company selling the prescription painkiller OxyContin. The business is facing hundreds of lawsuits in the US over its alleged role in the country's opioid crisis, which kills more than 100 people a day," wrote the *Guardian* newspaper in 2019: *Guardian*, 25 March 2019. For the background to the opioid crisis, see Patrick Radden Keefe, *Empire of Pain: The Secret History of the Sackler Dynasty* (London: Picador, 2021), awarded the 2021 Baillie Gifford prize for nonfiction.

51 *Observer*, 24 March 2019: the *Observer* is the Sunday version of the *Guardian*.

initial listings of the national inventories and have fed into museum knowledge and so into the cataloguing of collections. However, the main – and most radical – change for everyone working in museums with scientific instruments over the past four decades has been the advent of the desktop computer.

3 Digitisation of Museum Documentation: Information Technology Becomes the Curatorial Handmaid

Putting descriptions of instruments into a word processor is all very well, but retrieving that information in a meaningful format is another story. A paper given by Willem Hackmann in 1986 to the Scientific Instrument Society described a British experiment with early information classification on cards, then the mechanisation of information retrieval through an index. These cards had punched holes in them for retrieval. Interested people formed a group in Leicester in 1966 (Leicester University was the home of a post-graduate course in Museum Studies, for many years the only such course in the Western Hemisphere).[52] This was taken up by the UK Museums Association, which produced so-called IRGMA cards (Information Retrieval Group of the Museums Association), which subsequently became the MDA (the Museums Documentation Association), established in 1977. Relaunched in 2008 as the Collections Trust, it developed its Spectrum Standard in 1994, now in its fifth edition, which is used by the majority of UK museums for registration and description of objects of all disciplines.[53] The six main collections of scientific instruments, however, having investigated these, have decided on other systems of documentation, often with North American backing.

There is, of course, a fairly extensive literature about how computerisation came about in museums in, for instance, the United Kingdom, the United States, and The Netherlands – and meetings at the annual SIC conferences helped instrument curators keep in touch informally with what was going on.[54] Standards were discussed extensively, but although documentation

52 W.D. Hackmann, "Cataloguing Electrical Instruments: A Case Study," in Jon Darius, "Report on the Oxford Meeting," *Bulletin of the Scientific Instrument Society*, no. 11 (1986): 6–8: quote 7.

53 See "What We Do," Collections Trust, accessed 16 February 2023, https://collectionstrust.org.uk/what-we-do.

54 For instance, see David Williams, "A Brief History of Museum Computerization," in *A Guide to Museum Computing*, ed. D. Williams (Nashville, TN: American Association for State and Local History, 1987), 1–5; and Andrew Roberts, "The Changing Role of Information Professionals in Museums," MDA *Information* 5, no. 3 (2001): 15–17; both of these

personnel talked with curators (and indeed in the beginning might have been former curators themselves), they increasingly formed units of their own, often as a part of a "collections care department" along with conservators and technicians. An agreed terminology, even across a single language such as English, in a single subject such as historic instrumentation, proved difficult to implement. To illustrate, it is perhaps worth looking briefly at the example of the classification of sundials.

Two attempts to form some sort of sundial classification scheme were both published in 1953, one by the Belgian instrument historian Henri Michel and the other by the assistant curator at Oxford, Kathleen Higgins.[55] Margarida Archinard, curator of the collection in Geneva, wrote an article over fifty years later in which she explained that she would try to synthesise the two approaches "which are in fact quite similar and were sufficiently valid at the time, and propose a new approach to the problem based, for the first time, on principles directly related to astronomy and geometry."[56] Anthony Turner has, over his career, given the subject a lot of thought. His 1985 paper "Sundials: History and Classification" is worth revisiting.[57] Turner's catalogue of the collection of sundials at Florence, published in 2007, begins with this sundial classification, which he explains as follows:

> The classification here adopted … is largely based on the pioneering scheme developed some forty years ago by Kathleen Higgins[58] as it was revised by a committee of English museum curators and sun-dial specialists convened at the moment when revision of the scientific instrument catalogue of the British Museum was being undertaken.[59] With some further modifications this scheme was presented for discussion at the first

reprinted in *Museums in a Digital Age: Leicester Readers in Museum Studies*, ed. Ross Parry (Abingdon: Routledge, 2010), 15–21 and 22–27. For The Netherlands, see Trilce Navarrete, *A History of Digitization: Dutch Museums* (Amsterdam, 2014), accessed 3 March 2023 online at https://catalogus.boekman.nl/pub/p14-0752.pdf.

55 Henri Michel, *Catalogue des cadrans solaire du Musée de la Vie wallonne, pars 1 et 2* (Liège: Éditions du Musée de la Vie wallonne, 1953); Kathleen Higgins, "The Classification of Sundials," *Annals of Science* 9 (1953): 342–358.

56 Margarida Archinard, "Une classification des cadrans solaires," *Annals of Science* 64 (2007): 471–524, quote 471.

57 Anthony J. Turner, "Sundials: History and Classification," *History of Science* 27 (1989): 303–318; reprinted in Anthony J. Turner, *Of Time and Measurement: Studies in the History of Horology and Fine Technology* (Aldershot: Variorum, 1993), chapter II. Chapters VI, VIII and IX also concern aspects of the subject.

58 Higgins, "Classification of Sundials."

59 Ward, *Catalogue of European Scientific Instruments*, 149–150.

international conference devoted to the history of sun-dials in 1985 by the present writer.[60]

Nevertheless, even in English, such classifications have not stood still and there have been further discussions about how catalogues have been arranged and how they might have been arranged in a more perfect world.[61] Terminology remains tricky, sometimes disputed, and moves on. It certainly does not prevent or even hinder – at least not so the average catalogue user would notice – the further cataloguing of historic instrument collections.[62] Similar issues of terminology, as well as methods for linking across collections, are currently being explored by the Arts and Humanities Research Council-funded Towards a National Collection programme, a suite of projects that aims to forge connections and increase public engagement with cultural heritage in the UK.[63] Helped by the digitisation of information and of photography of objects, Anderson's 1981 "Disappearing Art Form" has been saved, indeed revolutionised, by the technology.

4 Dealers' and Collectors' Catalogues

Dealers' catalogues are produced to sell material to a fairly broad customer base, but they can also provide quality information to curators and scholars by showing examples of instruments that would otherwise be inaccessible. Collectors' catalogues may be seen as vanity projects, to some extent, but also provide the same service. Long before dealers produced and posted out their

60 Turner, "Sundials: History and Classification." Anthony J. Turner, *Istituto e Museo di Storia della Scienza: Catalogue of Sun-dials, Nocturnals and Related Instruments* (Florence: Guinti Editore, 2007), 8–9.

61 See David A. King, "Essay Review: Towards a History from Antiquity to the Renaissance of Sundials and Other Instruments for Reckoning Time by the Sun and Stars," *Annals of Science* 61 (2004): 375–388.

62 Sara J. Schechner, *Time of Our Lives: Sundials of the Adler Planetarium* (Chicago: Adler Planetarium, 2019), 38–41. This uses another form of classification – hierarchically based on the Sun's apparent motion being tracked (in terms of hour angle, altitude, azimuth, and combinations of these), the orientation of surfaces on which the Sun's motion is projected and the hour lines are drawn, the flat or curvy topography of the surface, and ultimately sub-groups defined culturally, stylistically, and by mechanical features – but this was some time in the making and thus was unknown to most readers until it was published.

63 See Towards a National Collection, https://www.nationalcollection.org.uk/.

catalogues of solely historical instruments in the mid to late twentieth century, booksellers occasionally sold instruments. For instance, in 1937, the then Royal Scottish Museum acquired an important Marshall-type microscope from E.P. Goldschmidt of London, offered because it was stamped on the leather body with the Royal coat-of-arms.[64] By this date, Goldschmidt had been joined by Ernst Weil (1891–1965), "one of the most respected and knowledgeable London book and instrument dealers of his day" who had fled with his family from Munich in 1933.[65]

In fact, some auction houses had produced instrument catalogues at the turn of the twentieth century, with important sales in 1911 of the Strozzi collection in Amsterdam and the Roussel collection in Paris.[66] By the early 1960s when the major London auctioneers were offering sales comprised entirely of scientific instruments – the word "historic" was understood – there was clearly a market eager to be supplied. Specialist dealers in instruments also emerged. For instance, Harriet Wynter, originally based in Brighton and dealing in European porcelain, moved to London in 1966 and began trading in scientific instruments. After retirement, she presented much of her archive to National Museums Scotland and wrote an autobiography that was published in 2010. Like many of the dealers, she joined the Scientific Instrument Society (founded 1983), and attended many of the SIC conferences.[67] She – together with her then business partner, Anthony Turner – produced her first catalogue in 1973: "I had been accumulating a great deal of stock since I began to buy scientific instruments in 1966, and I wanted to produce a catalogue, perhaps even a series of catalogues of instruments, something apparently never done in Britain before."[68] She bought her first personal computer in 1980, but by 1986 her publications had ceased. Other British dealers sent out less-polished

---

64 NMS.T.1937.131; from E.P. Goldschmidt & Co., Ltd., *Catalogue* (London: Goldschmidt, 1937), 60, item 209 and plate VII; see G.L'E. Turner, "Decorative Tooling on 17th and 18th Century Microscopes and Telescopes," *Physis* 8 (1966): 99–128; and reprinted in G.L'E. Turner, *Essays on the History of the Microscope* (Oxford: Senecio Publishing, 1980), 79–108.

65 William J.H. Andrewes, "The Legacy of David Wheatland," *Nuncius* 16 (2001): 687–701, see 692–693.

66 See A.J. Turner, "From Mathematical Practice to the History of Science: The Pattern of Collecting Scientific Instruments," *Journal of the History of Collections* 7 (1995): 135–150; Anthony J. Turner, "Paris, Amsterdam, London: The Collecting, Trade and Display of Early Scientific Instruments, 1830–1930," in *Scientific Instruments: Originals and Imitations, the Mensing Connection*, ed. Peter de Clercq (Leiden: Museum Boerhaave, 2000), 23–47.

67 Harriet Wynter, *The Price of Every Thinge* (Glasgow: The Grimsay Press, 2010); see 89–96 for London dealers.

68 Wynter, *The Price of Every Thinge*, 119.

catalogues, using Polaroid images, typescript, and photocopier – Christopher Sykes of Woburn Antiques, and Peter Delehar (1944–2022) of Portobello Road were two during the 1980s – but the properly printed and produced dealers' catalogues before the advent of the word processor came from overseas: those of Alain Brieux (1922–1983) in Paris,[69] Saul Moskowitz (1933–1991) of Marblehead, Massachusetts,[70] and in due course, David and Yola Coffeen of Hastings-on-Hudson, New York.

The Coffeens began their listing *Tesseract* in 1982 with *Catalog A*. In the early days, eager European museum curators who received their copies through the mail had to telephone the sleeping Coffeens to reserve particularly important items. David Coffeen had taken a career change from astronomer to dealer that year. With its large land mass, the United States has a tradition of buying goods from catalogues, for example, those of Sears or Montgomery Ward, and a few other instrument dealers were selling in this way, such as Saul Moskowitz from 1970. The Coffeens specialised in arcane early objects, which they researched extensively. David explains that there have been 113 issues since then (until March 2023), with a total of 5000 pages of instruments for sale. He does it all: acquisitions, research, writing, photography, and layout – and then goes off to an offset printing shop (these have changed with time). He remarks some landmark dates: 1982, the beginnings, saw a black-and-white product, using typewriter and Velox half-tones, cut-and-pasted by hand, using scissors and paste; 1988, still a black-and-white product, using a word processor (his first Mac) for text, plus Veloxes and scissors; 2004, black-and-white product, but now using a layout programme (Quark) for text plus digital photographs (no more scissors!); 2012, COLOUR at last, finally affordable and reliable. He notes that many customers prefer this, but sales have not increased. The advent of the word processor helped, as it is easy to make changes and corrections, and enabled more complex, faster layouts. On the downside, the learning time, the frequent software revisions and compatibility issues, may have meant that the IT revolution was more expensive than the scissors-and-paste method. It looks much more professional, however. By 1999, the Tesseract website was established, listing perhaps two hundred objects but not the catalogues. It was only in 2018 that the latest print catalogue was added to the website.[71] Along with the catalogues of the international auction houses, from about 1998, many dealers'

69 See Andrewes, "The Legacy of David Wheatland," 696–697.

70 See Andrewes, "The Legacy of David Wheatland," 698. Moskowitz's catalogues, dating from 1970 to 1987, *Historical Technology*, were made available in three volumes.

71 My thanks to David Coffeen for sending me speedy and comprehensive information about the changing state of his particular instrument catalogues, and the effect of the word processor on the method: DC to ADM-L, email 8 March 2023.

catalogues were placed on their websites – complete with photographs, descriptions, and more recently, conservation reports.[72]

A similar path has been taken with collectors' catalogues. In the past, private collectors during the sixteenth and seventeenth centuries might have had inventories, as in the first Renaissance *Kunstkammern* formed by European royalty.[73] Sometimes, manuscript inventories were formed upon the death of the collector[74] or sales catalogues of collections were published after that event.[75] In the early nineteenth century, a wealthy English eccentric named William Kitchiner (1778–1827) wrote a book entitled *Practical Observations on Telescopes, Opera-glasses and Spectacles* (1815), essentially a discussion of instruments in his own collection; he was a household name during his lifetime for another book, the best-selling *The Cook's Oracle* (1817). Subsequently, in 1825 he produced an enlargement of the earlier book, *The Economy of the Eyes: being the result of thirty years' experiments with Fifty-One Telescopes, of from One to Nine Inches in Diameter in the possession of William Kitchiner, M.D.*[76] At this distance in time, it is difficult to say whether Kitchiner was merely amusing himself with writing about his discussions with opticians or whether he had some other purpose in describing his telescopes.

Over the past forty years, collectors have produced catalogues of their collections for a number of reasons, and perhaps it would be invidious to try to attribute the motives behind them. It is enough, surely, to have private treasures shared with a wider public, especially if those treasures are eventually dispersed. One collector, Arthur Frank (1915–2005) came back from the Second World War to his father's Glasgow optical shop, and over time, became convinced that Britain's greatness stemmed from her role in providing

72 For the beginnings, see Desmond Squire, "Market Place: Buying & Selling Scientific Instruments on the Internet," *Bulletin of the Scientific Instrument Society*, no. 57 (1998), 34–36.

73 For example, see Joachim Menzhausen, "Elector Augustus' *Kunstkammer*: An Analysis of the Inventory of 1587," in *The Origins of Museums: The Cabinet of Curiosities in Sixteenth- and Seventeenth-Century Europe*, ed. Oliver Impey and Arthur MacGregor (London: House of Stratus, 2001), 91–99.

74 A manuscript inventory of the Earl of Ilay's London instrument collection was made in 1761 by George Adams senior; parts of it are described by A.D. Morrison-Low, "Two Enlightenment Collections of Scientific Instruments in Hanoverian Britain," *Bulletin of the Scientific Instrument Society*, no. 156 (2023): 2–11; the mss Inventory is held by the Bute Collection at Mount Stuart, Isle of Bute.

75 Ilay's collection was purchased by his nephew, John Stuart, third Earl of Bute, and after the latter's death in 1792 his instruments were sold at auction 5–7 February 1793: Morrison-Low, "Two Enlightenment Collections." This sale is examined in detail by Gerard L'E. Turner, "The Auction Sales of the Earl of Bute's Instruments 1793," *Annals of Science* 23 (1967): 213–242.

76 Anita McConnell, "Kitchiner, William," *Oxford Dictionary of National Biography* (2004), see https://doi.org/10.1093/ref:odnb/15690.

scientific instruments to the world, thus advancing the progress of science. His broad-based collection of instrumentation was offered, in parts, to various museums: and he, together with Dr R.H. Nuttall, described many of the items in catalogues and articles.[77]

Many collectors bring in acclaimed experts in their particular fields to catalogue their private collections. Before Seth G. Atwood (1917–2010) disposed of his collection through the salerooms between 1999 and 2004, he paid for some opulent volumes – hard-back, full colour, beautiful explanatory diagrams, and top-range scholarship – for sections of the material.[78] Sadly, the entire programme of catalogues was not completed, although the lavish auction catalogues went some way to make up for this.[79]Another collector, who invited a number of scholars in the field to help put on an exhibition and produce a catalogue of his collection – this time globes and planetaria – was Rudolf Schmidt (1924–2013). The venues were Christie's of London and the Museum Boerhaave in Leiden; both exhibitions borrowed material from outside the primary collection.[80] The well-known historian of New York architecture, Andrew Alpern, wrote in the catalogue of his collection of drawing instruments that:

> The primary tools of the architect of today are software programs and computer terminals. And as a logical outgrowth of that, the drawings produced by today's architects look very different from those drawn in decades and centuries past. … Practitioners brought up on CAD may have little concept of how of how those drawings in the past were produced.

---

77 A.D. Morrison-Low, "Obituaries: Arthur Frank (1915–2005)," *Bulletin of the Scientific Instrument Society*, no. 87 (2005): 36–38. See also Nuttall, *Microscopes from the Frank Collection.* Robert Nuttall taught chemistry at Strathclyde University and advised Frank on his purchases; subsequently he advised National Museums Scotland on their acquisitions of modern material from the Scottish universities.

78 A.J. Turner, *Time Measuring Instruments: Part 1, Astrolabes, Astrolabe-Related Instruments*, Time Museum Catalogue of the Collection, vol. 1, no. 1 (Rockford, IL: Time Museum, 1985); A.J. Turner, *Time Measuring Instruments: Part 3, Water-Clocks, Sand-Glasses, Fire-Clocks*, Time Museum Catalogue of the Collection, vol. 1, no. 3 (Rockford, IL: Time Museum, 1988); Anthony G. Randall, *The Time Museum Catalogue of Chronometers* (Rockford, IL: Time Museum, 1992).

79 The Time Museum was sold in four sales over five years by Sotheby's New York, and brought in over $18m: Part One, 2 December 1999; Part Two, 19 June 2002; Part Three, 30 October 2002; Part Four, 13–15 October 2004.

80 Tom Lamb and Jeremy Collins, eds., *The World in Your Hands: An Exhibition of Globes and Planetaria, from the Collection of Rudolf Schmidt* (London: Christie's, 1994). Petra Svatek, "Rudolf Schmidt," *Journal of the International Map Collectors' Society*, no. 135 (Winter 2013), 45.

> This collection of drawing instruments and related books, journals, trade catalogues, and other materials provides some of the answers.[81]

Alpern then goes on to describe how he put together his collection, starting in 1958 when he saw "several splendid sets of drawing instruments" in a teacher's home, inspiring him to visit antique fairs and infrequent local auctions. "When I discovered they were regularly held in London by Christie's, Sotheby's and Phillips, the die was cast. I bought in London at those sales, where I learned of the small handful of specialist dealers that offered drawing instrument sets and issued catalogues, including David Coffeen, Raymond Giordano, and the late Saul Moskowitz."[82]

Howard Dawes, a long-standing member of the Scientific Instrument Society although no longer collecting today, produced two themed volumes about aspects of his collection at the turn of the Millennium. The first, *Ideas in Brass*, used his books and instruments to illustrate a (mainly) English Enlightenment; while the second, *Instruments of the Imagination*, looked at drawing instruments in Britain from 1600 until 1850. Both volumes were written by Dawes, then digitally printed, with full colour photographs, and produced on demand.[83] Another long-standing member of the Scientific Instrument Society was Peter Louwman (1935–2023), who invited an eminent instrument historian to assist him to put together a catalogue of his wide-ranging collection of telescopes;[84] these are now housed in a gallery in a wing of the family automobile museum. Those attending the Scientific Instrument Commission's Symposium XXXVII in Leiden and Haarlem in September 2018 paid the collection a visit as part of the programme. Another telescope collection, recently assembled by the retired American academic, Edward D. Wolf, and discussed in a penetrating review by William Tobin, has been sold to the Beijing Planetarium and thus unlike many other private collections, will not be dispersed. Tobin states: "To Summarize: The Wolf Collection is important and extensive. Because of its numerous and excellent photographs *Wolf Telescopes* sets a new and exacting standard. It is a

---

81 [Andrew Alpern], *Catalogue of the Andrew Alpern Collections of Drawing Instruments at the Avery Architectural and Fine Arts Library* (New York: The Trustees of Columbia University in the City of New York, 2010), ix.

82 [Alpern], *Catalogue*, x.

83 Howard Dawes, *Ideas in Brass: A History of the Instruments of Science in Britain* (Worcestershire: Howard Dawes, 2002); Howard Dawes, *Instruments of the Imagination: A History of Drawing Instruments in Britain 1600–1850* (Worcestershire: Dawes Trust, 2009).

84 P.J.K. Louwman and H.J. Zuidervaart, *A Certain Instrument for Seeing Far: Four Centuries of Styling the Telescope, Illustrated by a Selection of Treasures from the Louwman Collection of Historic Telescopes* (Wassenaar: Louwman Collection of Historic Telescopes, 2013).

comprehensive record of the Collection and an unparalleled tool for the study of both the Collection itself and historic telescopes elsewhere."[85] Thus, with time, the quality of self-produced catalogues has continued to improve.

I conclude this very partial discussion with perhaps the most beautifully-produced catalogue of them all: Ralf Kern's *Wissenschaftliche Instrumente in ihrer Zeit*, produced in five volumes between 2010 and 2018.[86] This is so expensive – new or secondhand – that no public library in the United Kingdom appears to stock it.[87] Online it looks beautiful ... but out of reach. There is a moral here somewhere, but it does not concern the forty years of the Scientific Instrument Commission, whose work has ensured that historic scientific instrument collections and scholarship remains increasingly accessible, to curators and the public alike.

Bibliography

[Alpern, Andrew]. *Catalogue of the Andrew Alpern Collections of Drawing Instruments at the Avery Architectural and Fine Arts Library*. New York: The Trustees of Columbia University in the City of New York, 2010.

Anderson, R.G.W. "Instruments, Inventories and Catalogues." Pp. 145–164 in *Scientific Instruments and Museums*, edited by Maurice Dorikens. Turnhout: Brepols, 2002.

Anderson, R.G.W. "Introduction." Pp. 1–3 in *An Apparatus of Instruments: The Role of the Scientific Instrument Commission*, edited by R.G.W. Anderson and G.L'E. Turner. London: Scientific Instrument Commission of the International Union of the History and Philosophy of Science, 1993.

Anderson, R.G.W. "National Inventories of Scientific Instruments and the British Contribution." *Bulletin of the Scientific Instrument Society*, no. 4 (1984): 9–12.

Anderson, R.G.W. *The Mariner's Astrolabe*. Edinburgh: Royal Scottish Museum, 1972.

85 Edward D. Wolf, *Wolf Telescopes: A Collection of Historical Telescopes* (Trumansburg, NY: printed for the author, 2016); William Tobin, review of *Wolf Telescopes: A Collection of Historical Telescopes* by Edward D. Wolf, *Journal of Astronomical History and Heritage* 19, no. 3 (2016): 349–351.

86 Ralf Kern, *Wissenschaftliche Instrumente in ihrer Zeit*, 5 vols. (Cologne: Buchhandlung Walther Koenig, 2010–2018). Currently offered "as new" for £1297.34 for 5 vols. or individually for a three-figure sum each: https://www.abebooks.co.uk/servlet/SearchResults?an=RalfKern&cm_sp=SearchF-_-home-_-Results&ref_=search_f_hp&sts=t&tn=Wissenschaftliche (accessed 2 November 2024).

87 Or so I thought, until Richard Kremer pointed out that Volume 1 is available at Loughborough University Library: see https://vufind.lboro.ac.uk/Record/539250?sid=116728415 (accessed 2 November 2024).

Anderson, R.G.W. "The Museum Catalogue – A Disappearing Art Form?" Pp. 24–30 in *Museum Publications in Science, Technology and Medicine*, edited by [R.I.C. Charlton]. Oxford and London: Ashmolean Museum and Science Museum, 1982.

Anderson, R.G.W. "The Museum Curator as an Endangered Species." Pp. 27–29 in *The Management of Change in Museums: Proceedings of a Seminar held at the National Maritime Museums, Greenwich, on 22nd November 1982*, edited by Neil Cossons. London: National Maritime Museum, 1985.

Anderson, R.G.W. *The Playfair Collection and the Teaching of Chemistry at the University of Edinburgh 1713–1858*. Edinburgh: Royal Scottish Museum, 1978.

Andrewes, William J.H. "The Legacy of David Wheatland." *Nuncius* 16 (2001): 687–701.

Archinard, Margarida. "Une classification des cadrans solaires." *Annals of Science* 64 (2007): 471–524.

Bedini, Silvio A. "Obituary: Maria Luisa Righini Bonelli 1917–1981." *Annals of Science* 39 (1982): 327–328.

Bennett, Jim. "Cataloguing a Museum: Problems and Solutions." Pp. 165–169 in *Scientific Instruments and Museums*, edited by Maurice Dorikens. Turnhout: Brepols, 2002.

Bracegirdle, Brian. *A Catalogue of the Microscopy Collections at the Science Museum*. London: Little Imp Publications, CD-ROM, 2005.

Brenni, Paolo. "The Italian Scientific Instrument Heritage: An Impossible Inventory?" Pp. 191–197 in *Scientific Instruments and Museums*, edited by Maurice Dorikens. Turnhout: Brepols, 2002.

Clarke, T.N., A.D. Morrison-Low, and A.D.C. Simpson. *Brass & Glass: Scientific Instrument Making Workshops in Scotland*. Edinburgh: National Museums Scotland, 1989.

Clay, R.S., and T.H. Court. *The History of the Microscope*. London: C. Griffin and Co., Ltd., 1932; reprinted London: The Holland Press, 1985.

de Clercq, Peter. "The Scientific Instrument Commission and Its Bibliographies." *Bulletin of the Scientific Instrument Society*, no. 71 (2001): 38.

Dawes, Howard. *Ideas in Brass: A History of the Instruments of Science in Britain*. Worcestershire: Howard Dawes, 2002.

Dawes, Howard. *Instruments of the Imagination: A History of Drawing Instruments in Britain 1600–1850*. Worcestershire: Dawes Trust, 2009.

Delingpole, James. "Ouch! Is This the Direction Our Museums Have to Go?" *The Times*, 18 March 2006.

Dorikens, Maurice, ed. *Scientific Instruments and Museums*. Turnhout: Brepols, 2002.

Evans, Rand B. "Dividing the Effort: An Inventory of Historical Scientific Instruments in America." Pp. 171–177 in *Scientific Instruments and Museums*, edited by Maurice Dorikens. Turnhout: Brepols, 2002.

Gibbs, Sharon L., J.A. Henderson, and D.J. De Solla Price. *A Computerised Checklist of Astrolabes*. New Haven: Yale University, 1973.

Goldschmidt & Co., Ltd., E.P. *Catalogue*. London: Goldschmidt, 1937.

Gouk, Penelope. *The Ivory Sundials of Nuremberg 1500–1700*. Cambridge: Whipple Museum, 1988.

Guardian (London), 25 March 2019.

Hackmann, W.D. "Cataloguing Electrical Instruments: A Case Study." In "Report on the Oxford Meeting," by Jon Darius. *Bulletin of the Scientific Instrument Society*, no. 11 (1986): 6–8.

Hawkes, Nigel. "Scholarship at Risk, Say Unions. Job Cuts Will Turn Science Museum 'into Theme Park,'" *The Times* (London), 16 October 1993.

Higgins, Kathleen. "The Classification of Sundials." *Annals of Science* 9 (1953): 342–358.

Holbrook, Mary, R.G.W. Anderson, and D.J. Bryden, eds. *Science Preserved: A Directory of Scientific Instruments in Collections in the United Kingdom and Eire*. London: HMSO, 1992.

Holland, Julian. "Review: M. Dingley, *A Catalogue of Portable Microscopes*, self-published CD-ROM, 2002." *Bulletin of the Scientific Instrument Society*, no. 77 (2003): 39.

Keefe, Patrick Radden. *Empire of Pain: The Secret History of the Sackler Dynasty*. London: Picador, 2021.

Kern, Ralf. *Wissenschaftliche Instrumente in ihrer Zeit*, 5 vols. Cologne: Buchhandlung Walther Koenig, 2010–2018.

King, David A. "Essay Review: Towards a History from Antiquity to the Renaissance of Sundials and Other Instruments for Reckoning Time by the Sun and Stars." *Annals of Science* 61 (2004): 375–388.

Lamb, Tom, and Jeremy Collins, eds. *The World in Your Hands: An Exhibition of Globes and Planetaria, from the Collection of Rudolf Schmidt*. London: Christie's, 1994.

Louwman, P.J.K. and H.J. Zuidervaart. *A Certain Instrument for Seeing Far: Four Centuries of Styling the Telescope, Illustrated by a Selection of Treasures from the Louwman Collection of Historic Telescopes*. Wassenaar: Louwman Collection of Historic Telescopes, 2013.

McConnell, Anita. "Kitchiner, William." *Oxford Dictionary of National Biography* (2004).

Menzhausen, Joachim. "Elector Augustus' *Kunstkammer*: An Analysis of the Inventory of 1587." Pp. 91–99 in *The Origins of Museums: The Cabinet of Curiosities in Sixteenth- and Seventeenth-Century Europe*, edited by Oliver Impey and Arthur MacGregor. London: House of Stratus, 2001.

Michel, Henri. *Catalogue des cadrans solaire du Musée de la Vie wallonne, pars 1 et 2*. Liège: Éditions du Musée de la Vie wallonne, 1953.

Michel, Henri. "Notes sur l'inventaire mondial des instruments scientifiques d'intérêt historique." *Archives internationales d'histoire des sciences* 11 (1958): 394–401.

Morrison-Low, A.D. "Obituaries: Arthur Frank (1915–2005)." *Bulletin of the Scientific Instrument Society*, no. 87 (2005): 36–38.

Morrison-Low, A.D. "Two Enlightenment Collections of Scientific Instruments in Hanoverian Britain." *Bulletin of the Scientific Instrument Society*, no. 156 (2023): 2–11.

[National Maritime Museum]. *An Inventory of the Navigation and Astronomy Collections in the National Maritime Museum Greenwich*. London: National Maritime Museum, 1971.

Navarrete, Trilce. *A History of Digitization: Dutch Museums*. Amsterdam: Universiteit van Amsterdam, 2014.

Nuttall, R.H. *Microscopes from the Frank Collection 1800–1860*. Jersey: Arthur Frank, 1979.

Observer (London), 24 March 2019.

Price, Derek J. de Solla. "An International Checklist of Astrolabes." *Archives internationales d'histoire des sciences* 8 (1955): 243–263, 361–381.

Randall, Anthony G. *The Time Museum Catalogue of Chronometers*. Rockford, IL: Time Museum, 1992.

Roberts, Andrew. "The Changing Role of Information Professionals in Museums." *MDA Information* 5 no. 3 (2001): 15–17; and pp. 22–27 in *Museums in a Digital Age: Leicester Readers in Museum Studies*, edited by Ross Parry. Abingdon: Routledge, 2010.

Schechner, Sara J. *Time of Our Lives: Sundials of the Adler Planetarium*. Chicago: Adler Planetarium, 2019.

Simcock, A.V. *A Supplement to a Classified Bibliography on the History of Scientific Instruments*. Oxford: Scientific Instrument Commission, 1998.

Stimson, Alan. *The Mariner's Astrolabe: A Survey of Known Surviving Sea Astrolabes*. Utrecht: Hes Publishers, 1988.

Squire, Desmond. "Market Place: Buying & Selling Scientific Instruments on the Internet." *Bulletin of the Scientific Instrument Society*, no. 57 (1998): 34–36.

Tobin, William. Review of *Wolf Telescopes: A Collection of Historical Telescopes* by Edward D. Wolf. *Journal of Astronomical History and Heritage* 19 no. 3 (2016): 349–351.

Turner, A.J. "From Mathematical Practice to the History of Science: The Pattern of Collecting Scientific Instruments." *Journal of the History of Collections* 7 (1995): 135–150.

Turner, A.J. *Istituto e Museo di Storia della Scienza: Catalogue of Sun-dials, Nocturnals and Related Instruments*. Florence: Guinti Editore, 2007.

Turner, A.J. "Paris, Amsterdam, London: The Collecting, Trade and Display of Early Scientific Instruments, 1830–1930." Pp. 23–47 in *Scientific Instruments: Originals and Imitations, The Mensing Connection*, edited by Peter de Clercq. Leiden: Museum Boerhaave, 2000.

Turner, A.J. "Sundials: History and Classification." *History of Science* 27 (1989): 303–318. Reprinted as Chapter II in Turner, Anthony J. *Of Time and Measurement: Studies in the History of Horology and Fine Technology*. Aldershot: Variorum, 1993.

Turner, A.J. *Time Measuring Instruments: Part 1, Astrolabes, Astrolabe-Related Instruments*, Time Museum Catalogue of the Collection, vol. 1, no. 1. Rockford, IL: Time Museum, 1985.

Turner, A.J. *Time Measuring Instruments: Part 3, Water-Clocks, Sand-Glasses, Fire-Clocks.* Time Museum Catalogue of the Collection, vol. 1, no. 3. Rockford, IL: Time Museum, 1988.

Turner, G.L'E. "A Decade in the Study of Scientific Instruments." Pp. 5–14 in *An Apparatus of Instruments: The Role of the Scientific Instrument Commission*, edited by R.G.W. Anderson and G.L'E. Turner. London: Scientific Instrument Commission of the International Union of the History and Philosophy of Science, 1993.

Turner, G.L'E. "Decorative Tooling on 17th and 18th Century Microscopes and Telescopes." *Physis* 8 (1966): 99–128. Reprinted pp. 79–108 in Turner, G.L'E. *Essays on the History of the Microscope*. Oxford: Senecio Publishing, 1980.

Turner, G.L'E. "Do Museums of Science and Technology have a Publishing Policy?" Pp. 3–9 in *Museum Publications in Science, Technology and Medicine*, edited by [R.I.C. Charlton]. Oxford and London: Ashmolean Museum and Science Museum, 1982.

Turner, G.L'E. "Editor's Page: Collecting and Collectors." *Bulletin of the Scientific Instrument Society*, no. 19 (1988): 1–2.

Turner, G.L'E. "Recent Advances in the History of Scientific Instruments." *Bulletin of the Scientific Instrument Society*, no. 19 (1988): 3–6.

Turner, G.L'E. "The Auction Sales of the Earl of Bute's Instruments 1793." *Annals of Science* 23 (1967): 213–242.

Turner, G.L'E. *The Great Age of the Microscope: The Collection of the Royal Microscopical Society through 150 Years*. Bristol and New York: Adam Hilger, 1989.

Turner, G.L'E., and D.J. Bryden. *A Classified Bibliography on the History of Scientific Instruments*. Oxford: Scientific Instrument Commission, 1997.

Turner, G.L'E., and T.H. Levere. "Van Marum's Scientific Instruments in Teyler's Museum, Leyden." Vol. IV, Part II, pp. 129–401 in *Martinus van Marum: Life and Work*, edited by E. Lefebvre and J.G. De Bruijn. Leiden: Noordhoff International Publishing, 1973.

Ward, F.A.B. *A Catalogue of European Scientific Instruments in the Department of Medieval and Later Antiquities of the British Museum*. London: British Museum, 1981.

Williams, David. "A Brief History of Museum Computerization." Pp. 1–5 in *A Guide to Museum Computing*, edited by D. Williams. Nashville, TN: American Association for State and Local History, 1987. Reprinted pp. 15–21 in *Museums in a Digital Age: Leicester Readers in Museum Studies*, edited by Ross Parry. Abingdon: Routledge, 2010.

Wolf, Edward D. *Wolf Telescopes: A Collection of Historical Telescopes*. Trumansburg, NY: printed for the author, 2016.

Wyka, Ewa. "The World Inventory of Historical Scientific Instruments – the Polish Contribution." *Kwartainik Historii Nauikii Techniki* 65, no. 1 (2020): 81–99.

Wynter, Harriet. *The Price of Every Thinge*. Glasgow: The Grimsay Press, 2010.

CHAPTER 8

Forty Years of Permanent and Temporary Exhibitions: A Curator's Meditation

Giorgio Strano

The Oxford English Dictionary explains that "to exhibit" is the action "to show publicly for the purpose of amusement or instruction, or in a competition." This particular meaning is, however, the last in a list of ten. All the other meanings refer to juridical, bureaucratic, political, economic, and religious actions.

The peculiar collocation of "our" meaning at the end of the list might refer to the huge quantity of documents that museum curators must preliminarily "exhibit" to the various authorities in order to receive permission, support, and funding to refurbish a permanent exhibition, create a temporary exhibition, or to loan or borrow items to be displayed in an exhibition.[1]

In the following sections, I will leave aside all the other meanings of "to exhibit." I will occasionally deal with the "amusement or instruction" components of a permanent or temporary exhibition, as well as with the "competition" component. Displays may, in fact, receive awards from national and international institutions, and for different motives.

I will not even try to be exhaustive. A reasoned list of permanent and temporary exhibitions of scientific instruments in the last forty years would surely exceed the length limits of this chapter. I will mostly deal with institutions connected to the Scientific Instrument Commission, but, in any case, I will make rather arbitrary selections and therefore apologise, in advance, to all those who will not see any mentions of their exhibitions.

1 Permanent Exhibitions

The creation of a permanent exhibition usually involves a relevant number of professionals with specific skills: architects, curators, conservators, engineers, art-transporters, graphic designers, climate experts, showcase makers, educators, security agents, lighting experts, not to mention electricians and other

1 I consulted the second edition of *The Oxford English Dictionary* (Oxford: Clarendon Press, 1989). More recent editions might have relocated "our" meaning.

 | DOI:10.1163/9789004737587_010

workers if architectural modifications are necessary. In the case of museums of historical scientific instruments, the crucial members of any exhibition project are the historians of science and technology. They are the only individuals able to assess the importance of the items to be displayed and to select them from among the many available objects. A conspicuous part of a museum's collections is, in fact, usually "condemned" to remain in storage until a positive conjuncture occurs.

All the individuals involved in an exhibition project should supposedly cooperate. Joining forces would, in fact, appear as an obvious precondition to anyone; but museologists very well know that some professionals often try to impose their point of view above everyone else's. Usually, the most delicate relationship is the one between the historians of science and technology (who would exhibit an object for its cultural importance) and the architects (who praise the aesthetic value of the objects, to say nothing of their showcases). Museologists should act as mediators, but are seldom involved. In fact, they usually intervene only after things are done, by praising or criticising the exhibition.

Over the last forty years, the Istituto e Museo di Storia della Scienza (Institute and Museum of the History of Science) in Florence, nowadays Museo Galileo, was the epicentre of a remarkable renewal. The project was cultivated for more than a decade without, however, securing what was essential to its completion: enough money. Once suitable funds were granted in 2008 by public and private sponsors, six historians and philosophers of science and scientific instruments, coordinated by Paolo Galluzzi (1942–), collaborated with another two dozen professionals to refurbish eighteen exhibition rooms. The re-opening of the museum in June 2010 was the occasion to host the XXIX Symposium of the Scientific Instrument Commission (Florence, 4–9 October 2010). The complexity of the renovation prompted the publication of a volume dedicated to its thorough explanation.[2]

It should be noted that, originally, the Museo Galileo's collections were located elsewhere. Over a span of four and a half centuries, they had been gathered or scattered in various grand-ducal residences and institutions.[3]

2 Filippo Camerota, ed., *Displaying Scientific Instruments: From the Medici Wardrobe to the Museo Galileo* (Trezzano sul Naviglio: Goppion, 2012).

3 Before their present site, the collections were placed across four Florentine buildings: Palazzo Vecchio, the Uffizi Galleries, Palazzo Pitti, and Palazzo Torrigiani. They were either privately or publicly displayed; see Filippo Camerota, "The Medici Collection of Mathematical Instruments: History and Museography," in *European Collections of Scientific Instruments, 1550–1750*, ed. Giorgio Strano, Stephen Johnston, Mara Miniati, and Alison Morrison-Low (Leiden: Brill, 2007), 129–148; Filippo Camerota and Mara Miniati, eds., *The Medici and the Sciences: Instruments and Machines in the Grand-Ducal Collections* (Florence: Giunti, 2009),

Their present whereabouts, removed from such "original" places, allowed a relative freedom in designing the permanent exhibition. The relevant constraints were, in fact, the walls of Palazzo Castellani (where the museum has resided since 1930), the amount of funds available for the renovation, and the inventiveness of the scientific committee. Under these conditions, the new exhibition realised the curatorial ideas in an "artificial" environment (a historical building, anyway), employing a design that took advantage of state-of-the-art showcases, dedicated lighting, and interactive information devices.[4] The museum visitors' experience was therefore optimised in both visibility and accessibility.[5]

Over the last four decades, there have been other conspicuous refurbishments of permanent displays. Unlike the Florence case, the next two examples occurred in the same building where the collections had originally been gathered or formed. These examples highlight how different levels of integration between the collections and the building may impact a permanent exhibition and its visitors.

The nucleus of scientific instruments of the Mathematisch-Physikalischer Salon in Dresden was assembled by Augustus I of Saxony (1526–1586) in his

67–69, 87, 93–96, 105 and 139; Filippo Camerota, "The Mathematical Rooms," in *Displaying Scientific Instruments*, 3–17; Simone Contardi, "The Museum of Physics and Natural History," in *Displaying Scientific Instruments*, 19–35; Mara Miniati, "From the Museum of Physics to the Museum of Ancient Instruments," in *Displaying Scientific Instruments*, 37–55.

4 The Museo Galileo received three awards: ICOM Italia, Museums of the Year 2010, First Prize for "Best Management;" British Society of History of Science: Great Exhibitions Competition for 2010, First Prize; European Museum Academy Award 2011.

5 Several museums host collections that were originally formed or held elsewhere. Here are a few examples: The National Museum Boerhaave, Leiden, founded in 1928, was established in 1931 in a building that started as a convent and was then transformed into the hospital where Hermann Boerhaave taught his pupils. After a momentous renovation, the Museum was nominated European Museum of the Year 2019–2020 by the European Museum Forum of the European Council; see: Willem Otterspeer *et al.*, *75 Jaar Museum Boerhaave* (Leiden: Museum Boerhaave, 2006), 21–71. The Whipple Museum of the History of Science, Cambridge (UK), began with the donation of more than one thousand scientific instruments collected by Robert S. Whipple (1871–1953). The museum resides in the old building of the former Perse School, and is part of the University of Cambridge; see: Joshua Nall, Liba Taub and Frances Willmoth, eds., *The Whipple Museum of the History of Science: Objects and Investigations to Celebrate the 75th Anniversary of R.S. Whipple's Gift to the University of Cambridge* (Cambridge: Cambridge University Press, 2019), 1–4. The Whipple Museum is presently undergoing renovations. The Collection of Historical Scientific Instruments (CHSI) at Harvard University, Cambridge (USA) grew around a nucleus of Harvard's earliest scientific instruments collected by David P. Wheatland (1898–1993). The instruments, moved from place to place over the years, finally found their way to the Harvard Science Center, where some are displayed in permanent and temporary exhibition galleries; see: David P. Wheatland, *The Apparatus of Science at Harvard 1765–1800* (Cambridge, MA: Harvard University Press, 1968), 7–9.

Royal Palace. In 1728, the collection was moved to its present location, the Zwinger Palace, completed by order of Augustus II "the Strong" (1670–1733). The collection grew in quantity over the years but, sadly, the building was gravely damaged in February 1945 by the British-American aerial bombing of Dresden. Reconstruction started under the Soviet Union's political control of East Germany. But a full recovery, restoration, and valorisation of the surviving scientific instruments was not completed until 2013.[6]

The location inside the "original" building posed some constraints on the refurbishment. As a matter of fact, the newly installed Mathematisch-Physikalischer Salon combines areas with modern construction and educational tools (the terrestrial and celestial globes section)[7] and areas which, although modernised, are designed to recall an eighteenth-century cabinet of physics (the pavilion dedicated to instruments of the Enlightenment).[8] This combination was made possible and "inspired" by the circumstance that the bombing deleted relevant pieces of information about the pre-existing display, therefore partially removing any obligations to preserve it. The current Mathematisch-Physikalischer Salon therefore embodies curatorial and architectural ideas in a relatively original building (the Zwinger) via a partially "artificial" design. The permanent exhibition makes use of state-of-the-art showcases, dedicated lighting, and modern methods of information delivery, but also contains a memento of the historical display. This museological solution facilitates accessibility to the rooms; there is no limit to the number of visitors allowed depending on the way the objects are displayed. The only exception is the Enlightenment pavilion, where the instruments, placed outside showcases, could be touched or damaged if visitors were to enter there in large numbers (*Figs. 8.1a, 8.1b*).[9]

---

6 Klaus Schillinger, "The Development of the Collections of Scientific Instruments of the Mathematical-Physical Salon between the Years 1750 and 1850 and Their Use for Research and Teaching," in *Proceedings of the Eleventh International Scientific Instrument Symposium Held at Bologna University, Italy, 9–14 September* 1991, ed. Giorgio Dragoni, Anita McConnell, and Gerard L'E. Turner (Bologna: Grafis Edizioni, 1994), 101–106; Peter Plassmeyer, Wolfram Dolz, and Michael Korey, *Mathematisch-Physikalischer Salon – Zwinger Meisterwerke* (Munich: Deutscher Kunstverlag, 2020), 7–13.

7 Plassmeyer *et al., Meisterwerke*, 115–139.

8 Plassmeyer *et al., Meisterwerke*, 53–77.

9 The British Museum in London was a similar case. The museum building was bombed on 10 May 1941. After reconstruction, no attempt to reorganise the collections in their former fashion was made. See: Silke Ackermann, "'Of Sufficient Interest …, but not of Such Value …': 260 Years of Displaying Scientific Instruments in the British Museum," in *Scientific Instruments on Display*, ed. Silke Ackermann, Richard L. Kremer, and Mara Miniati (Leiden: Brill, 2014), 77–93, and especially 86–93.

FIGURE 8.1
(*a*) Scientific instruments in the upper storey of the south-western pavilion of the Zwinger in Dresden after a renovation in the 1920s
© MATHEMATISCH-PHYSIKALISCHER SALON, STAATLICHE KUNSTSAMMLUNGEN DRESDEN. PHOTO: ARCHIVE MPS
(*b*) The current display in the same upper hall following the latest renovation, completed in 2013
©MATHEMATISCH-PHYSIKALISCHER SALON, STAATLICHE KUNSTSAMMLUNGEN DRESDEN. PHOTO: PETER MÜLLER

The relationship between the original building and collections led to a different outcome in the case of the Cabinet of Physics of the Fondazione Scienza e Tecnica (FST) in Florence. Renovation of the permanent exhibition started in 1983, and from the beginning it involved the late Paolo Brenni (1954–2021). The renovation process was very slow because of limited annual funding of the FST. The refurbished Cabinet of Physics opened to the public only in 2007.

Preconditions for the refurbishment were the existence of most of the scientific and technological instruments, the original showcases, and their placing inside the original building of the Istituto Tecnico Toscano (today's I.I.S.-I.T.L. Salvemini Duca d'Aosta). The collections had been formed from the early 1850s, under the initial directorship of Filippo Corridi (1806–1877). After the official inauguration in 1857, the collections expanded until the 1910s. Between the First and Second World Wars, they became obsolete and increasingly neglected.[10]

Rather than seeking a "new" display, the curators and architects of the FST chose to pursue a very accurate reconstruction of the Cabinet of Physics and its adjoining rooms. The work was "facilitated" by historical photographs showing the instruments both inside and outside the old showcases. The renovation focused on the restoration of the scientific instruments – an activity in which Brenni excelled – and the showcases. Concessions to modern display practices resulted in upgrading the showcases with LED lights and introducing labels with QR codes to give access to information on specific objects.

The refurbished Cabinet of Physics offers a full immersion experience in a nineteenth-century space. Its organisation, however, implies a number of restrictions. Some of the exhibited instruments have been placed, as they were in the past, outside the showcases. Their number, however, is larger than that revealed by historical iconography. To prevent damage caused by inopportune touching or manipulation, visitors can access the Cabinet only in guided groups.

This type of permanent exhibition makes tangible the curators' and architects' ideas about historical display – i.e., about preserving the overall layout of the historical showcases and exhibition rooms in the original building. On the one hand, this concept represents a very special case; like the restoration of a single scientific instrument, a historical cabinet can be "restored" as a whole.[11]

10 Paolo Brenni, "The Physics Cabinet of the Istituto Tecnico Toscano," in *Cabinets of Experimental Philosophy in Eighteenth-Century Europe*, ed. Jim Bennett and Sofia Talas (Leiden: Brill, 2013), 215–241, see 221–241.

11 Among similar examples, the Oval Room of the Teylers Museum in Haarlem (The Netherlands) could be regarded as the preservation, since 1784, of a historical display

On the other hand, such a permanent exhibition is a luxury; it is more expensive than it is lucrative. And, in fact, in order to survive economically, the FST must rely on not only admission fees but other income.

As a final example we turn to a museum residing in the same building where it originated but including a state-of-the-art permanent exhibition making no reference to any historical displays. To overcome the problem of admitting a potentially large number of visitors, a modern exhibition satisfies conservation and information exigencies. It allows better preservation of the collections and richer information for the public on the significance of the displayed objects. Although spectacular *in situ*, the large number of exhibited items in the Cabinet of Physics of the FST may, for many visitors, recall a storage facility. In fact, it was originally the stores where the lecturers and their assistants gathered the instruments for class demonstrations. The "stores-look," typical of numerous early twentieth-century museums, could be overwhelming for the many visitors who, usually, are neither collectors nor scientific instrument experts.

The History of Science Museum (HSM) of Oxford provides an example of a state-of-the-art permanent exhibition located in the same place – the Old Ashmolean Building on Broad Street – where a research institution was created by Elias Ashmole (1617–1692) in 1683. Over the centuries, the institution modified its vocation and the nature of the collections. Scientific instruments became dominant in 1919 for the interests and activities of Robert W.T. Gunther (1869–1940) and Lewis Evans (1853–1930).[12] Besides the original nucleus, the collections were, and still are, enlarged by donations and acquisitions. During the last century, the permanent exhibition, which had been opened as a public

---

in the historical showcases of the original building; see: Martin Weiss, "'Monuments of Science': How the Teyler Museum's Instrument Collection Became Historical," in *Cabinets of Experimental Philosophy*, 195–214; Eric Jorink, "The First Museum in the Netherlands? The Establishment of Teyler's Oval Room in Historical Perspective (c. 1600–1800)," in *Teyler's Foundation in Haarlem and Its "Book and Art Room" of 1779: A Key Moment in the History of a Learned Institution*, ed. Ellinoor Bergvelt and Debora Meijers (Leiden: Brill, 2021), 87–108. Very close to the case of the FST are the Cabinet of Physics of the University of Coimbra, on which see: Emelinda Ramos Antunes and Catarina Pires, "O Gabinete de Física da Universidade de Coimbra," in *Coleções Científicas Luso-Brasileiras: Patrimônio a ser Descoberto*, ed. Marcus Granato and Marta C. Lourenço (Rio de Janeiro: Museu de Astronomia e Ciências Afins-MAST/MCT, 2010), 157–210; and the Chemical Laboratory of the Polytechnic School of Lisbon, restored in 2007, on which see: Marta C. Lourenço, "O Museu de la Ciência da Universidade de Lisboa: Patrimônio, Coleções e Pesquisa," in *Coleções Científicas Luso-Brasileiras*, 257–276.

12 A.V. Simcock, *The Ashmolean Museum and Oxford Science, 1683–1983* (Oxford: Museum of the History of Science, 1984), 1–3.

museum in 1925, was refurbished several times, passing from wooden cupboards filled with dozens of items – the "stores-look" – to a more rationally elegant display. This was based upon the selection of a limited number of important items, attention to conservation issues, and a practical method of providing information. The fact that the largest part of the collections had to remain in storage was understood as an opportunity to change the permanent exhibition in a "physiological" turnover. The appearance of "new" objects on display is not only a consequence of scholarly research on the collections but also a way to revitalise the interest of the media and visitors. Indeed, since the late Jim Bennett (1947–2023) became its director in 1994, the HSM has become a place of continued reordering and renovation (*Fig. 8.2*).

FIGURE 8.2
The late Jim Bennett in the top gallery of the History of Science Museum in Oxford
COURTESY OF THE HISTORY OF SCIENCE MUSEUM, UNIVERSITY OF OXFORD. PHOTO: KEIKO IKEUCHI

2 *Horror Vacui* versus White Cubes

Permanent exhibition renovations are sensitive elements of a museum's life. They may indicate vitality, but also crises. Renovations are fashion dependent and, as such, are subject to external influences. One of the most dangerous outcomes of these times is, in fact, a historically grounded museum becoming excessively open to inputs from the public and some political groups. Of course, politicians and visitors bring money, which is essential to survive. There are, however, limits to which an audience may be satisfied; otherwise, a historical scientific museum risks transforming itself into an institution whose purposes are totally disconnected from its collections.

The first "odd" trends emerged in the late 2010s in the UK with renovations at the Natural History Museum (NHM) of London, a world-leading institution in biodiversity preservation and documentation. The museum directorship sympathised with concerns of the public and took to heart suggestions from political groups to emphasise in programming the effects of human impact on the biosphere. Species conservation, environmental protection, sustainability, and so on, became relevant topics for the permanent exhibition, and some areas of the NHM were dedicated to them. Since space inside historic buildings does not expand by itself, part of the collections was not relocated to other rooms but rather moved to new stores at Harwell, about seventy-five kilometres outside London. Furthermore, curators and taxonomists lost relevance. As they retired, other more useful professionals were hired: communication managers, marketing experts, and so on.

One of the NHM curators criticised this trend as a betrayal of the museum's mission: preservation and documentation. Millions of specimens gathered in the past are vital for natural history but only if they are preserved and made accessible for study. They represent the past flora and fauna of particular geographical areas and can tell stories about past natural history collecting. This is what the public should learn by visiting the NHM. Millions of specimens can, however, tell very little about the difficulties that species and the environment are suffering (except, perhaps, by contrast with what we have lost). In no way can those specimens outline strategies to save the biosphere. Rather than responding to planetary emergencies, a historical museum neglecting its roots and giving up its unique mission may tragically descend into irrelevance. Many other institutions, which do not possess huge historical collections, can explain the human impact on the biosphere much better and with greater resonance than the NHM.[13] And, in fact, the NHM directorship has partially turned around already.

13 Fred Naggs, "The Tragedy of the Natural History Museum, London," *Megataxa* 7, no. 1 (2022): 85–112.

Does the SIC community, with no direct concern for natural history, have anything to fear from this example? The *horror vacui* of the past, expressed by permanent exhibitions crowded with too many objects of the same type has, museologically speaking, given way to reasoned selections of objects that better satisfy visitors' "amusement or instruction," as *The Oxford English Dictionary* suggests. While a collector may jump for joy in front of dozens of, let's say, microscopes in a "stores-looking" exhibition (*Fig. 8.3a*), the general public and today's students undoubtedly receive more benefit from a calibrated and state-of-the-art display (*Fig. 8.3b*). But does this museological strategy represent the ultimate step by history of science permanent exhibitions to balance internal (preservation and study) and external (desiderata from the public) forces?

We might say that it all began by introducing personal computer monitors in the museum rooms. The original idea was to take advantage of the new affordable technology to vary the "standard" information delivery system which, until several decades ago, consisted of printed (i.e., static) labels crowded with text and pictures placed next to the exhibited objects. Monitors offered the opportunity to catch visitors' attention with colour, movement, sound, and streams of information.

Shortly thereafter, other new technologies paved excellent ways to enhance the public's comprehension of historical scientific instruments. While it would be "odd" in an art museum to combine a painting or a statue with a monitor – visitors would like to look at them with no interference from the outside – a scientific instrument has much to tell that is usually silenced by a static presentation. The instrument-and-monitor combination made it possible to see details, components, and even functioning, either in reality (by filming the instrument)[14] or in the virtual world (by 3D models).[15] In more advanced versions, the monitor became a touch-screen or an audio- or video-guide, providing visitors with the opportunity to virtually interact with the informational apparatus and the instrument itself.

Then other perspectives emerged. Why don't we make more informative museums by introducing who we are, our mission, our staff, our publications, our forthcoming projects, and so on? The monitors (either interactive or not) multiplied but, of course, with a couple of drawbacks at least. Technicians are

---

14 See the filmed demonstrations prepared by Paolo Brenni using the instruments of the Fondazione Scienza e Tecnica, https://www.youtube.com/@florencefst.

15 See the 3D animations of some instruments of the Museo Galileo; e.g., "Multimedia: Mechanical paradox," Virtual Museum, Museo Galileo, accessed 13 September 2024 https://catalogue.museogalileo.it/multimedia/MechanicalParadox.html.

FIGURE 8.3 Museo Galileo, Florence: (*a*) Partial view of the room "Microscopes," before 2009, with about eighty items on display in four shop-style showcases. (*b*) Partial view of the room "Measuring Natural Phenomena," January 2010, with a selection of about fifteen microscopes plus other items in a large state-of-the-art showcase

indispensable to fix breakdowns quickly (and, incidentally, by multiplying the monitors, the breakdowns multiply too). The space of such information areas cannot contain many objects (if any); part of the permanent exhibition must be relocated. Once again, as space doesn't expand for itself inside a building, a number of items must go to the stores.

Who is the target audience of the "object-less" information areas? Of course, it is not the usual visitors, who basically look annoyed at "institutional" monitors and pass by quickly. We might suspect that the informative areas are intended to target the local administrators and the political class. By visiting a museum, both of these audiences are impressed by auto-referential claims about local hyper-activity. These claims "exhibit" that public money is positively spent and, implicitly, ask for more funds.

Unfortunately, such "object-less" informational areas resemble the so-called "white cubes" of many modern art museums. The similarity could be considered a valid justification to their creation. There is, however, a momentous difference. An informational area is a permanent exhibition by itself while a white cube is a functional empty space designed to periodically host temporary exhibitions.[16]

Besides avoiding the confusion between these two museological entities, people involved in a museum refurbishment should accurately weigh the pros and cons of (too many) informational tools. The pros to attract public money, and the consequent creation of "object-less" areas, should not exceed the cons of minimising the presence of originals or the opportunity to make them tell their own stories. Such stories – the relevant ones – are hardly any story just conceived to address a public's present concerns. As an extreme consequence, the public might understand that there is something wrong in the information strategy and become dissatisfied with what the museum offers.

3 Temporary Exhibitions

A permanent exhibition is to a novel as a temporary exhibition is to a short story. Like a novel, a permanent exhibition conveys multiple narratives which can be singled out and elaborated by the museum's educational staff according to its visitors' age, education, and interests. A museum of scientific instruments

16 A typical "white cube" is the MA XXI – National Museum of Twentieth-Century Arts in Rome. See MA XXI, Museo nazionale delle arti del XXI secolo, home page, accessed 26 September 2024, https://www.maxxi.art/en/. The MA XXI does not possess any collections. On white cubes in general see: Patricia Falguières, "Politics for the White Cube: The Italian Way," *Grey Room* 64 (2016): 6–39.

contains narratives on the history of the collections, on the discoveries made with a given set of the objects, on the society that used and took advantage of those objects, as well as many others. For example, a group of children might be recipients of anecdotes about scientists and discoveries, while a group of engineers would prefer a technical narrative on how precision was achieved or improved over time. Both groups would linger in front of different highlights featured within the many objects on display.

On the other hand, like a short story, a temporary exhibition expounds upon a single narrative which is usually revealed by its title and subtitle. The narrative of *The Astrolabe, East and West* was unlikely to focus on new geographical discoveries. Such a topic might be mentioned in a section but will never form a narrative. The exhibition will instead "employ" the objects on display to outline the astronomical ideas behind astrolabe making and design, its use for astronomical, astrological, and religious calculations, its being a status symbol for some social classes, and the circulation of instruments and designs in the Mediterranean world.[17] The narrative of *Galileo's Telescope* will not focus on theories of the structure of Saturn's rings, even if the planet was carefully observed with that instrument. It will, instead, outline the origin of the telescope, its making, improvement, use, and its role in changing our understanding of the cosmos.[18]

The dichotomy of novel/short story also applies to the relationship of permanent and temporary exhibitions in time. A permanent exhibition lasts for years (with limited adjustments). Visitors can come back more than once to see and understand various narratives outlined by the objects on display. Consequently, a permanent exhibition must endure the stresses of huge numbers of visitors; convey narratives which remain current or are somewhat "neutral" to everyday scientific, social, and political life; present itself in a way that remains captivating over the years. Such requirements imply the adoption of robust materials and the use of information strategies which are not at risk of becoming obsolete too soon. Because of this, a permanent display has to be more "classic" or "traditional" than a temporary one. Wrongly chosen design, materials or information strategies would require intermittent remodelling of the (very expensive) permanent display.

---

17 Museum of the History of Science, Oxford University, *The Astrolabe, East and West*, 29 October 2005–12 March 2006, online catalogue, accessed 25 September 2024, https://www.mhs.ox.ac.uk/astrolabe/exhibition/.

18 Giorgio Strano, ed., *Galileo's Telescope: The Instrument that Changed the World* (Florence: Giunti, 2008); "Galileo's Telescope," Museo Galileo, accessed 2 February 2025, https://brunelleschi.imss.fi.it/telescopiogalileo/index.html.

Temporary exhibitions need not be "classic." They will last a few months at maximum. Therefore, they can, for example, rely upon basic showcases which, even if awkwardly complicated to deal with, will be opened just a couple of times over their life span (*Fig. 8.4*). Informational strategies could be extremely innovative and appeal to the public in exotic ways. In fact, if the lamp of some extremely performative projector fails in six months, or if the brilliant colours of a grandiose interactive TV monitor fade in the same time interval, it doesn't matter much. Additional care would only be necessary for travelling exhibitions, which require more robust modular solutions, capable of being packed, transported, and unpacked many times, and adaptable to the rooms of (almost) any possible venue.

From a positive viewpoint, temporary exhibitions could be a laboratory to experiment with display and communication strategies. If successful, such strategies could be transferred to permanent exhibitions; if not, they can be

FIGURE 8.4
A curious small triangular showcase with mirror and interlocking opening glass with screws. From the exhibition, *Copernicus and his World*, The Royal Castle, Warsaw, 25 April–30 July 2023
PHOTO: BY AUTHOR

discarded when the temporary exhibition closes. More importantly, a temporary exhibition of a private collection could form the nucleus of a future museum, as was the case of the Whipple Museum of the History of Science in Cambridge (UK), the Collection of Historical Scientific Instruments (CHSI) at Harvard University in Cambridge (USA), and the Museo Galileo in Florence.[19] Out of many, these three examples suggest that, sometimes, the boundaries between "temporary" and "permanent" could be fuzzy. This is particularly true for small exhibitions conceived to save collections of decommissioned scientific instruments owned by research or educational institutions.[20]

From a less positive viewpoint, the ephemeral connotation of a temporary exhibition means that whatever display and information solutions are used, they will last only a few months. At the end, the objects on display will return to their usual whereabouts and the informational apparatus will disappear (thrown away, with all due respect to the environment). Any memories of the exhibition will be entrusted to a catalogue (if any), a website (if any), a leaflet (if any), or only to the cameras and brain-cells of its curators and visitors. The idea to preserve the informational apparatus for some future re-display usually results in costs for storing voluminous materials and interferes with the important "guests" of the museum stores – the original instruments. And usually a corresponding loss of interest and "natural" decay result in the final destruction of the temporary exhibition materials.

A remarkable difference between permanent and temporary exhibitions is their "protagonists." A permanent exhibition is built around the objects that a museum possesses; the collections come first. On the contrary, by being based

19 On the Whipple Museum and the CHSI, see n. 5 above. The turning point of the present Museo Galileo was the *Prima Esposizione Nazionale di Storia della Scienza* (First National Exhibition of the History of Science) held in Florence between May and November 1929; see: Francesco Barreca, "The First National Exhibition of the History of Science (1929)," in *Displaying Scientific Instruments*, 95–107; Marco Beretta, "Andrea Corsini and the Creation of the Museum of the History of Science in Florence (1930–1961)," in *Scientific Instruments on Display*, 1–36. See also "The First National Exposition of History of Science – 1929," Museo Galileo, accessed 26 September 2024, https://exhibits.museogalileo.it/esposizione1929/index.html.

20 See, for instance, the Richard A. Paselk Scientific Instrument Museum, located in Humboldt State University (HSU) at Arcata, California (USA): Richard A. Paselk, "The Display of Twentieth-Century Instruments at the Humboldt State University," in *Scientific Instruments on Display*, 148–158; museum website, accessed 27 September 2024, https://scimus.humboldt.edu/. This exhibition is "permanent" because it is placed inside a public institution, it includes old instruments which belong to that institution, and it is meant to last over the years. But it is also "temporary," because it is essentially a one-person project, confined to a single room, and, as such, dependent on the fickle purposes of the HSU directorship.

on a specific topic (a particular historical actor, a time period, a scientific endeavour, a particular instrument, and so on) a temporary exhibition is a combination of objects selected from the host institution and from other museums and private collectors. The objects may also be combined with books, manuscripts, and paintings from public and private libraries, which help to outline the narrative in a symbiotic relationship. If the selection is poor, the exhibition is poor or even wrong. It is therefore not by chance that a rough "unit" to measure the quality of a temporary exhibition is the number of originals exhibited or, as an alternative, the number of originals on loan.

Loans are a "dramatic" component of curatorial work but often make the difference in attracting the public. Initially temporary exhibitions were, in fact, meant to offer a unique opportunity to the people of a specific place to admire, gathered under a single roof, a number of selected artefacts which, if not on loan, would require a tour around the world to be seen. One such scientific instrument exhibition of this type was *Galileo: Images of the Universe from Antiquity to the Telescope* (Florence, 13 March–30 August 2009). The exhibition assembled in the white cube of the Strozzi Palace about fifty items from the Museo Galileo and another 170 loans (*Fig. 8.5*).[21]

Such exhibitions are, of course, extremely expensive because of the great labour required by any single original loan (art transport, "nail-to-nail" insurance, courier expenses, etc.). Recently, this labour has become overwhelmingly complicated due to: (1) the formidable increase of bureaucratic documentation required to move any historical object outside the museum where it usually resides and to temporarily export/import it; (2) the increasing attention of the international community to the physical materials of which a historical object is made (such materials could be listed in the *Convention on International Trade in Endangered Species of Wild Fauna and Flora* (CITES) or in a number

21 Noteworthy loans were the Farnese Atlas, Museo Archeologico Nazionale, Naples; Mūsa's spherical astrolabe, HSM, Oxford; Philipp Immser's astronomical clock, Technisches Museum für Industrie und Gewerbe, Vienna; Joost Bürgi's and Erasmus Habermel's sextants, Národní Technické Muzeum, Prague; Ole Rømer's planetarium and eclipsareon, Bibliothèque nationale de France, Paris. See: Paolo Galluzzi, ed., *Galileo: Images of the Universe from Antiquity to the Telescope* (Florence, Giunti, 2009), 140 (III.1.7), 195 (IV.3.2), 322 (VI.4.1), 337 (VI.5.14 and VI.5.15), 403–404 (VIII.4.1 and VIII.4.2); exhibition website, accessed 27 September 2024: https://brunelleschi.imss.fi.it/galileopalazzostrozzi/index.html. On a national (Italian) level, the temporary exhibition *AStrum 2009: Astronomy and Instruments* (Rome, 16 October 2009–16 January 2010), gathered a selection of important original objects owned by Italian astronomical observatories. Exhibition catalogue: Ileana Chinnici, ed., *AStrum 2009: Astronomy and Instruments: Italian Heritage Four Hundred Years after Galileo* (Livorno: Sillabe, 2009).

FIGURE 8.5 A view of the temporary exhibition, *Galileo: Images of the Universe from Antiquity to the Telescope*, Strozzi Palace, Florence, 13 March–30 August 2009 © MUSEO GALILEO, FLORENCE. PHOTO: SABINA BERNACCHINI

of national and regional laws and regulations);[22] (3) the increasing attention to materials which, at different exposure levels, may endanger human and environmental health and life (mercury, lead, asbestos, radioactive substances, chemicals); and (4) the increasing attention to materials and objects which are connected to Western imperialism and colonialism over the last five centuries. Any loan made by neglecting international, national, or regional laws can have serious consequences for the objects (up to seizure and confiscation) and for those who grant the loan.

One way to avoid such complications would be the creation of temporary exhibitions primarily based on reconstructions, replicas, interactive tools, and digital projections (in the style of a mini science centre). Such ventures are less expensive (but for making replicas and interactive tools which, nevertheless,

22 Scientific instruments often contain – in what proportion it does not matter – materials derived from protected animals (ivory, coral, some types of animal skin, etc.) and plants (rosewood, some types of mahogany, some types of walnut wood, etc.). See the CITES official website: https://cites.org/eng, and in particular the *Washington Convention* text: https://cites.org/eng/disc/text.php and its *Appendices*: https://cites.org/eng/app/appendices.php, all accessed 25 September, 2024. Local law and regulations are just a mess.

can be re-used) and relatively easy to set up, especially without anxious museum curators underfoot. When not educationally intended, however, they may look like exhibitions just made for the sake of making of exhibitions (not a joke!). From a curatorial point of view, however, despite their attractiveness, these "science centre displays" do not satisfy the original idea of an exhibition: to bring objects together from all over the world and display the *real things*.

4 Centripetal and Centrifugal Forces

A temporary exhibition can also be created as a transitory section or "appendix" to a museum. In this case, a portion of the many objects confined to the stores can rise to visibility or, as an alternative, the importance of one or more originals may be outlined by special information and education apparatuses.[23] Such "internal" exhibitions usually emphasise one of the narratives conveyed by the permanent exhibition and, in any case, can serve the purpose of crystallising public attention around a new experience. Every modern museum has, in fact, at least one area for temporary exhibitions. It was, for instance, Jim Bennett who devoted a room of the HSM to temporary exhibitions and Paolo Galluzzi who transformed the Museo Galileo cellars into a multi-functional floor for conferences, educational activities, and, of course, exhibitions.

Internal temporary exhibitions may or may not include loans. Whatever the case, they exert centripetal forces on the public, pulling the visitors towards the museum. History of science temporary exhibitions of more recent years have tended, however, to occupy "external" white cubes. As such, they exert centrifugal forces, pulling the visitors away from the museum and its permanent exhibitions. In designing "external" exhibitions, the organisers should accurately weigh the pros and cons. If the exhibition is placed in another city, the effect of the centrifugal forces is reduced; if it is placed in the same city, however, a general visitor who already entered a history of science site will hardly repeat the experience by entering another. This is also true for school trips which, usually, include one scientific museum per year (at least in Italy). The centrifugal forces will therefore split the visitors between two venues.

---

23 See, for example: Michael Korey and Samuel Gessner, eds., *The Wondrous Course of the Planets: A Heavenly Machine for Elector August of Saxony* (Dresden: Mathematisch-Physikalischer Salon, 2022); exhibition website, accessed 30 September 2024: https://mathematisch-physikalischer-salon.skd.museum/en/ausstellungen/the-wondrous-course-of-the-planets/. What began as "temporary" seems, however, to have become "permanent."

It is not by chance that the noteworthy "centrifugal" temporary exhibitions organised by the Museo Galileo were either cautiously set up in art-history museums[24] or during the limited time span of 2008–2010. In fact, both *Galileo: Images of the Universe* and *The Medici and Science*, respectively located at the Strozzi Palace and the Pitti Palace in Florence,[25] were open to the public when the Museo Galileo was closed for renovation.

5 Conclusions

Evolution in the contents and style of permanent and temporary exhibitions is a natural social process. As any social process, it may produce benefits as well as dangers. The outcomes of any single step in the process – producing success or failure – cannot be understood in advance or on the spot, but only in retrospect.

In this chapter, I have tried to draw a rough taxonomy of permanent museum displays. In particular, I focused on the drawbacks and benefits deriving from the relationship between a permanent exhibition and the building that hosts it. A historic building may serve the display as a constraint or an opportunity, whereas a modern building, which today validates the collections, may become a constraint or a resource for the future. By looking at the story of their institutions, museum curators can follow the evolving tension between collections and their "containers" over a span of decades or even centuries. If successful, any new permanent display of the past will just represent another step in the evolution process. If not, the collections become at risk of being stored elsewhere or dispersed.

From the end of the nineteenth century, this evolution process has been complicated by the appearance of temporary exhibitions. In the last forty years, temporary exhibitions have become a conspicuous part of curatorial work both directly (by creating them) or indirectly (by managing temporary loans). Sometimes, temporary exhibitions are opportunities to ameliorate the permanent display and bring visitors into a museum; at other times they disperse creative energies and bring the visitors elsewhere. The multiplication of

24 For example: *Scienziati a Corte*, Uffizi Gallery, Florence, 18 March–18 June 2001; exhibition catalogue: Paolo Galluzzi, ed., *Scienziati a Corte: L'Arte della Sperimentazione nell'Accademia Galileiana del Cimento (1657–1667)* (Livorno: Sillabe, 2001).

25 On *Galileo: Images of the Universe*, see n. 21. On *The Medici and Science*, see n. 3 above; exhibition website: "The Medici and Science: Instruments and Machines in the Grand-Ducal Collections," Museo Galileo, accessed 30 September 2024, https://brunelleschi.imss.fi.it/mediciscienze/index.html.

temporary exhibitions could also be seen as a matter of concern by itself, as it might cause a surplus of conflicting cultural offers to tourists visiting a particular locality.

Over the last forty years, the Scientific Instrument Commission symposia have introduced many case-studies of both permanent and temporary exhibitions, among which I have selected only a few examples. Speakers at these symposia also expounded upon strategies to increase the impact of permanent and temporary exhibitions on the public, especially by facilitating the comprehension of the role of historic scientific instruments in human life. Debate within the Commission also singled out a number of critical cases where economic or political problems (which are often inextricably interconnected) could "materially" endanger museums and collections. Less discussed, however, has been the question of how economic and political influences might redirect a museum toward a mission that could be unrelated to the nature of its collections. Concerning the "irrelevance" resulting from such influences – giving up a museum's or a collection's specific identity and becoming just one among many – the Scientific Instrument Commission will surely have something to say.

Bibliography

Ackermann, Silke. "'Of Sufficient Interest ..., but not of Such Value ...': 260 Years of Displaying Scientific Instruments in the British Museum." Pp. 77–93 in *Scientific Instruments on Display*, edited by Silke Ackermann, Richard L. Kremer, and Mara Miniati. Leiden: Brill, 2014.

Barreca, Francesco. "The First National Exhibition of the History of Science (1929)." Pp. 95–107 in *Displaying Scientific Instruments: From the Medici Wardrobe to the Museo Galileo*, edited by Filippo Camerota. Trezzano sul Naviglio: Goppion, 2012.

Beretta, Marco. "Andrea Corsini and the Creation of the Museum of the History of Science in Florence (1930–1961)." Pp. 1–36 in *Displaying Scientific Instruments: From the Medici Wardrobe to the Museo Galileo*, edited by Filippo Camerota. Trezzano sul Naviglio: Goppion, 2012.

Brenni, Paolo. "The Physics Cabinet of the Istituto Tecnico Toscano." Pp. 215–241 in *Cabinets of Experimental Philosophy in Eighteenth-Century Europe*, edited by Jim Bennett and Sofia Talas. Leiden: Brill, 2013.

Camerota, Filippo, ed. *Displaying Scientific Instruments: From the Medici Wardrobe to the Museo Galileo*. Trezzano sul Naviglio: Goppion, 2012.

Camerota, Filippo. "The Mathematical Rooms." Pp. 3–17 in *Displaying Scientific Instruments: From the Medici Wardrobe to the Museo Galileo*, edited by Filippo Camerota. Trezzano sul Naviglio: Goppion, 2012.

Camerota, Filippo. "The Medici Collection of Mathematical Instruments: History and Museography." Pp. 129–148 in *European Collections of Scientific Instruments, 1550–1750*, edited by Giorgio Strano, Stephen Johnston, Mara Miniati, and Alison Morrison-Low. Leiden: Brill, 2007.

Camerota, Filippo, and Mara Miniati, eds. *The Medici and the Sciences: Instruments and Machines in the Grand-Ducal Collections*. Florence: Giunti, 2009.

Chinnici, Ileana, ed. *AStrum 2009: Astronomy and Instruments: Italian Heritage Four Hundred Years after Galileo*. Livorno: Sillabe, 2009.

Contardi, Simone. "The Museum of Physics and Natural History." Pp. 19–35 in *Displaying Scientific Instruments: From the Medici Wardrobe to the Museo Galileo*, edited by Filippo Camerota. Trezzano sul Naviglio: Goppion, 2012.

Falguières, Patricia. "Politics for the White Cube: The Italian Way." *Grey Room* 64 (2016): 6–39.

Galluzzi, Paolo, ed. *Galileo: Images of the Universe from Antiquity to the Telescope*. Florence, Giunti, 2009.

Galluzzi, Paolo, ed. *Scienziati a Corte: L'Arte della Sperimentazione nell'Accademia Galileiana del Cimento (1657–1667)*. Livorno: Sillabe, 2001.

Jorink, Eric. "The First Museum in the Netherlands? The Establishment of Teyler's Oval Room in Historical Perspective (c. 1600–1800)." Pp. 87–108 in *Teyler's Foundation in Haarlem and Its "Book and Art Room" of 1779: A Key Moment in the History of a Learned Institution*, edited by Ellinoor Bergvelt and Debora Meijers. Leiden: Brill, 2021.

Korey, Michael, and Gessner, Samuel, eds. *The Wondrous Course of the Planets: A Heavenly Machine for Elector August of Saxony*. Dresden: Mathematisch-Physikalischer Salon, 2022.

Lourenço, Marta C. "O Museu de la Ciência da Universidade de Lisboa: Patrimônio, Coleções e Pesquisa." Pp. 257–276 in *Coleções Científicas Luso-Brasileiras: Patrimônio a ser Descoberto*, edited by Marcus Granato and Marta C. Lourenço. Rio de Janeiro: Museu de Astronomia e Ciências Afins-MAST/MCT, 2010.

Miniati, Mara. "From the Museum of Physics to the Museum of Ancient Instruments." Pp. 37–55 in *Displaying Scientific Instruments: From the Medici Wardrobe to the Museo Galileo*, edited by Filippo Camerota. Trezzano sul Naviglio: Goppion, 2012.

Naggs, Fred. "The Tragedy of the Natural History Museum, London." *Megataxa* 7, no. 1 (2022): 85–112.

Nall, Joshua, Liba Taub, and Frances Willmoth, eds. *The Whipple Museum of the History of Science: Objects and Investigations to Celebrate the 75th Anniversary of R.S. Whipple's Gift to the University of Cambridge*. Cambridge: Cambridge University Press, 2019.

Otterspeer, Willem, *et al. 75 Jaar Museum Boerhaave*. Leiden: Museum Boerhaave, 2006.

Paselk, Richard A. "The Display of Twentieth-Century Instruments at the Humboldt State University." Pp. 148–158 in *Scientific Instruments on Display*, edited by Silke Ackermann, Richard L. Kremer, and Mara Miniati. Leiden: Brill, 2014.

Plassmeyer, Peter, Wolfram Dolz, and Michael Korey. *Mathematisch-Physikalischer Salon – Zwinger Meisterwerke*. Munich: Deutscher Kunstverlag, 2020.

Ramos Antunes, Emelinda, and Catarina Pires. "O Gabinete de Física da Universidade de Coimbra." Pp. 157–210 in *Coleções Científicas Luso-Brasileiras: Patrimônio a ser Descoberto*, edited by Marcus Granato and Marta C. Lourenço. Rio de Janeiro: Museu de Astronomia e Ciências Afins-MAST/MCT, 2010.

Schillinger, Klaus. "The Development of the Collections of Scientific Instruments of the Mathematical-Physical Salon between the Years 1750 and 1850 and Their Use for Research and Teaching." Pp. 101–106 in *Proceedings of the Eleventh International Scientific Instrument Symposium Held at Bologna University, Italy, 9–14 September 1991*, edited by Giorgio Dragoni, Anita McConnell, and Gerard L'E. Turner. Bologna: Grafis Edizioni, 1994.

Simcock, A.V. *The Ashmolean Museum and Oxford Science, 1683–1983*. Oxford: Museum of the History of Science, 1984.

Strano, Giorgio, ed. *Galileo's Telescope: The Instrument that Changed the World.* Florence: Giunti, 2008.

Weiss, Martin. "'Monuments of Science': How the Teyler Museum's Instrument Collection Became Historical." Pp. 195–214 in *Cabinets of Experimental Philosophy in Eighteenth-Century Europe*, edited by Jim Bennett and Sofia Talas. Leiden: Brill, 2013.

Wheatland, David P. *The Apparatus of Science at Harvard 1765–1800*. Cambridge, MA: Harvard University Press, 1968.

CHAPTER 9

Education through Instruments: What We've Learned (and Achieved) over the Last Forty Years

Jean-François Gauvin

The Scientific Instrument Commission (SIC) and its international membership have long encouraged educational initiatives in all of its pursuits. It welcomes students to its activities and motivates its members to reach out to learners of all ages. As a community of teachers, curators and experts, we have spared no effort over the last forty years in researching and publishing, collecting, digitising, assessing, saving and preserving (sometimes breaking!), buying and donating, restoring, communicating, exhibiting, and teaching with scientific instruments. Such undertakings led to innovative ways of engaging students and the wider public. The latter often became integral participants and partners in our understanding of the material culture of science. I'm convinced that what a scientific instrument is "now" is defined by these all-encompassing activities, which thus offer a versatile ground for education, communication, and interaction.[1] The recently edited volume by Elizabeth Cavicchi and Peter Heering in the Brill Series Scientific Instruments and Collections splendidly illustrates this point by laying out the rich experiences, challenges, and perspectives gained from the increased pedagogical attention given to historical scientific instruments in the formal (classrooms) and non-formal (e.g., museums) education of students.[2] Such attention is a testament to the significant progress accomplished since the SIC started holding annual symposia in the early 1980s.

The SIC's first *Newsletter* of May 1980 delineated three goals for the new organisation: to stimulate the production of inventories; to stimulate the production of catalogues; and to facilitate the exchange of information among research workers. Nothing about teaching or communication with the non-expert public was mentioned. That moment apparently called for exploring, cataloguing, documenting, researching, preserving, and networking with and about these

1 Liba Taub, "What Is a Scientific Instrument, Now?," *Journal of the History of Collections* 31 (2019): 453–467.

2 Elizabeth Cavicchi and Peter Heering, eds., *Historical Scientific Instruments in Contemporary Education*, Scientific Instruments and Collections, 9 (Leiden: Brill, 2022).

 | DOI:10.1163/9789004737587_011

(too often) forgotten or neglected treasure troves that had been collected and stored in universities, colleges, high schools, industrial sites, museums, and private collections. The founding members of the reconstituted SIC sought to know and document what was "out there" before beginning to think about how to bring these collections into play for teaching, exhibiting, and public outreach; such a sequence is typical in the formation of research fields. A quick survey of paper titles presented at the annual SIC symposia between 1980 and 2000 finds few presentations dealing with scientific instruments and education. Yet at the same time scholars such as Gudrun Wolfschmidt at the Hamburg Observatory, Falk Rieß at Oldenburg, Mara Miniati at the Istituto e Museo di Storia della Scienza, or Deborah Warner at the Smithsonian Institution were already devoting considerable time and effort to create educational curricula and public outreach programmes around exhibitions and instrument collections. We might thus ask how welcoming was the SIC to educators and educational initiatives during its first twenty years? What did it mean to be an "instrument scholar" in the 1980s or the 1990s? Could it be that education was not valued at the same level as the "hardcore" history of instrumentation or collections? Might we dare say that education, as such, was not initially recognised as a legitimate and creative research process in instrument studies?

From the turn of the century, however, we see a sharp and significant rise in the number of SIC symposia presentations specifically addressing pedagogical topics. Semester-length seminars and individual classes were introduced at Cambridge, Oxford, Harvard, Dartmouth, MIT, Florence, Rio de Janeiro, and Flensburg, to name but a few places. Although still not "mainstream," this approach to scholarly engagement with instruments was no longer invisible. This shift not only occurred within universities, observatories, museums, and the SIC symposia but also at international conferences, often organised by SIC members. I am thinking, for instance, of the two "Scientific Instrument Collections in the University" (SICU) symposia held in the United States in 2004 (Hanover, New Hampshire) and 2007 (Oxford, Mississippi), where participants explored education in multiple sessions;[3] of the 2009 Regensburg conference, "Learning by Doing: Experiments and Instruments in the History of Science Teaching," which provided relevant historical perspectives and concrete examples of instrument use in education;[4] of the first week-long "Reading Artifacts" seminar held in Ottawa in 2011 at the Canada Science and Technology

3 Selected papers from the SICU 2007 Symposium were published in *Rittenhouse* 22, no. 1–2 (2008), with some specifically discussing educational uses of instruments.

4 Peter Heering and Roland Wittje, eds., *Learning by Doing. Experiments and Instruments in the History of Science Teaching* (Stuttgart: Franz Steiner Verlag, 2011).

Museum, which demonstrated hands-on what it meant to be closely engaged with historical objects;[5] and of the Ninth International Conference for the History of Science in Science Education held in Flensburg, Germany in 2012 on the theme "Enabling Scientific Understanding through Historical Instruments and Experiments in Formal and Non-Formal Learning Environments."[6]

Another pivotal initiative was the still very successful YouTube channel launched by Paolo Brenni and Anna Giatti in 2009. Their simple didactic approach to scores of historical instruments (kept in two separate Italian collections) focused on how they were initially used for pedagogical demonstrations without providing exhaustive commentaries. It was left to the audience (from students to teachers, museum curators, experts, and the public) to make their own assessment of what they saw. Always eager to talk about and demonstrate how instruments worked, Brenni and Giatti understood early that this social media platform was an ideal way to reach a maximum number of people around the world; the Fondazione Scienza e Tecnica's channel @florencefst currently has over 4200 subscribers and half a million views.[7]

In what follows, I shall divide this short and necessarily incomplete review into three parts to survey a field that is now receiving significant support by the SIC community. In the first section, I consider how instruments were used over the last centuries to teach science and train students in producing scientific knowledge – a reminder that instruments were always a part of past scientific education. Now in "historical" collections, these objects teach today's inquirers about the nature of science and its history. In the second section, I discuss the epistemic value of replicating experiments and reconstructing instruments in training today's students and the wider public to view science as a fact-producing activity requiring specific skills only acquired through hands-on practice. The last section concerns museums, where historical instruments can be put to good use in exhibitions and in motivating playful activities

5 A team that carefully studied an Eötvös torsion balance during the week published their results: Katharine Anderson, Mélanie Frappier, Elizabeth Neswald, and Henry Trim, "Reading Instruments: Objects, Texts and Museums," *Science & Education* 22 (2013): 1167–1189.

6 It resulted in a publication: Peter Heering, Stephen Klassen, and Don Metz, eds., *Enabling Scientific Understanding through Historical Instruments and Experiments in Formal and Non-Formal Learning Environments*, Flensburg Studies on the History and Philosophy of Science in Science Education, 2 (Flensburg: Flensburg University Press, 2013). Other papers were also published in the journal *Science & Education*.

7 Paolo Brenni, "Filming Nineteenth-Century Physics Demonstrations with Historical Instruments," in *Historical Scientific Instruments in Contemporary Education*, eds. Cavicchi and Heering, 34–49. See Paolo Brenni's demonstration videos, Fondazione Scienza e Tecnica, https://www.youtube.com/@florencefst/; and at "Adminstrator Liceo Sarpi," Liceo Paolo Sarpi, https://www.youtube.com/@administratorliceosarpi7144/.

to showcase the past and more recent nature of science. These three sections, taken together, suggest that instruments perform different functions over their existence; they possessed an original, pedagogically active life, became historical and were reconsidered in a museological context, and finally can be replicated to make them pedagogical once again in a second, modern "active life." In the conclusion, I will consider as a new frontier of research and application several sensitive areas of investigation, such as decolonisation, that might ensure that the pedagogical tools developed by the original thinkers and educators in the SIC and elsewhere remain relevant in the coming years.

1 Teaching with Scientific Instruments, Now and Then

In 2017, a newly minted architect from MIT, Francesca Liuni, now assistant professor at the Rhode Island School of Design (RISD) in Providence, contacted me *à propos* recreating on a smaller scale her idea for a museum that would bring the viewer's body into relationship with an astrolabe (usually about 15–20 cm in diameter). This concept, which arose in a course taught at MIT by Elizabeth Cavicchi, and eventually became an award-winning master's thesis, challenged the museum visitor's experience and the conventional way of thinking about exhibits. The project turned into one of my favourite collaborations. I found the money to erect two large metal rings and spent a few evenings with Francesca building two 8 × 16 foot frames to support her renditions of the astrolabe's three-dimensional stereographic projection and geometry of solid angles (*Fig. 9.1*). Physically immersed in the mathematical dimensions of an astrolabe, students, faculty, and visitors entering the Department of the History of Science at Harvard University encountered first-hand the powerful alliance of history, mathematics, science, and art crystallised in this classic scientific instrument invented more than 1000 years ago.[8]

A simple instrument can open a vast horizon of science teaching moments, from elementary schools to university programmes. Furthermore, when you introduce historical, cultural, philosophical, and technological perspectives to basic physics principles and mathematical equations you discover a complex and multifaceted apparatus. This was the idea behind the International Pendulum Project (IPP) conducted in the early 2000s by forty academic researchers and numerous school teachers from sixteen countries.

8 Elizabeth Cavicchi, "Historical Scientific Instruments in Exploratory Teaching and Learning," in *Historical Scientific Instruments in Contemporary Education*, eds. Cavicchi and Heering, 174–178.

FIGURE 9.1
(*a*) Partial view of an exhibit, *Syntax of an Astrolabe*, in the foyer of the Department of the History of Science, Harvard University, 2017
PHOTO: MAUREEN TON
(*b*) The Ṣafavid-dynasty astrolabe, circa 1620, that inspired Francesca Liuni's project. It was in the small case in the middle of the exhibit
COLLECTION OF HISTORICAL SCIENTIFIC INSTRUMENTS, HARVARD UNIVERSITY, DW0596

By demonstrating how to develop, in the liberal arts tradition, a well-designed science course based on a simple instrument such as the pendulum, the project's instigators sought ways "for students to learn important components of scientific knowledge, key features of scientific method, and important aspects of the interplay between science and its social and cultural context."[9] By opening a course with simple objects, one introduces the notion of instrumentation and encourages critical thinking about what an instrument of science is. As Simon Werrett argues, in looking at the thrifty practices of early modern savants using kitchen utensils to cleaning tools, "[s]cientific instruments might perhaps be better thought of as *processes* rather than as fixed, static objects. ... 'Instrument' should be thought of as a practice, rather than a thing."[10] In other words, simple objects can demonstrate that instruments are situated and contextualised – not typified, epitomised, and presupposed.

Historically, educators periodically call for science teaching reforms and these invariably rest upon scientific instrumentation. Already in the eighteenth century, abbé Jean-Antoine Nollet, among others, made it salient and fashionable to learn about the "*physique amusante*" while creating a notable and lucrative instrument-making business based on his *Leçons de physique expérimentale*. After the French Revolution, Nollet's style of demonstration lectures carried into the newly founded *Écoles centrales*. In the second half of the nineteenth century, physics laboratories in Britain were primarily developed as *teaching* (not research) spaces, focusing on the processes of precision measurement. These laboratories were not only meant for aspiring physics teachers but were also considered "as the optimum medium for a liberal education in rational and accurate reasoning." In nineteenth-century Germany, the physics seminar established by Franz Neumann at Königsberg University in East Prussia combined mathematical and experimental physics. Its quantitative and practical educational methods trained the new professional "physicists" as well as secondary school teachers. At the same time, in Spain the creation of a national secondary school system was undertaken simultaneously with the establishment of cabinets of physics, chemistry, and natural history (a development not unlike that occurring in French Catholic schools in Québec during the same period, see *Fig. 9.2*). The focus was put on science and technology, to which instruments were crucial for teaching experimental science

9 Michael R. Matthews, Colin Gauld, and Arthur Stinner, "The Pendulum: Its Place in Science, Culture and Pedagogy," *Science & Education* 13 (2004): 261–277, quote 274. Two full numbers of this journal were dedicated to the project's results.

10 Simon Werrett, *Thrifty Science: Making the Most of Materials in the History of Experiment* (Chicago: The University of Chicago Press, 2019), 82 (emphasis original).

FIGURE 9.2 *Cabinet de physique* of Université Laval, circa 1895. In the early nineteenth century, this cabinet belonged to the Séminaire de Québec. Such cabinets, while usually smaller, were found in most *séminaires* in the province of Québec between 1830 and 1930. Jules-Ernest Livernois, *Le Musée de physique de l'Université Laval, Québec*, from the album, *Maisons d'éducation de la province de Québec*, circa 1895
MUSEE NATIONAL DES BEAUX-ARTS DU QUEBEC, GIFT OF YVES BEAUREGARD (2014.115.06). PHOTO: MNBAQ

and disciplined reasoning and, by the end of the century, for conducting what came to be called "original" research. Outside Europe, similar patterns are likewise detectable in large states such as Brazil and Japan, and in smaller settings, such as the flourishing Greek community in Istanbul.[11]

---

11 Lewis Pyenson and Jean-François Gauvin, eds., *The Art of Teaching Physics: The Eighteenth-Century Demonstration Apparatus of Jean-Antoine Nollet* (Sillery, Quebec: Septentrion, 2002). Graeme Gooday, "Precision Measurement and the Genesis of Physics Teaching Laboratories in Victorian Britain," *British Journal for the History of Science* 23 (1990): 25–51, quote 26. Kathryn M. Olesko, *Physics as a Calling: Discipline and Practice in the Könisburg Seminar for Physics* (Ithaca: Cornell University Press, 1991). Josep Simon and Mar Cuenca-Lorente, "Science Education and the Material Culture of the Nineteenth-Century Classroom: Physics and Chemistry in Spanish Secondary Schools," *Science & Education* 21 (2012): 227–244. Pedro Ruiz-Castell, "Scientific Instruments for Education in Early Twentieth-Century Spain," *Annals of Science* 65 (2008): 519–527. Reginaldo Alberto Meloni, "Tools for Teaching Physics and Chemistry in Secondary Schools: The Case of the Colégio Culto à Ciência, Brazil, 1899–1902," in *Scientific Instruments in the History of Science: Studies in Transfer, Use and Preservation*, eds. Marcus Granato and Marta C. Lourenço

In the United States, scientific educational reforms also came with lots of instruments. In the nineteenth century, antebellum colleges embraced the general teaching of the natural sciences, as observed *inter alia* by the increased number of colleges' cabinets of instruments and the rise of new buildings specially designed for laboratories and museums (often dubbed the "apparatus" of an institution). The rise of the American instrument-making industry closely followed this sustained educational reform just before the Civil War (*Fig. 9.3*).[12] Antebellum colleges maintained a positive relationship with the growing scientific establishment. In fact, "the increase in equipment and professors, the improved quality of textbooks, and the pervasive concern for the perpetuation of the scientific spirit proceeded virtually unhindered" during that period.[13] By the 1870s, a "New Education" programme was being pushed by science reformers, especially Daniel Coit Gilman from the Sheffield Scientific School at Yale College. Applied science (and engineering), with its growing *armamentarium*, should not be interpreted as competing against the classics (Latin and Greek), according to its proponents, but perceived as another way to mould well-rounded individuals.[14] With the emergence of land-grant universities and the move of laboratory-based science teaching into high schools

(Rio de Janeiro: Museu de Astronomia e Ciência Afins, 2014), 247–258. Tetsuo Isozaki, "The Organisation and the Recontextualization of *Rika* (School Science) Education in the Second Half of the Nineteenth Century in Japan," *Science & Education* 23 (2014): 1153–1168. Panagiotis Lazos, George N. Vlahakis, and Constantine Skordoulis, "Instruments and Laboratories in the Schools of the Greek Community of Istanbul, 1850–1960," in *Scientific Instruments Between East and West*, eds. Neil Brown, Silke Ackermann, and Feza Günergun (Leiden: Brill, 2019), 168–186.

12 Deborah J. Warner, "Commodities for the Classroom: Apparatus for Science and Education in Antebellum America," *Annals of Science* 45 (1988): 387–397.

13 Stanley M. Guralnick, "Sources of Misconception on the Role of Science in the Nineteenth-Century American College," *Isis* 65 (1974): 352–366, quote 365. Several scholars besides Guralnick reassessed in the mid-1970s the antebellum scientific stance; see Sally Gregory Kohlstedt, "Reassessing Science in Antebellum America," *American Quarterly* 29 (1977): 444–453. For a standard history of the period, Daniel J. Kevles, *The Physicists: The History of a Scientific Community in Modern America* (Cambridge, MA: Harvard University Press, 1987), 3–24.

14 Jean-François Gauvin, "The Apparatus of Yale College's Sheffield Scientific School and the Promise of a 'New Education' (ca 1870)," in *Scientific Instruments as Cultural Artefacts*, eds. Paola Bertucci and Alexi Baker (New Haven: Yale University Press, forthcoming). See also Steven Turner, "Chicago Scientific Instrument Makers, 1871–1918, Part 1: The School Market," *Rittenhouse* 19 (2005): 65–128; Steven Turner, "Chicago Scientific Instrument Makers, 1871–1918, Part 2: The University and Research Market," *Rittenhouse* 20 (2006): 1–64.

FIGURE 9.3 Double-beam axial electromagnetic reciprocating engine by Daniel Davis, Jr, Boston, circa 1848. Davis was a prominent instrument maker in Boston before the Civil War
COLLECTION OF HISTORICAL SCIENTIFIC INSTRUMENTS, HARVARD UNIVERSITY, 1997-1-1801

after the Civil War, scientific instrument making in the USA boomed by 1900. And most of the apparatus being produced was intended for pedagogical purposes.

The goal of teaching science to form individual character was reawakened a century later, with a twist, when Gerald Holton created with F. James Rutherford and Fletcher G. Watson the Project Physics Course – also known as the Harvard Project Physics. Developed in the 1960s after the shock of Sputnik, the Project's objective was to rethink the pedagogical approach to the teaching of physics (for eleventh and twelfth graders in US high schools) using a humanistic approach (in contrast to the Physical Science Study Committee programme developed at MIT). Built from the ground up, the Harvard Project Physics (dubbed the "New Physics" by some, though Holton never liked the term) required a vast array of novel curriculum materials: instruction booklets, projectable transparencies, 8-mm short film loops, teacher training films, *and* a wealth of laboratory materials and equipment specially designed and manufactured by Damon Engineering, Inc., a local Massachusetts firm whose products were distributed by Holt, Rinehart and Winston, Inc. from New York

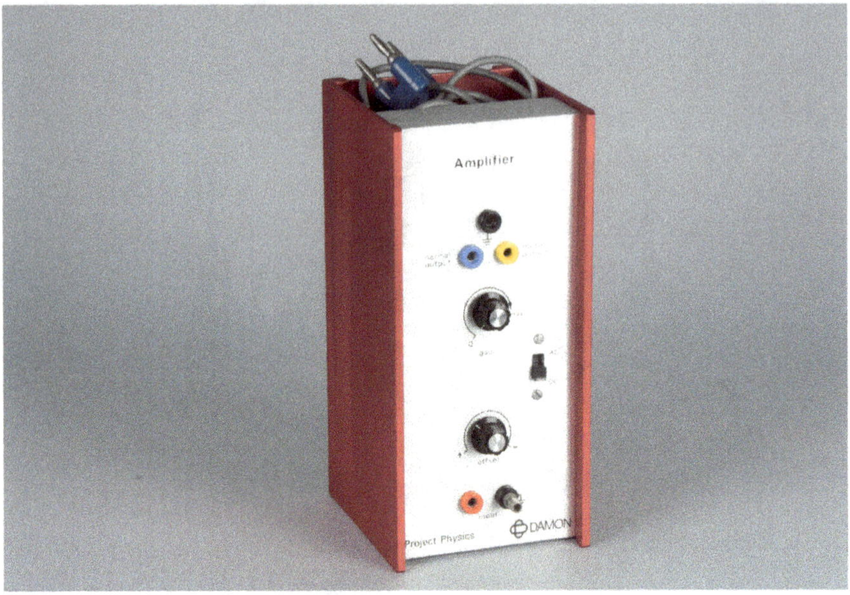

FIGURE 9.4 AC-DC amplifier by Damon Engineering, Inc., Needham, Massachusetts, 1962–1972, designed for Harvard Project Physics
COLLECTION OF HISTORICAL SCIENTIFIC INSTRUMENTS, HARVARD UNIVERSITY, 2006-1-0087

City (*Fig. 9.4*). Most of these laboratory devices were unsophisticated, inexpensive, and robustly built to withstand the wear-and-tear inflicted by hundreds of eager (and sometimes indifferent) students.[15]

These examples, stretching over three centuries, only scratch the surface of the creative and didactic use of instruments geared toward science training and education.[16] Over the last forty years, SIC members have contributed considerably to the discovery, history, inventorying, and preserving of such apparatus. These efforts led, sometimes directly, to experiments in historical teaching. In Italy, for example, scholars and educators have featured combined standard laboratory exercises with formal visits to museum collections, or have employed high-school students to help document historical scientific

15 Gerald Holton, "The Project Physics Course, Then and Now," *Science & Education* 12 (2003): 779–786. The best analysis of the Harvard Project Physics is David Meshoulam, "'Teaching Physics as One of the Humanities': The History of (Harvard) Project Physics, 1961–1970" (PhD Diss., University of Wisconsin-Madison, 2014).

16 It is important to remember here that these school collections were not always used to their full potential. Paolo Brenni, "The Evolution of Teaching Instruments and Their Use Between 1800 and 1930," *Science & Education* 21 (2012): 191–226.

instruments.[17] In France, similar activities have been organised with the help of the Association de sauvegarde et d'étude des instruments scientifiques et techniques de l'enseignement (ASEISTE), founded in 2004, which mobilises six universities and sixty secondary schools (*lycées*) around a common purpose: to preserve, document, educate, and communicate about science with historical collections.[18] In Brazil, the Museum of Astronomy and Related Sciences (MAST) in Rio de Janeiro has developed new educational methods to teach about science and technology using its own extensive historical collections and advances new statistical techniques of evaluation to test the effectiveness of the latter.[19]

Although some educators have advocated since the early 2000s the introduction of "microcomputer-based laboratories" to assist in teaching fundamental skills of laboratory work, many others (including this author) believe that encouraging the operation of (historical) scientific instruments catalyses pedagogical and didactic elements (we might speak of tacit knowledge here) not found elsewhere in a standard curriculum.[20] The *materiality* of the instruments is at the core of these initiatives. Too often forgotten or neglected by science education specialists and authorities, material practices compose a natural dyad with social practices (as Steven Shapin and Simon Schaffer's *Leviathan and the Air-Pump* convincingly argued four decades ago). Through these historical objects, if they were made available, students would receive hands-on experience that would stay with them long after they leave the laboratory. According to philosopher of science Brandon Boesch, skills thus acquired exist in a sort of "life cycle" between learning, practising, and transmitting.

17 Ornella Pantano and Sofia Talas, "Physics Thematic Paths: Laboratorial Activities and Historical Scientific Instruments," *Physics Education* 45 (2010): 140–146. F. Marcon, "Between School and Universities: The Study of the Historical Scientific Instruments Collection of the Liceo Maffei in Verona," *Il Nuovo Cimento* 40C (2017): 1–6.

18 Françoise Khantine-Langlois, Alfonso San-Miguel, and Pierre Lauginie, "Examples of the Use in Education of Historical Physics Instruments at Secondary School and University Level in France Supported by ASEISTE," in *Historical Scientific Instruments in Contemporary Education*, eds. Cavicchi and Heering, 122–138. Two special numbers (44 and 45) of the journal *In Situ. Revue des patrimoines* discuss the scientific heritage of nineteenth- and twentieth-century collections in French *lycées*.

19 See the education section on the MAST website: Museu de Astronomia e Ciencias Afins – MAST, accessed 10 August 2023, https://www.gov.br/mast/pt-br.

20 Ricardo Trumper, "The Physics Laboratory – A Historical Overview and Future Perspectives," *Science & Education* 12 (2003): 645–670.

Skill transmittance in science education is nothing short of "a valuable pathway to developing a more complete understanding of scientific practice."[21]

2 Replicating Experiments, Giving Instruments Agency?

To some, replicating historical experiments captures the spirit of eighteenth-century itinerant lecturer-demonstrators. It catches the humanistic endeavour of science and provides "a romantic view of the world around us."[22] As we briefly saw above, bringing together the "histories of use" with "tacit skills" is controversial, but has become familiar to scholars, educators and curators in recent years, in formal and non-formal pedagogical settings.[23] And it is here to stay.[24] We usually attribute the first formal initiatives to Falk Rieß, trained in physics, philosophy, and didactics in Germany, who developed in the 1980s innovative pedagogical programmes in his laboratory at the Carl von Ossietzky University of Oldenburg. Future physics teachers learned more than physics; they explored the history of physics by performing experiments. In this practical training they replicated iconic historical instruments and experiments such as Tycho Brahe's sextant, Charles-Augustin Coulomb's torsion balance, Moritz Hermann von Jacobi's electric motor, and a prize-winning demonstration of the Magnus effect. This method provided pupils and students a novel way of looking at how to better train future teachers *and* how to study the history of science. It opened new vistas on the historical practice of science.[25]

21 Catherine Milne and Kathryn Scantlebury, eds., *Material Practice and Materiality: Too Long Ignored in Science Education* (Cham: Springer, 2019). Brandon Boesch, "Skill Transmittance in Science Education. Studying the Skills of Scientific Expertise," *Science & Education* 28 (2019): 45–61, quote 59.

22 Don Metz and Art Stinner, "A Role for Historical Experiments: Capturing the Spirit of the Itinerant Lecturers of the 18th Century," *Science & Education* 16 (2007): 613–624, quote 614.

23 Conferences such as the one held at the Science Museum in 2017 have acquired their *lettres de noblesse*. Tim Boon, ed., "A Symposium on Histories of Use and Tacit Skills," *Science Museum Group Journal* 8 (2017), https://journal.sciencemuseum.ac.uk/article/histories-of-use-and-tacit-skills/.

24 Marieke M.A. Hendriksen, "Rethinking Performative Methods in the History of Science," *Berichte zur Wissenschaftsgeschichte* 43 (2020): 313–322.

25 Falk Rieß, "History of Physics in Science Teacher Training in Oldenburg," *Science & Education* 9 (2000): 399–402. Peter Heering, Falk Rieß, and Christian Sichau, eds., *Im Labor der Physikgeschichte: Zur Untersuchung historischer Experimentalpraxis* (Oldenburg: Universität Oldenburg Verlag, 2000). To the chagrin of Rieß, the motivation barrier to study physics was not solved by 2000. Rieß, "Problem with German Science Education," *Science & Education* 9 (2000): 327–331.

Despite the achievements of this method, critics of experimental reconstructions in the mid-1990s saw it as "relics (or revivals) of positivist approaches." For proponents of a "strong program" in the growing field of sociology of scientific knowledge, themselves being challenged by practising scientists during the so-called Science Wars, the reconstruction of historical experiments revealed more about the historians conducting said experiments than it did about the original experimenters.[26] None of these criticisms deterred the scholars and educators invested in this epistemic and pedagogical course of action. This controversy, to my knowledge, never was a methodological issue at the SIC – even though there were competing approaches and results between replicationists.[27] Uncovering a relevant "dynamic relationship between library and laboratory," historians and teachers sought in experimental reproductions a way to "bridge the unavoidable gap that exists between the actions and ideas of historical actors and the textual descriptions or artefactual residues of those actions and ideas that have come down to us."[28]

The value of this approach for teaching the nature of science is well documented and continues to be studied across the globe. Results have varied. Although students appear to enjoy the reconstruction of experiments and instruments, it is not always clear whether they acquire a better grasp of the physics (or chemistry) or of the apparatus's original historical and sociological contexts. They do, however, gain a sense of wonder about nature and discover that scientific activities and practices are very much social, materiality driven, and not the sole domain of disembodied "geniuses." To ensure that such pedagogical projects work well, teachers require adequate training and forums where they can share their achievements and compare difficulties encountered regarding various types of experiments and devices. For students and educators alike, new scientific and technical questions arise, and the former often find it easier to challenge authority after performing hands-on experiments and confronting the materiality and vicissitudes of building an instrument. All this is not only true for students and educators who reenact and put into practice these ideas, but also for scholars in Germany, England, Turkey, Norway, and the United States who develop and improve this approach by

26 To understand how this war started, see Alan D. Sokal and Jean Bricmont, *Fashionable Nonsense: Postmodern Intellectuals' Abuse of Science* (New York: St Martins Press, 1999).

27 One example is Alberto A. Martínez, "Replication of Coulomb's Torsion Balance Experiment," *Archives for the History of Exact Sciences* 60 (2006): 517–563, which arrives at a conclusion different from that of Peter Heering on this topic.

28 Hjalmar Fors, Lawrence M. Principe, and Otto Sibum, "From the Library to the Laboratory and Back Again: Experiment as a Tool for Historians of Science," *Ambix* 63 (2016): 85–97, quote 88.

simultaneously studying its pedagogical benefits and examining the epistemology of knowledge production.[29]

Since the founding of the SIC, scholars from all horizons have replicated instruments and reworked experiments to valuable heuristic, pedagogical, and intellectual results.[30] These reproductions and reworkings, "which aimed at understanding more deeply the historical actors, their experiments and ideas and their origins," have laid the foundation to a transformative encounter with past practices and scientific knowledge.[31] In recent years, Pamela Smith's major "Making and Knowing Project" at Columbia University, Sven Dupré's interdisciplinary "Artechne – Technique in the Arts 1500–1950" at Utrecht University and the University of Amsterdam, William Newman's decades-long reworking of Newton's alchemical experiments, Peter Heering and Cibelle Celestino Silva's reevaluation of eighteenth-century Leiden jar operational accounts, and Tom Everett's reconstruction and reworking of the ear phonautograph originally built in 1874 by Alexander Graham Bell and Clarence J. Blake at the Canada Science and Technology Museum, to name but a few projects, have opened up perspectives for a fruitful exchange between textual knowledge, hands-on practices, and museum artefacts.[32] These practices can,

29 Peter Heering, "Make-Keep-Use: Bringing Historical Instruments Into the Classroom," *Interchange* 46 (2015): 5–18. Hasok Chang, "How Historical Experiments Can Improve Scientific Knowledge and Science Education: The Cases of Boiling Water and Electrochemistry," *Science & Education* 20 (2011): 317–341. Zeynep Neslihan Alisir and Serhat Irez, "The Effect of Replicating Historical Scientific Apparatus on High School Students' Attitudes Towards Science and Their Understanding of Nature of Science," *Science & Education* 29 (2020): 1201–1234. Per-Odd Eggen, Lise Kvittingen, Annette Lykknes, and Roland Wittje, "Reconstructing Iconic Experiments in Electrochemistry: Experiences from a History of Science Course," *Science & Education* 21 (2012): 179–189. Elizabeth Cavicchi, "Historical Experiments in Students' Hands: Unfragmenting Science Through Action and History," *Science & Education* 17 (2008): 717–749.

30 My definition of a reconstructed instrument is based on Klaus Staubermann, ed., *Reconstructions: Recreating Science and Technology of the Past* (Edinburgh: National Museums of Scotland, 2011). For historical context, Deborah Warner, "Replicating Historic Scientific Instruments," *Bulletin of the Scientific Instrument Society*, no. 151 (2021): 10–13. Warner focuses on replications made for exhibitions and not for hands-on experimental practices.

31 Fors, Principe, and Sibum, "From the Library to the Laboratory and Back Again," 88. For more on restoration and scientific instruments, Jean-François Gauvin, *Instruments of Knowledge: Finding Meaning in Objects, Habits, and Museums*, Nuncius Series, 12 (Leiden: Brill, 2023), chapter 5.

32 Pamela H. Smith and the Making and Knowing Project, "Historians in the Laboratory: Reconstruction of Renaissance Art and Technology in the Making and Knowing Project," *Art History* 39 (2016): 210–233; and "The Making and Knowing Project," accessed 28 May 2020, https://www.makingandknowing.org/. Smith, *From Lived Experience to the*

I would argue, strengthen our trust in science by concretely "demystifying" how it was performed in the past and, to a large extent, still is today.[33] Museums can play a much bigger role in engaging the public (and the scholarly community) with the creation of a non-formal setting where historical instruments are not simply idealised and artfully displayed behind vitrines but are instead actively presented as performing agents in scientific endeavours.[34]

3 Museums

The modern museum never was unenthusiastic or indifferent to pedagogical methods, including those described in the previous section.[35] Several scholars in our field have tested the practicality and value of using historical and/or replicated instruments in science museums, in the context of wider public interactions (i.e., mostly families), special students and school programme engagements, and continuing teacher education. Some institutions, like the Lorentz Lab at the Teylers Museum in Haarlem, have transformed historic laboratories into creative and interactive science education spaces; others have experimented with putting real scientists on display by moving their functional laboratory (apparatus and related paraphernalia) *into* the museum. These initiatives have often been met with success, although the cost

Written Word: Reconstructing Practical Knowledge in the Early Modern World (Chicago: The University of Chicago Press, 2022). On the pedagogical role of this initiative, Tianna Helena Uchacz, "Reconstructing Early Modern Artisanal Epistemologies and an 'Undisciplined' Mode of Inquiry," *Isis* 111 (2020): 606–613. Sven Dupré's online project: "Artechne – Technique in the Arts, 1500–1950," Utrecht University, accessed 28 February 2023, https://artechne.wp.hum.uu.nl/. William Newman, *Newton the Alchemist: Science, Enigma, and the Quest for Nature's "Secret Fire"* (Princeton: Princeton University Press, 2019). Peter Heering and Cibelle Celestino Silva, "Re-Examining the Early History of the Leiden Jar: Stabilization and Variation in Transforming a Phenomenon Into a Fact," *History of Science* 56 (2018): 314–342. Tom Everett, "Writing Sound with a Human Ear: Reconstructing Bell and Blake's 1874 Ear Phonautograph," *Science Museum Group Journal* 12 (2019), http://journal.sciencemuseum.ac.uk/browse/issue-12/writing-sound-with-a-human-ear/.

33 Naomi Oreskes, *Why Trust Science?* (Princeton: Princeton University Press, 2019).

34 On "beautiful" instruments, Richard Dunn, "'More Artistic than Scientific': Exhibiting Instruments as Decorative Arts in the Victoria and Albert Museum," in *Scientific Instruments on Display*, eds. Silke Ackermann, Richard L. Kremer, and Mara Miniati (Leiden: Brill, 2014), 61–76. On the concept of art in science museums, Catia Rodrigues Barbosa, "L'esthétique dans les musées des sciences et des techniques," *La Lettre de l'OCIM* 95 (2004): 27–30.

35 Robert G.W. Anderson, "'What is Technology?': Education Through Museums in the Mid-Nineteenth Century," *British Journal for the History of Science* 25 (1992): 169–184.

for organising and operating these participatory projects makes it difficult to maintain them.[36] From the reproduction of scary and marvellous images using an eighteenth-century solar microscope to the "shocking" electrical entertainment of Benjamin Franklin ("Come, Gentlemen, – who'll be next, – that's it, go-o-od, Line of Fops, all hold hands, Line of Fops, how many have we now …") and the down-to-earth mechanics of Galileo, science museum and science centre visitors are asked to "play" with historical exhibits.[37] As I have previously discussed when looking at simple devices crafted in the 1960s at Harvard and dubbed "quantum toys," play possesses epistemic virtues and is an essential and legitimate part of learning.[38] It is often said that Frank Oppenheimer, who founded the Exploratorium in San Francisco, got angry whenever an exhibit could not be made to "misbehave." Experimental failures and exhibit breakdowns were all part of having fun with science.[39] Though adults may

---

36 Peter Heering, "Educational Experiences in Re-enacting Historical Experimental Procedures," in *Historical Scientific Instruments in Contemporary Education*, eds. Cavicchi and Heering, 226–242. Jürgen Teichmann, "Historical Experiments and Science Education. From Conceptual Planning of Exhibitions to Continuing Education for Teachers," *Interchange* 46 (2015): 31–43. Trienke M. van der Spek, "The Lorentz Lab: Reviving the Scientific History of Teylers Museum with Working Replica," in *Historical Scientific Instruments in Contemporary Education*, eds. Cavicchi and Heering, 243–262. Morgan Meyer, "Researchers on Display: Moving the Laboratory into the Museum," *Museum Management and Curatorship* 26 (2011): 261–272.

37 Peter Heering, "The Enlightened Microscope: Re-enactment and Analysis of Projections with Eighteenth-Century Solar Microscopes," *British Journal for the History of Science* 41 (2008): 345–367. David J. Rhees, "Sparks and Shocks: Replicas of Historical Instruments in Museum Education," *Interchange* 46 (2015): 45–56. Jürgen Teichmann, "Studying Galileo at Secondary School: A Reconstruction of His 'Jumping Hill' Experiment and the Process of Discovery," *Science & Education* 8 (1999): 121–136. Franklin's imaginary quote comes from Thomas Pynchon, *Mason & Dixon* (New York: Henry Holt and Company, 1997), 294.

38 Jean-François Gauvin, "Playing with Quantum Toys: Julian Schwinger's Measurement Algebra and the Material Culture of Quantum Mechanics Pedagogy at Harvard in the 1960s," *Physics in Perspective* 20 (2018): 8–42. These toys were adapted and replicated, based on my article, by Fernand Brunschwig for STEMteachersNYC (2022). For a classic interpretation of play, J. Huizinga, *Homo ludens: A Study of the Play-Element in Culture* (London: Routledge & Kegan Paul, 1949). Gerard L'E. Turner, "Presidential Address: Scientific Toys," *British Journal for the History of Science* 20 (1987): 377–98. The role of material objects is also of significant importance. Maarit Alasuutari, Marleena Mustola, and Niina Rutanen, eds., *Exploring Materiality in Childhood: Body, Relations and Space* (Abingdon: Routledge, 2021).

39 K.C. Cole, *Something Incredibly Wonderful Happens: Frank Oppenheimer and His Astonishing Exploratorium* (Chicago: The University of Chicago Press, 2009). See also one among many recent case studies from the National Science Museum in Thailand: Peeranut Kanhadilok and Mike Watts, "Youth at Play: Some Observations from a Science Museum," *International Journal of Adolescence and Youth* 22 (2017): 179–194.

sometimes feel unwelcome in science museums,[40] research has shown how significant the pedagogy of play is towards learning science. From John Locke's *Some Thoughts Concerning Education* (1693) to Jean-Jacques Rousseau's *Émile* (1762) and, more recently, to Friedrich Froebel's Kindergärten and Maria Montessori's Casa dei Bambini, educators have long-since discerned a core concept – i.e., "the more natural and unstructured the environment is, the richer is children's play as an educational activity."[41] And this is true whether a museum sets up an engineering or a virtual environment.[42]

"Science objects," according to Samuel Alberti, "can be *fun*."[43] Together with labels ("A good label is poetry"), spatial planning, specially designed educational activities, digital interactions, and social media networking, today's science exhibits, Alberti asserts, should present "science [a]s a cultural activity, embedded within other things we do." The best exhibits will instantiate the "*joy* of science" and hook visitors using surprising objects chosen to demonstrate that science is a "contingent, messy and sometimes controversial activity."[44] This, however, is seldom an easy task since the "persistence of heroic and genius narratives" is always lurking in museums (and expected by visitors), as Rebekah Higgitt cogently illustrates by describing a major exhibit focused on the eighteenth-century quest to solve the longitude problem at sea.[45] Interactive displays might not be the answer here. A study at the Museo Galileo, for instance, explains that exhibit visitors tend to focus more on interactive models than on reading labels and seizing a richer understanding of the phenomenon. However, when only labels are available, it is important to

---

40 Winifred Kehl, "I Have a Confession: I'm an Adult. How Can Science Museums Welcome Grown-Ups?," *Museum* (July–August 2013): 46–51. Steven Conn, *Do Museums Still Need Objects?* (Philadelphia: University of Pennsylvania Press, 2010), 138–171. The chapter is entitled "Where Have All the Grown-Ups Gone?"

41 Roberto Farné, "Pedagogy of Play," *Topoi* 24 (2005): 169–181, quote 173. On the role of games to apprehend and make sense of culture, social connection, politics, and economic change, see Silvia Lindtner and Paul Dourish, "The Promise of Play: A New Approach to Productive Play," *Games and Culture* 6 (2011): 453–478.

42 Pirko Tõugu, *et al.*, "Connecting Play Experiences and Engineering Learning in a Children's Museum," *Journal of Applied Developmental Psychology* 53 (2017): 10–19. Jonathan P. Rowe, *et al.*, "Play in the Museum: Design and Development of a Game-Based Learning Exhibit for Informal Science Education," *International Journal of Gaming and Computer-Mediated Simulations* 9 (2017): 96–113.

43 Samuel J.M.M. Alberti, *Curious Devices and Mighty Machines: Exploring Science Museums* (London: Reaktion Books Ltd, 2022), 155 (emphasis original). See also Alberti, "Objects and the Museum," *Isis* 96 (2005): 559–571.

44 Alberti, *Curious Devices and Mighty Machines*, 196.

45 Rebekah Higgitt, "Challenging Tropes: Genius, Heroic Invention, and the Longitude Problem in the Museum," *Isis* 108 (2017): 371–380.

carefully consider the theory of cognitive ergonomics, i.e., the meaning of symbols and cues familiar to people, in creating visual interpretations that accompany the text.[46] Producing a meaningful and educational science exhibition is a difficult enterprise, especially when you aim at making regular scientific objects "fun."

The "joy of science," however, should not be achieved at the expense of historical objects. In 2005, Jim Bennett observed the "baleful influence" science centres have had "on the esteem accorded to historical objects." The educational authority granted to science centres, who foster a playful environment championing a public understanding of science, led to "new ambitions" that "demanded clear and decontextualised presentation of 'the science'"; objects in this context seemed more than ever "ambiguous and contingent." Historic scientific instruments have slowly become troublesome, epistemically ambivalent, and no longer "fun" or strangely beautiful and mysterious:

> Insufficiently malleable to the new mission, they retained too much of their own agenda, derived in large part from their preexisting 'biographies.' Galleries full of them seemed to raise too many questions where what was wanted were answers. Better to start afresh, to engage the visitor with newly created objects and present them in galleries that could manage and direct the visitor's experience of learning 'the science.'[47]

Striking a balance between science museums' and science centres' approaches to education requires a specialised staff to successfully relay the "joy of science" by means of instruments.[48]

Between 1996 and 2006, to that end, the University of Oxford offered a master's degree entitled "History of Science: Instruments, Museums, Science, Technology," largely based on the collections of the History of Science Museum. This one-year course, believed at the time to be one of a kind (to my knowledge such a course has not since been reproduced), focused on "the role of instruments in the history of science and technology from ancient times to the early twentieth century. It also covered collecting and the place of museums in the history of science. Students had opportunities to become involved

---

46 Débora Amaral Taveira Mello, *et al.*, "A Model-Based Analysis of the Museo Galileo Interactive Area," *Science & Education* 32 (2023): 1197–1219.

47 Jim Bennett, "Museums and the History of Science: Practitioner's Postcript," *Isis* 96 (2005): 602–608, quote 606. For an overview of the field of museum studies in general at the same time, Randolph Starn, "A Historian's Brief Guide to New Museum Studies," *American Historical Review* 110 (2005): 68–98.

48 Peter Heering, "Science Museums and Science Education," *Isis* 108 (2017): 399–406.

with the Museum's programmes of documentation, display, and exhibitions."[49] Although the cohorts were usually small (five to ten, sometimes fewer), several alumni pursued careers in science museums or as faculty – and today are SIC members, as we have seen in the recent tribute given to Jim Bennett, founder of the Oxford course, at the 2024 Ottawa SIC symposium. Such training aims likewise at looking beyond the materiality, science, technology, and use of scientific instruments. Through their careful study, we learn about not only the history of scientific disciplines and their methods, but also topics like science and religion, economics and capitalism, consumption, exploration and navigation, colonialism, power and politics, gender issues, social and class distinctions, standardisation, fashions and local styles, knowledge transfer, social construction of scientific units, epistemology, recycling, art history, workshop collaborations, etc. The study and methods of investigation for instruments need not differ from those applied to history, art, anthropology or archaeology collections in museums.[50]

Over the last twenty years, a growing number of university collections have been integrated into the teaching and research of academic programmes in the history of science, humanities, and engineering. Museums in the UK, through collaborations with universities and the Arts and Humanities Research Council (AHRC), have developed original and distinctive PhD projects highlighting the history of instrumentation and scientific collections – which profited several current SIC members. The early 2000s also saw the creation of communities around university collections, such as the European network UNIVERSEUM (2000) and the International Council of Museums Committee for University Museums and Collections (ICOM-UMAC, 2001), which seek "to contribute to society, for the benefit of all, by sustaining the continued development of university museums and collections as essential resources devoted to research, education, and the preservation of cultural, historic, natural and scientific heritage."[51] Nonetheless, despite these positive and impactful initiatives over the last twenty years, scientific instrumentation is too often absent in the "standard" edited volumes organised around themes of the museum and the collection.[52]

49 The quote is found here: http://www.mhs.ox.ac.uk/web-archive/museum-graduate-course/, accessed 28 March 2023.

50 Laurel Thatcher Ulrich, Ivan Gaskell, Sara J. Schechner, Sarah Anne Carter, and with photographs by Samantha van Gerbig, *Tangible Things: Making History Through Objects* (Oxford: Oxford University Press, 2015).

51 "International Council of Museums Committee for University Museums and Collections," ICOM, accessed 23 January 2024, http://umac.icom.museum/fr/.

52 See, for instance, Sandra Dudley, ed., *Museum Objects: Experiencing the Properties of Things* (New York: Routledge, 2012); and Bruce A. Campbell and Christian Baars, eds., *The*

Enlisting more SIC members dedicated to promoting education with historical instruments and their replicas is certainly a good strategy to mitigate the "science centre trend" that threatens to overwhelm science museums.[53] Better documenting scientific instrument collections increases collaborations between museum curators, educators, and historians of science and may require non-trivial changes in collecting practices and proactive research.[54] Science education has been an active field of research for decades. What experts and teachers, such as Flora Paparou and Chryssanthi Pikoula, who presented their classroom work at the SIC meeting in Athens (2022), have provided over the last twenty years are innovative ways to include historic scientific instruments into the pedagogical equation. Everybody wins by pursuing the research on, and the experiments with, historic scientific instruments in both formal and non-formal environments.

4 Conclusion

When former SIC president Jim Bennett in 2020 received the highest distinction awarded by the History of Science Society, the Sarton Medal, it was regarded by most as an acknowledgment of Jim's formidable contribution to bridging museum practices, history of science, and the study of its material culture. As a curator and director of major science collections in the UK, Bennett understood the necessary balance required for research, curating, formal teaching, and educating the public about instruments. Within the latter two activities, several SIC members were also recognised by the History of Science Society with its Joseph H. Hazen Education Prize, recognising outstanding teaching practices with in the history of science. Created in 1998, its first recipient was Gerald Holton – whom I mentioned above regarding the Harvard Project Physics. Since then, four members of the SIC community have been awarded this prize: Liba Taub (1999), Falk Rieß (2004), Sara Schechner

Curation and Care of Museum Collections (New York: Routledge, 2019). In the latter, only natural history is treated.

53 A recent exhibition at the Science Museum in London demonstrates how far museums have come over the past thirty years: *Science City 1550–1800: The Linbury Gallery*. On the same topic, tackled in a more formal way, Jim Bennett and Rebekah Higgitt, "London 1600–1800: Communities of Natural Knowledge and Artificial Practices," *British Journal for the History of Science* 52 (2019): 183–196. A collection of papers is devoted to this exhibit in the *Science Museum Group Journal* 15 (2021), accessed on 28 March 2023, https://journal.sciencemuseum.ac.uk/issue/spring2021/.

54 Marta C. Lourenço and Samuel Gessner, "Documenting Collections: Cornerstones for More History of Science in Museums," *Science & Education* 23 (2014): 727–745.

(2008), and Marvin Bolt (2017). The Paul Bunge Prize, awarded by the German Chemical Society and the German Bunsen Society for Physical Chemistry for "outstanding research on all aspects of the history of scientific instruments", has gone to several members of the SIC (including, again, Jim Bennett). These international prizes demonstrate how well the SIC, over the past twenty years, has emphasised the study *and* pedagogical treatment of scientific instruments. And the prizes certainly do not suggest that nothing significant happened on the educational front between 1980 and 2000; Falk Rieß began his laboratory in the 1980s and Dominique Bernard received in 2019 the prestigious Paul Doistau-Émile Blutet Prize awarded by the Académie des sciences for his work going back decades at the Université de Rennes. Yet as a nascent community, the SIC's first twenty years were dedicated primarily to establishing the field of scientific instrument studies by finding, saving, and documenting collections around the world. By roughly 2000 the field had grown strong enough to support a variety of initiatives and approaches that add value to scholarship and to the public understanding of science. This is in large part due to the strong leadership of the SIC officers over the last forty years and the willingness of SIC members themselves to experiment.

What will the next forty years be like? The 2022 SIC symposium in Athens may provide hints. As the study of instruments, collections, and educational practices has become mainstream, several members of our community are exploring difficult and controversial themes that were long marginalised. Sessions at Athens entitled "Scientific Instruments and Women (Parts 1 & 2)" and "Scientific Instruments Studies – Decolonised (Parts 1 & 2)" illustrate the effort deployed by our community to look beyond the usual narrative and tropes in order to foreground systematic racism and misogynous inclinations in science and technology and its instrumentarium. Similar conversations are already taking place, for instance, in the Society for the History of Technology (SHOT) where scholars, following the police shootings of Black citizens in American cities, and the subsequent uprisings, have questioned not only the scarcity of such topics in the history of technology, but also the lack of recognition and sometimes ill-treatment received by scholars of colour who do tackle these sensitive issues.[55] The same is true in the history of science, where scholars ask for a more "global" view of science in order to fight against the "universalism" concept of science that for most, obviously, originates in "the

55 Chandra Bhimull, Gabrielle Hecht, Edward Jones-Imhotep, Chakanetsa Mavhunga, Lisa Nakamura, and Asif Siddiqi, "Systemic and Epistemic Racism in the History of Technology," *Technology and Culture* 63 (2022): 935–952.

West."[56] Our collective studies of instrument collections and original pedagogical methods are here highly significant since science museums and science centres tended in the past to exclude disenfranchised groups who simply did not see their own traditions and cultures in these non-formal educational settings.[57] The project described by Silke Ackermann and Nicola Bird at the SIC Ottawa symposium in 2024, MultakaOxford, showcases the substantial role our community can play in these difficult times.[58] Based at the History of Science Museum in Oxford since 2017, the presentation described the initiative as "a multi-award-winning inclusive volunteering programme" that champions "collaborative practice to bring diverse representation and narratives to science, history, culture, and faith, enriching our understanding of contemporary society." There is no better place than the SIC to generate and enter such conversations and to figure out relevant pedagogical and didactic solutions that will transform the way we grapple with science and technology so they become more inclusive and connected to one and all. The SIC's future is bright and relevant to the whole of society.

Acknowledgments

I am indebted to the editors of this volume and to Elizabeth Cavicchi, who have provided numerous comments and suggestions to make this essay better. I take full responsibility for its final form and content.

Bibliography

Alasuutari, Maarit, Marleena Mustola, and Niina Rutanen, eds. *Exploring Materiality in Childhood: Body, Relations and Space*. Abingdon: Routledge, 2021.

Alberti, Samuel J.M.M. *Curious Devices and Mighty Machines: Exploring Science Museums*. London: Reaktion Books Ltd., 2022.

56 James Delbourgo, "The Knowing World: A New Global History of Science," *History of Science* 57 (2019): 373–399. Geert J. Somsen, "A History of Universalism: Conceptions of the Internationality of Science from the Enlightenment to the Cold War," *Minerva* 46 (2008): 361–379.

57 Emily Dawson, "'Not Designed for Us': How Science Museums and Science Centers Socially Exclude Low-Income, Minority Ethnic Groups," *Science Education* 98 (2014): 981–1008.

58 For more on this initiative, "MultakaOxford," History of Science Museum, University of Oxford, accessed on 7 January 2025, https://www.hsm.ox.ac.uk/multaka-oxford.

Alberti, Samuel J.M.M. "Objects and the Museum." *Isis* 96 (2005): 559–571.

Anderson, Katharine, Mélanie Frappier, and Elizabeth Neswald, and Henry Trim. "Reading Instruments: Objects, Texts and Museums." *Science & Education* 22 (2013): 1167–1189.

Anderson, Robert G.W. "'What is Technology?': Education Through Museums in the Mid-Nineteenth Century." *British Journal for the History of Science* 25 (1992): 169–184.

Bennett, Jim. "Museums and the History of Science: Practitioner's Postcript." *Isis* 96 (2005): 602–608.

Bennett, Jim, and Rebekah Higgitt. "London 1600–1800: Communities of Natural Knowledge and Artificial Practices." *British Journal for the History of Science* 52 (2019): 183–196.

Bhimull, Chandra, Gabrielle Hecht, Edward Jones-Imhotep, Chakanetsa Mavhunga, Lisa Nakamura, and Asif Siddiqi. "Systemic and Epistemic Racism in the History of Technology." *Technology and Culture* 63 (2022): 935–952.

Boesch, Brandon. "Skill Transmittance in Science Education. Studying the Skills of Scientific Expertise." *Science & Education* 28 (2019): 45–61.

Boon, Tim, ed. "A Symposium on Histories of Use and Tacit Skills." *Science Museum Group Journal* 8 (2017), https://journal.sciencemuseum.ac.uk/article/histories-of-use-and-tacit-skills/.

Brenni, Paolo. "Filming Nineteenth-Century Physics Demonstrations with Historical Instruments." Pp. 34–49 in *Historical Scientific Instruments in Contemporary Education*, edited by Elizabeth Cavicchi and Peter Heering. Leiden: Brill, 2022.

Brenni, Paolo. "The Evolution of Teaching Instruments and Their Use Between 1800 and 1930." *Science & Education* 21 (2012): 191–226.

Campbell, Bruce A. and Christian Baars, eds. *The Curation and Care of Museum Collections*. New York: Routledge, 2019.

Cavicchi, Elizabeth. "Historical Experiments in Students' Hands: Unfragmenting Science Through Action and History." *Science & Education* 17 (2008): 717–749.

Cavicchi, Elizabeth. "Historical Scientific Instruments in Exploratory Teaching and Learning." Pp. 158–180 in *Historical Scientific Instruments in Contemporary Education*, edited by Elizabeth Cavicchi and Peter Heering. Leiden: Brill, 2022.

Cavicchi, Elizabeth, and Peter Heering, eds. *Historical Scientific Instruments in Contemporary Education*. Scientific Instruments and Collections, 9. Leiden: Brill, 2022.

Chang, Hasok. "How Historical Experiments Can Improve Scientific Knowledge and Science Education: The Cases of Boiling Water and Electrochemistry." *Science & Education* 20 (2011): 317–341.

Cole, K.C. *Something Incredibly Wonderful Happens: Frank Oppenheimer and His Astonishing Exploratorium*. Chicago: The University of Chicago Press, 2009.

Conn, Steven. *Do Museums Still Need Objects?* Philadelphia: University of Pennsylvania Press, 2010.

Dawson, Emily. "'Not Designed for Us': How Science Museums and Science Centers Socially Exclude Low-Income, Minority Ethnic Groups." *Science Education* 98 (2014): 981–1008.

Delbourgo, James. "The Knowing World: A New Global History of Science." *History of Science* 57 (2019): 373–399.

Dudley, Sandra, ed. *Museum Objects: Experiencing the Properties of Things*. New York: Routledge, 2012.

Dunn, Richard. "'More Artistic than Scientific': Exhibiting Instruments as Decorative Arts in the Victoria and Albert Museum." Pp. 61–76 in *Scientific Instruments on Display*, edited by Silke Ackermann, Richard L. Kremer, and Mara Miniati. Leiden: Brill, 2014.

Eggen, Per-Odd, Lise Kvittingen, Annette Lykknes, and Roland Wittje. "Reconstructing Iconic Experiments in Electrochemistry: Experiences from a History of Science Course." *Science & Education* 21 (2012): 179–189.

Everett, Tom. "Writing Sound with a Human Ear: Reconstructing Bell and Blake's 1874 Ear Phonautograph." *Science Museum Group Journal* 12 (2019), http://journal.sciencemuseum.ac.uk/browse/issue-12/writing-sound-with-a-human-ear/.

Farné, Roberto. "Pedagogy of Play." *Topoi* 24 (2005): 169–181.

Fondazione Scienza e Tecnica. "florencefst." https://www.youtube.com/@florencefst/.

Fors, Hjalmar, Lawrence M. Principe, and Otto Sibum. "From the Library to the Laboratory and Back Again: Experiment as a Tool for Historians of Science." *Ambix* 63 (2016): 85–97.

Gauvin, Jean-François. *Instruments of Knowledge: Finding Meaning in Objects, Habits, and Museums*. Nuncius Series, 12. Leiden: Brill, 2023.

Gauvin, Jean-François. "Playing with Quantum Toys: Julian Schwinger's Measurement Algebra and the Material Culture of Quantum Mechanics Pedagogy at Harvard in the 1960s." *Physics in Perspective* 20 (2018): 8–42.

Gauvin, Jean-François. "The Apparatus of Yale College's Sheffield Scientific School and the Promise of a 'New Education' (ca 1870)." In *Scientific Instruments as Cultural Artefacts*, edited by Paola Bertucci and Alexi Baker. New Haven: Yale University Press, forthcoming.

Gooday, Graeme. "Precision Measurement and the Genesis of Physics Teaching Laboratories in Victorian Britain." *British Journal for the History of Science* 23 (1990): 25–51.

Guralnick, Stanley M. "Sources of Misconception on the Role of Science in the Nineteenth-Century American College." *Isis* 65 (1974): 352–366.

Heering, Peter. "Educational Experiences in Re-enacting Historical Experimental Procedures." Pp. 226–242 in *Historical Scientific Instruments in Contemporary Education*, edited by Elizabeth Cavicchi and Peter Heering. Leiden: Brill, 2022.

Heering, Peter. "Make-Keep-Use: Bringing Historical Instruments into the Classroom." *Interchange* 46 (2015): 5–18.

Heering, Peter. "Science Museums and Science Education." *Isis* 108 (2017): 399–406.

Heering, Peter. "The Enlightened Microscope: Re-enactment and Analysis of Projections with Eighteenth-Century Solar Microscopes." *British Journal for the History of Science* 41 (2008): 345–367.

Heering, Peter, and Cibelle Celestino Silva. "Re-Examining the Early History of the Leiden Jar: Stabilization and Variation in Transforming a Phenomenon into a Fact." *History of Science* 56 (2018): 314–342.

Heering, Peter, and Roland Wittje, eds. *Learning by Doing. Experiments and Instruments in the History of Science Teaching*. Stuttgart: Franz Steiner Verlag, 2011.

Heering, Peter, Falk Rieß, and Christian Sichau, eds. *Im Labor der Physikgeschichte: Zur Untersuchung historischer Experimentalpraxis*. Oldenburg: Universität Oldenburg Verlag, 2000.

Heering, Peter, Stephen Klassen, and Don Metz, eds. *Enabling Scientific Understanding through Historical Instruments and Experiments in Formal and Non-Formal Learning Environments*. Flensburg Studies on the History and Philosophy of Science in Science Education, 2. Flensburg: Flensburg University Press, 2013.

Hendriksen, Marieke M.A. "Rethinking Performative Methods in the History of Science." *Berichte zur Wissenschaftsgeschichte* 43 (2020): 313–322.

Higgitt, Rebekah. "Challenging Tropes: Genius, Heroic Invention, and the Longitude Problem in the Museum." *Isis* 108 (2017): 371–380.

History of Science Museum, University of Oxford. "MultakaOxford." Accessed on 7 January 2025. https://www.hsm.ox.ac.uk/multaka-oxford.

Holton, Gerald. "The Project Physics Course, Then and Now." *Science & Education* 12 (2003): 779–786.

Huizinga, J. *Homo ludens: A Study of the Play-Element in Culture*. London: Routledge & Kegan Paul, 1949.

Isozaki, Tetsuo. "The Organisation and the Recontextualization of *Rika* (School Science) Education in the Second Half of the Nineteenth Century in Japan." *Science & Education* 23 (2014): 1153–1168.

Kanhadilok, Peeranut, and Mike Watts. "Youth at Play: Some Observations from a Science Museum." *International Journal of Adolescence and Youth* 22 (2017): 179–194.

Kehl, Winifred. "I Have a Confession: I'm an Adult. How Can Science Museums Welcome Grown-Ups?" *Museum* (July–August 2013): 46–51.

Kevles, Daniel J. *The Physicists: The History of a Scientific Community in Modern America*. Cambridge, MA: Harvard University Press, 1987.

Khantine-Langlois, Françoise, Alfonso San-Miguel, and Pierre Lauginie. "Examples of the Use in Education of Historical Physics Instruments at Secondary School and University Level in France Supported by ASEISTE." Pp. 122–138 in *Historical Scientific Instruments in Contemporary Education*, edited by Elizabeth Cavicchi and Peter Heering. Leiden: Brill, 2022.

Kohlstedt, Sally Gregory. "Reassessing Science in Antebellum America." *American Quarterly* 29 (1977): 444–453.

Lazos, Panagiotis, George N. Vlahakis, and Constantine Skordoulis. "Instruments and Laboratories in the Schools of the Greek Community of Istanbul, 1850–1960." Pp. 168–186 in *Scientific Instruments Between East and West*, edited by Neil Brown, Silke Ackermann, and Feza Günergun. Leiden: Brill, 2019.

Liceo Paolo Sarpi. "Administrator Liceo Sarpi." https://www.youtube.com/@administratorliceosarpi7144/.

Lindtner, Silvia, and Paul Dourish. "The Promise of Play: A New Approach to Productive Play." *Games and Culture* 6 (2011): 453–478.

Lourenço, Marta C., and Samuel Gessner. "Documenting Collections: Cornerstones for More History of Science in Museums." *Science & Education* 23 (2014): 727–745.

Marcon, F. "Between School and Universities: The Study of the Historical Scientific Instruments Collection of the Liceo Maffei in Verona." *Il Nuovo Cimento* 40C (2017): 1–6.

Martínez, Alberto A. "Replication of Coulomb's Torsion Balance Experiment." *Archives for the History of Exact Sciences* 60 (2006): 517–563.

Matthews, Michael R., Colin Gauld, and Arthur Stinner. "The Pendulum: Its Place in Science, Culture and Pedagogy." *Science & Education* 13 (2004): 261–277.

Meloni, Reginaldo Alberto. "Tools for Teaching Physics and Chemistry in Secondary Schools: The Case of the Colégio Culto à Ciência, Brazil, 1899–1902." Pp. 247–258 in *Scientific Instruments in the History of Science: Studies in Transfer, Use and Preservation*, edited by Marcus Granato and Marta C. Lourenço. Rio de Janeiro: Museu de Astronomia e Ciência Afins, 2014.

Meshoulam, David. *"Teaching Physics as One of the Humanities": The History of (Harvard) Project Physics, 1961–1970*. PhD dissertation, University of Wisconsin-Madison, 2014.

Metz, Don, and Art Stinner. "A Role for Historical Experiments: Capturing the Spirit of the Itinerant Lecturers of the 18th Century." *Science & Education* 16 (2007): 613–624.

Meyer, Morgan. "Researchers on Display: Moving the Laboratory into the Museum." *Museum Management and Curatorship* 26 (2011): 261–272.

Milne, Catherine, and Kathryn Scantlebury, eds. *Material Practice and Materiality: Too Long Ignored in Science Education*. Cham: Springer, 2019.

Neslihan Alisir, Zeynep, and Serhat Irez. "The Effect of Replicating Historical Scientific Apparatus on High School Students' Attitudes Towards Science and Their Understanding of Nature of Science." *Science & Education* 29 (2020): 1201–1234.

Newman, William. *Newton the Alchemist: Science, Enigma, and the Quest for Nature's "Secret Fire."* Princeton: Princeton University Press, 2019.

Olesko, Kathryn M. *Physics as a Calling: Discipline and Practice in the Könisburg Seminar for Physics*. Ithaca: Cornell University Press, 1991.

Oreskes, Naomi. *Why Trust Science?* Princeton: Princeton University Press, 2019.

Pantano, Ornella, and Sofia Talas. "Physics Thematic Paths: Laboratorial Activities and Historical Scientific Instruments." *Physics Education* 45 (2010): 140–146.

Pyenson, Lewis, and Jean-François Gauvin, eds. *The Art of Teaching Physics. The Eighteenth-Century Demonstration Apparatus of Jean-Antoine Nollet.* Sillery, Quebec: Septentrion, 2002.

Pynchon, Thomas. *Mason & Dixon.* New York: Henry Holt and Company, 1997.

Rhees, David J. "Sparks and Shocks: Replicas of Historical Instruments in Museum Education." *Interchange* 46 (2015): 45–56.

Rieß, Falk. "History of Physics in Science Teacher Training in Oldenburg." *Science & Education* 9 (2000): 399–402.

Rieß, Falk. "Problem with German Science Education." *Science & Education* 9 (2000): 327–331.

Rodrigues Barbosa, Catia. "L'esthétique dans les musées des sciences et des techniques." *La Lettre de l'OCIM* 95 (2004): 27–30.

Rowe, Jonathan P., *et al.* "Play in the Museum: Design and Development of a Game-Based Learning Exhibit for Informal Science Education." *International Journal of Gaming and Computer-Mediated Simulations* 9 (2017): 96–113.

Ruiz-Castell, Pedro. "Scientific Instruments for Education in Early Twentieth-Century Spain." *Annals of Science* 65 (2008): 519–527.

"Science in the City." Special issue, *Science Museum Group Journal* 15 (2021), https://journal.sciencemuseum.ac.uk/issue/spring2021/.

Simon, Josep, and Mar Cuenca-Lorente. "Science Education and the Material Culture of the Nineteenth-Century Classroom: Physics and Chemistry in Spanish Secondary Schools." *Science & Education* 21 (2012): 227–244.

Smith, Pamela H. *From Lived Experience to the Written Word: Reconstructing Practical Knowledge in the Early Modern World.* Chicago: The University of Chicago Press, 2022.

Smith, Pamela H., and the Making and Knowing Project. "Historians in the Laboratory: Reconstruction of Renaissance Art and Technology in the Making and Knowing Project." *Art History* 39 (2016): 210–233.

Sokal, Alan D., and Jean Bricmont. *Fashionable Nonsense: Postmodern Intellectuals' Abuse of Science.* New York: St Martins Press, 1999.

Somsen, Geert J. "A History of Universalism: Conceptions of the Internationality of Science from the Enlightenment to the Cold War." *Minerva* 46 (2008): 361–379.

Starn, Randolph. "A Historian's Brief Guide to New Museum Studies." *American Historical Review* 110 (2005): 68–98.

Staubermann, Klaus, ed. *Reconstructions: Recreating Science and Technology of the Past.* Edinburgh: National Museums of Scotland, 2011.

Taub, Liba. "What is a Scientific Instrument, Now?" *Journal of the History of Collections* 31 (2019): 453–467.

Taveira Mello, Débora Amaral, *et al.* "A Model-Based Analysis of the Museo Galileo Interactive Area." *Science & Education* 32 (2023): 1197–1219.

Teichmann, Jürgen. "Historical Experiments and Science Education. From Conceptual Planning of Exhibitions to Continuing Education for Teachers." *Interchange* 46 (2015): 31–43.

Teichmann, Jürgen. "Studying Galileo at Secondary School: A Reconstruction of His 'Jumping Hill' Experiment and the Process of Discovery." *Science & Education* 8 (1999): 121–136.

"The Making and Knowing Project." https://www.makingandknowing.org/.

Tõugu, Pirko, *et al.* "Connecting Play Experiences and Engineering Learning in a Children's Museum." *Journal of Applied Developmental Psychology* 53 (2017): 10–19.

Trumper, Ricardo. "The Physics Laboratory – A Historical Overview and Future Perspectives." *Science & Education* 12 (2003): 645–670.

Turner, Gerard L'E. "Presidential Address: Scientific Toys." *British Journal for the History of Science* 20 (1987): 377–98.

Turner, Steven. "Chicago Scientific Instrument Makers, 1871–1918, Part 1: The School Market." *Rittenhouse* 19 (2005): 65–128.

Turner, Steven. "Chicago Scientific Instrument Makers, 1871–1918, Part 2: The University and Research Market." *Rittenhouse* 20 (2006): 1–64.

Uchacz, Tianna Helena. "Reconstructing Early Modern Artisanal Epistemologies and an 'Undisciplined' Mode of Inquiry." *Isis* 111 (2020): 606–613.

Ulrich, Laurel Thatcher, Ivan Gaskell, Sara J. Schechner, Sarah Anne Carter, and with photographs by Samantha van Gerbig. *Tangible Things: Making History Through Objects*. Oxford: Oxford University Press, 2015.

Utrecht University. "Artechne – Technique in the Arts, 1500–1950." https://artechne.wp.hum.uu.nl/.

Van der Spek, Trienke M. "The Lorentz Lab: Reviving the Scientific History of Teylers Museum with Working Replica." Pp. 243–262 in *Historical Scientific Instruments in Contemporary Education*, edited by Elizabeth Cavicchi and Peter Heering. Leiden: Brill, 2022.

Warner, Deborah J. "Commodities for the Classroom: Apparatus for Science and Education in Antebellum America." *Annals of Science* 45 (1988): 387–397.

Warner, Deborah J. "Replicating Historic Scientific Instruments." *Bulletin of the Scientific Instrument Society*, no. 151 (2021): 10–13.

Werrett, Simon. *Thrifty Science: Making the Most of Materials in the History of Experiment*. Chicago: The University of Chicago Press, 2019.

CHAPTER 10

Instruments and New Media

Pedro M.P. Raposo and Floor Anna Koeleman

Beginning in the late 1960s, the digitisation of museum collections and their associated data has added new dimensions to how collections are managed, accessed, and activated.[1] This massive production and availability of digital assets and resources as it applies to scientific instruments – much of the hard work of which was carried out with gathering impetus from the late 1980s – has opened new possibilities not only for research communities interested in the subject, but also for engaging broader audiences with differing interests, motivations, and background knowledge.[2] Here, new media such as listservs, social networks, video and video-conferencing platforms, digital-exhibition repositories, and crowdsourcing hubs, which have evolved with the Internet since the late 1990s, play a very important role as they have already gained substantial traction among the community of curators and scholars engaged with instruments. Another important dimension is now being added by Artificial Intelligence (AI) and Large Language Models (LLMs), which are increasingly dominating debates about electronic devices, computing, digital technologies, and their impact on all aspects of human life.

This chapter presents an overview of how these new media and evolving technologies are already being used by individuals, groups, and institutions engaged with the preservation and study of historical scientific instruments, including the SIC. Our exploration of this theme is intended to highlight how these new media are shaping the curation of scientific instrument collections, opening up new research avenues, and offering opportunities for broader

1 Metropolitan Museum of Art, *Computers and Their Potential Applications in Museums: A Conference Sponsored by the Metropolitan Museum of Art, April 15, 16, 17, 1968* (New York: Arno Press, 1968). The discussions of the late 1960s may have been aspirational, but they were critical in shaping the vision for digitisation, even if implementation took decades to become a practical reality.

2 For examples, see David Williams, "Brief History of Museum Computerisation," in a *Guide to Museum Computing*, ed. D. Williams (Nashville, TN: American Association for States and Local History, 1987), 1–5; John Burnett and Ian Morrison, *Wimps, Worms and Winchesters: A Guide to Documentation in Museums* (Edinburgh: National Museums Scotland, 1989), especially "Case Studies: National Museums Scotland," 28–31; Trilce Navarrete, *A History of Digitisation: Dutch Museums* (Amsterdam: Boekman, 2014).

 | DOI:10.1163/9789004737587_012

engagement with the field, while generating new problems and challenges. It is framed by two major issues concerning *standards* and their fluid and mutable nature and the challenge of *translation*.

These technological changes in information technologies have evinced the impossibility of attaining overarching, universal, timeless standards with which to apprehend, describe, interpret, and communicate about scientific instruments, a problem that had already become evident with the rise of electronic databases in the late 1980s. These databases made scientific instrument collections much more accessible and discoverable and provided for a more centralised and efficient management of museum collections. Data for instrument collections increasingly became shared online.

Prominent online collections include those of the History of Science Museum of the University of Oxford, the Collection of Historical Scientific Instruments of Harvard University, the Science Museum London, the Smithsonian National Museum of American History, and the Yale Peabody Museum.[3] Information about instruments can thus be found and compared more easily. This has been significant since previously instrument collections were readily accessible only to local visitors and only a small fraction of the objects in museum collections were typically on display at any given time.

However, online catalogues *per se* did not solve all problems for researchers. Stewards and managers of all kinds of collections struggle with the challenges of describing their materials with metadata entailing enough breadth and depth to provide for flexible searchability and discovery. Additionally, description standards may vary significantly. Glossaries, thesauri, and works such as the detailed indices of the virtual Museo Galileo, *Epact: Scientific Instruments of Medieval and Renaissance Europe*, and the 1998 book *Instruments of Science: An Historical Encyclopedia* provide valuable references and may have contributed to a significant degree of standardisation.[4] Still, challenges persist. Honouring historical terminology and structure is arguably an untenable endeavour. There is no inherent or ahistorical categorisation of these artefacts, which were perceived and classified in different ways throughout history and

3 Online catalogue, History of Science Museum, University of Oxford, https://hsm.ox.ac.uk/collections-online#/search; Waywiser online catalogue, Collection of Historical Scientific Instruments, Harvard University, http://waywiser.fas.harvard.edu/collections; Online catalogue, Science Museum Group, https://collection.sciencemuseumgroup.org.uk/; "Collections Search Center," Smithsonian Institution, https://collections.si.edu/search/; and History of Science and Technology online catalogue, Yale Peabody Museum, https://peabody.yale.edu/explore/collections/history-science-technology.

4 "Virtual Museum" online catalogue, Museum Galileo, https://catalogue.museogalileo.it/index/IndepthIndex.html; *Epact*, http://www.mhs.ox.ac.uk/epact/glossary.php; Robert Bud and Deborah J. Warner, eds., *Instruments of Science: An Historical Encyclopedia* (New York: Garland Publishing, 1998).

across varied geographical contexts. Whatever description is applied to them today will be an arbitrary imposition reflecting contemporary concerns about the place, role, and meaning of these historical artefacts. This is shown, for example, in Adam Mosley's analysis of "Sundials and Other Cosmographical Instruments," which highlights the fluidity of historical categories as well as the agency of historians, curators, and archivists.[5]

Another issue that keen students and explorers of instruments have to deal with is that of *translation*. The accessibility of databases, either paper-based or digital, is often limited by language and other constraints. The act of translation entails both enrichment and loss. The process of transferring something over from one cultural body to another inevitably entails alteration.[6] Instrument databases involve various practices of translation: turning a scientific instrument into data (i.e. images, labels, and other descriptors); converting concepts and terms from the past into modern-day equivalents; and, of course, translating between different languages. With English as the current *lingua franca* among the Western research community (and beyond), national and regional nuances as expressed through language might well be lost.

One additional dimension of translation has to do with the interoperability and complementarity of databases. Consider the database of Scientific Instrument Makers, Observations and Notes (SIMON), comprising records on individual instrument makers and firms from Great Britain and Ireland.[7] A similar project is the *Dictionary of Precision Instrument-Makers and Related Craftsmen*, which has a different geographic scope, namely, France and Switzerland.[8] Such projects have their own idiosyncrasies and may not be completely compatible – if they are compatible at all.[9]

Moreover, how can the knowledge and curation of instruments be translated between different groups and audiences and contribute to a shared

5 Adam Mosley, "'Sundials and Other Cosmographical Instruments': Historical Categories and Historians' Categories in the Study of Mathematical Instruments and Disciplines," in *The Whipple Museum of the History of Science: Objects and Investigations, to Celebrate the 75th Anniversary of R.S. Whipple's Gift to the University of Cambridge*, ed. J. Nall, L. Taub, and F. Willmoth (Cambridge: Cambridge University Press, 2019), 55–82.

6 Sven Dupré, "Introduction: Science and Practices of Translation," *Isis* 109 (2018): 302–307, see 303.

7 *Tools of Knowledge*, accessed 3 July 2024, https://toolsofknowledge.org/.

8 Denis Beaudoin, Paolo Brenni, and Anthony Turner, *The Dictionary of Precision Instrument-Makers and Related Craftsmen* (France & Switzerland, 1430–1960), last modified July 2018, https://bibnum.explore.psl.eu/s/dictionarypim/page/prolegomena.

9 For SIMON, rather than relying on existing thesauri, a new ontology was created, the Scientific Instrument Makers and Events Ontology (SIMEOn); see "Database progress – and a new model," accessed 3 July 2024, https://toolsofknowledge.org/2023/05/24/database-progress-and-a-new-model/.

understanding of instruments? This is by no means a unidirectional process; rather, it is a multidirectional one. We should not underestimate the fact that the topic of scientific instruments and their history is a highly specialised and complex one. In fact, it is no less abstruse to scholars and academics not attuned to hands-on research and to the study of material culture than it is to general audiences. And the latter can play an important role in asking new questions, generating insights on the cultural significance of certain objects for specific groups and geographical regions, and performing research tasks that do not require specialised knowledge. This chapter provides several examples of how communication and engagement around instruments can be advanced through the forms of new media brought into focus here.

Since one of the major issues of translating the materiality of instruments into the digital realm is how to replicate their three-dimensionality, we begin by reflecting on how 3D modelling has affected the field. Then we address the relationship between historical instruments, social media networks, and crowdsourced research platforms and initiatives, given the importance of these technologies not only in how scholars communicate and collaborate among themselves, but also in how they engage broader, non-expert audiences with their field. In a similar vein, we then provide an overview of how the use of video and video conferencing platforms can serve both the purposes of broader outreach and scholarly exchange. While many of the technologies involved here were in full swing before the COVID-19 pandemic, this major world event and its impact on how such technologies have been used and perceived by those engaged with instruments and their history cannot be overlooked. Finally, we offer a reflection on and preliminary results from simple experiments with AI, namely with ChatGPT, as applied to instruments, since this kind of AI-powered tool is quickly gaining a prominent, albeit contested and problematic space in scholarly life. We conclude by reflecting on the losses and gains entailed by the adoption of these new technologies, and by pointing out open questions that are certain to generate discussion and debate among those who seek to advance innovative ways of studying and presenting scientific instruments.

1 Virtual Instruments: From 2D to 3D Modelling and New Possibilities for Research

A single image seldom suffices to fully illustrate 3D objects whose complexity is, in many cases, significantly reduced and distorted when transposed to a two-dimensional plane. Scale constitutes another issue, as does colour, even if to a lesser extent compared to collections of paintings, for example. Many museums have now gone through several cycles of photographic documentation of their collections. Tempting as it might be to dismiss the products of such undertakings as outdated pictorial material devoid of curatorial significance, they might offer relevant historical insights on scientific instruments as museum objects and on the different choices (lighting, composition, colour and tonality) made over time to photographically document them. A digitised black-and-white photograph, for example, might render certain contrasts more visible than a twenty-first century digital image will do. These different pictorial representations are not neutral; they shape our view of the depicted instruments.[10] Photography as applied to museum objects goes back to the early days of photography itself. Early techniques and media gave way to glass plates, which were then replaced by photographic film, and more recently, by electronic image sensors. These developments are illustrated, for example, by the online documentation of the *Instrument of the Primum Mobile* (1568) by Ignazio Danti in the online archive of Museo Galileo, the current home of this Florentine instrument. It includes a digital version of a broken photographic plate dating to 1929 as well as scanned film photographs from 1960 and 1983 plus a digital photograph from 2005 (*Fig. 10.1*).[11]

One way to overcome some of the limitations of two-dimensional depictions is to create three-dimensional models of scientific instruments. There are several processes available to create such models. One is photogrammetry, which consists essentially of taking a number of photographs from all angles and combining them with a computer programme into a single, stitched model. The quality of the finalised model depends on the quality of the photography. Shiny materials reflect light and can act as mirrors, which is usually undesirable. That constitutes one of the most common problems with this method for metallic instruments.

Another option is LIDAR (Light Detection and Ranging) scanning, which relies on the detection of distances between the scanning device and the scanned object. The resulting point cloud models can be converted to three-dimensional surfaces in order to recreate textures and outward appearances. While there are physical features of scientific instruments that cannot be captured in a digital model (for example, their weight), three-dimensional models allow for a simulated experience of using the instrument. Operations such as

10 Koeleman studied this topic at the summer seminar, "The Art Historical Image in the Digital Age," funded by the Samuel H. Kress Foundation and hosted by the Kunsthistorisches Institut in Florenz – Max-Planck-Institut (KHI) in Florence from 26 June to 7 July 2023. The seminar was led by Emily Pugh and David Ogawa, to whom she is indebted for raising sensitivity to the materiality of photography archives in relation to digitisation projects.

11 "Primo mobile," Museo Galileo, accessed 5 July 2024, https://opac.museogalileo.it/imss/search?collection_data=14&s=25&o=score&q=primo+mobile&v=l&h=any_bc.

FIGURE 10.1 Ignazio Danti, instrument of the *primum mobile*, 1568, brass, 27.9 × 17.5 cm. Photographs, *from top left to bottom right*, taken in 2005, 1983, 1960, and 1929
MUSEO GALILEO, 2643

rotating the model can unlock different angles and perspectives, which may or may not be feasible with the actual object depending on the type of device and its condition.

In some cases, a digital model might not only recreate the experience of using the instrument but also emulate the adjustments and replacements that must take place for use in specific circumstances. Such is the case of the astrolabe. This is exemplified by the online interactive astrolabe by Alexander Boxer, where the effect of three-dimensionality may be less relevant than the scope of functionality. Boxer's website offers a versatile digital version of the instrument that allows the user to easily adjust date/epoch and latitude. It is further complemented with a visual effect to simulate day and night on the tympan of the virtual astrolabe.[12] This digital model recreates visual but not haptic experiences of the original.

Other kinds of instruments will benefit from the effect of three-dimensionality. Importantly, the complementary practice of annotation lends itself to storytelling, with annotations placed in a certain manner providing a narrative

12 "Astrolabe," accessed 3 July 2024, https://alexboxer.com/astrolabe/.

of the simulated object.[13] For instance, the app *Behind the Stars*, developed by the Staatliche Kunstsammlungen in Dresden, offers a step-by-step introduction to a nocturnal, a sundial, a globe clock, and an astrolabe.[14] Making these digital models workable implies making design choices that, during the process of translating a scientific instrument into a data model, may imply the loss of certain features and decorative details.[15] However, their value for education and for engaging a broader audience with instruments should not be underestimated.

2 Digital Outreach and Engagement: Social Media and Crowdsourcing

During the SIC Annual General Meeting at Scientific Instrument Symposium XXXVII in Leiden, one of the authors of this chapter (Pedro Raposo) took the podium at the Rijksmuseum Boerhaave to propose a social media presence for SIC on Facebook and Twitter (now X), an idea he had been developing in collaboration with SIC members Taha Yasin Arslan and Emily Margolis. Although advancing a social media presence for a decades-old organisation in 2018 might not sound like a particularly bold undertaking, it would expand the SIC's presentation of scientific instruments and their history beyond the reach of the Rete mailing list, the annual symposia, and similar face-to-face events.

Rete is hosted by the University of Oxford and managed by the History of Science Museum "as a service to the historical instrument community."[16] Started in 1996, it provides a good example of how mailing lists powered by the Listserv technologies, developed from the late 1980s onward,[17] have fostered communication, exchange, and collaboration between experts in various fields

13 See for example the 3D model of the travel theodolite in the collection of ETH Zürich via ETHZ_IFGP_0214, accessed 5 July 2024, http://doi.org/10.21264/ethz-a-000013490. The 3D model was made using photogrammetry. For more technical metadata, see "ETH: Travel theodolite, 1899 / Reise-Theodolit," Sketchfab, accessed 5 July 2024, https://skfb.ly/oDpVZ.

14 "Behind the Stars," Mathematisch-Physikalischer Salon, accessed 3 July 2024, https://mathematisch-physikalischer-salon.skd.museum/vermittlung/behind-the-stars/.

15 This was the topic of Frederik Nehm's talk "Behind the Stars: An app to teach basic astronomy through interactive early-modern instruments" at the Scientific Instrument Symposium XLI in Athens of 2022.

16 "Mailing list (Rete)," History of Science Museum, University of Oxford, accessed 26 July 2024, https://www.hsm.ox.ac.uk/mailing-list.

17 See David Allne Grier and Mary Campbell, "A Social History of Bitnet and Listserv, 1985–1991," *IEEE Annals of the History of Computing* 22 (2020): 32–41. For Rete's "birth," see Sara J. Schechner's chapter about "The Scientific Instrument Commission in the 1990s."

and special interest groups. They dramatically improved the way scholars from different regions, countries and continents share information, and disseminate their work, communication that previously relied mainly on attendance at conferences and scholarly publications in print. These mailing lists also conferred a stronger collaborative dimension on scholarly work, which, in the era of the traditional academic paper, tended to be more individualistic and to develop in a certain isolation (especially in the humanities).

Rete remains a vital forum for instrument curators, researchers, and enthusiasts, and a prime communication channel for the SIC. As is the case with other scholarly mailing lists, subscribers and users are expected to abide by rules of professionalism and mutual respect, and the kinds of information, inquiries, and announcements that can be posted are clearly delimited.

This general sense of efficiency and scholarly decorum conveyed by Rete plus the perception that it was meeting the basic communication needs of a global historical instrument community underpinned the reservations expressed by several attendees of the 2018 Leiden Symposium to the idea of creating a social media presence for the SIC. Some participants pointed out the perils of sharing data, personal or other, on social media, expressing concerns with privacy, copyright, improper use of shared content, and trolling. Generational differences about patterns of scholarly exchange and socialisation were surely also at play. These concerns were legitimate. The year before, it had been revealed that Facebook (now part of the Meta company) had allowed the British consultancy firm Cambridge Analytica to harvest data of eighty-seven million users that was subsequently used in political advertising. Distrust of Facebook and its attitudes towards the platform's users stood understandably high, but in the end a majority of the attendees of the 2018 Symposium agreed that the SIC should launch a one-year experiment with a social media presence on Facebook and Twitter/X.

Accounts were then created on Facebook, Twitter/X, and eventually, also on Instagram (in July 2021). A monthly cycle of posts was implemented, in four major categories: Instrument of the Month; Mystery Instrument (in which followers were invited to guess what instrument was pictured in a shared photograph, with the correct answer being subsequently posted); Featured Instrument Collection Database (or other online resource); and Instrumental People (short biographies of historical figures of varied backgrounds and fields of study who had designed, used, traded, or promoted scientific instruments). By December 2023, the SIC accounts had garnered 1,000 followers on Facebook, about 900 on Twitter/X, and 386 on Instagram. But since January 2024 SIC social media has been in pause due to lack of volunteers to manage the accounts.

Other institutions and organisations holding instrument collections use social media to share information about their collections and events, for which they maintain dedicated accounts or groups. Examples active on Facebook currently include the Harvard Collection of Historical Scientific Instruments, which maintains its own page; the History of Science and Technology at the Yale Peabody Museum public group; and the Oxford History of Science Museum, which often shares content relating to instruments in its collections.

Although follower and engagement numbers for these social media outlets might look modest, we should again be mindful of how complex and seemingly arcane the topic of scientific instruments and their history can be. In that regard, the aesthetic qualities of many instruments kept in museum and private collections (which, as Jim Bennett would point out, was likely the primary reason why they were deemed worth preserving for posterity) present an advantage for outreach and wider engagement, particularly on platforms that favour pictorial content and limited text. For example, instruments have had a prominent place in the popular "History of Astronomy" account on Twitter/X, a personal endeavour of Voula Saridakis (Museum of Science and Industry, Chicago). With content typically consisting of visually appealing historical astronomy imagery and instruments with a very short caption, the account had gained a noteworthy number of 28,500 followers as of June 2024.

However, liking and resharing an image of a visually compelling instrument such as an astrolabe or a highly embellished diptych sundial on social media might not fully represent significant engagement with instruments and their cultural contexts. A possibility to create more direct engagement if not with instruments proper, at least with their digital surrogates, resides in crowdsourced platforms such as Zooniverse.

Between March 2020 and February 2021, the Adler Planetarium, one of the host institutions of Zooniverse (together with the University of Oxford), implemented the project Digital Historical Skies, in which over 2,500 volunteers tagged constellations on segmented digital images of star charts and celestial globes from the Adler's collections.[18] The goal was to produce metadata that would then make it easier for users to find images of specific constellations in the Western astronomical tradition, including both extant constellations (those that are part of the list of eighty-eight constellations adopted by the International Astronomical Union in 1922)[19] as well as constellations that fell out of use in celestial cartography. Volunteers were asked

18 Jessica BrodeFrank, Samantha Blickhan, Adriana Guzman Diaz, and Nick Lake, "A Place for Everything: Adapting Content across Platforms and Audiences – A Series of Case Studies at the Adler Planetarium," in "Revisiting Museums," special issue, *Museological Review* 25 (2021): 26.

19 "The 88 IAU Constellations," International Astronomical Union (IAU), accessed 23 November 2024, https://www.iau.org/public/themes/constellations/#n2.

to follow a well-defined workflow and to observe each image attentively. The project generated 7,663 identifications considered to have an accuracy of 98%. Additionally, many volunteers used a forum feature to pose questions to the project team about specific constellations, charts, and globes and to discuss these matters with other users. Although the project was not strictly focused on instruments as they are usually understood, it nevertheless hints at the potential of crowdsourcing to engage worldwide audiences with their study, provide a forum for meaningful conversations about their history and cultural significance, and generate useful metadata for museum collections.

3 Staying Connected around Instruments: Video Conferences, Platforms, and Tours

The COVID-19 pandemic made 2020 a challenging year for the entire world, and the scientific instrument community was no exception. The annual SIC symposium has typically boasted a busy social programme of instrument collections tours, including private collections that might otherwise be difficult to access and visits to other institutions and sites of historical and scientific interest. This robust social component has reinforced the collegiality of SIC symposia over the years, constituting one of its distinctive aspects compared to other academic conferences in the field of history of science and technology.[20]

Issues relating to the environmental impacts and carbon footprints of travelling had already permeated discussions about where to host the SIC symposia. Although video conferencing is also not free from environmental impacts, in the circumstances of the pandemic it offered a solution for maintaining a degree of live interaction among scholars around the globe. And it had the advantage of enabling the participation of those who might not be able to travel to the conference site in regular circumstances. But like many other digital platforms and services, it also posed questions of privacy and security.

The 2020 XXXIX SIC Symposium, jointly organised by the Science Museum, London, and the Royal Museums, Greenwich, and originally meant to be co-hosted in person by these two institutions, was held virtually, with pre-recorded presentations followed by live Q&A sessions. The importance of informal social interactions among participants was not forgotten. The programme included virtual social hours and even a virtual pub quiz on mystery objects. The format proved successful and inspired the SIC Board to explore similar ways not only for keeping the SIC community connected, but also to expand its reach.

20 See the chapter by Richard L. Kremer in this volume.

The Board decided to launch a series of virtual tours of instrument collections from around the world, following a format similar to that of presentations in the virtual SIC symposium, i.e., pre-recorded videos followed by a Q&A with a guest speaker. Asynchronous video had already been used for educational demonstrations of scientific instruments on the YouTube platform, most notably by the prominent SIC member and former President Paolo Brenni (1954–2021), who between 2010 and 2017 contributed dozens of videos to the channel of the Fondazione Scienza e Tecnica – Museo de Firenze.[21] The videos currently are organised by six categories – Acoustics, Electrodynamics and Electromagnetism, Mechanics, Optics, Electrostatics, and Thermology – and feature lively hands-on demonstrations of historical apparatus such as a thunder house (a model to demonstrate the destructive effect of lightning bolts in the absence of a proper conductor), Tyndall's apparatus for boiling water with friction, or a Soleil-Duboscq saccharimeter (a device to determine the saccharine strength of a sugar solution). With glove-clad hands, and sometimes aided by a collaborator, Brenni performs the demonstrations with the idiosyncratic expert confidence for which he was well-known. Without voice narration (sometimes complemented with subtitles) or any special soundtrack, the videos highlight the sounds produced by the handling of the instruments, echoing through the rooms of the Fondazione's instrument-filled, glass-doored cabinets. The atmosphere is at once solemn and dramatic, often exuding a strong note of spectacle that in many cases would not have been strange to the original performances of the featured devices (*Fig. 10.2*).

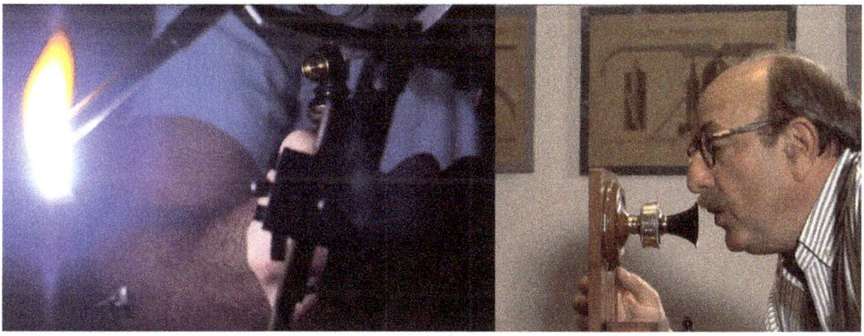

FIGURE 10.2 YouTube video still of the "Speaking arc and singing arc." *On right*, operator Paolo Brenni whistles into the microphone, while Roland Wittje, *on left*, regulates the length of the arc
COURTESY OF THE FONDAZIONE SCIENZA E TECNICA

21 Paolo Brenni's demonstration videos of working apparatus, Fondazione Scienza e Tecnica, final version 17 August 2015, http://www.youtube.com/@florencefst.

For the SIC virtual tour series, previously recorded demonstrations (or slide presentations) were used by speakers to circumvent the challenges and perils of live demonstrations and gallery walkthroughs. The series, comprising a total of six virtual tours with an average attendance of about eighty participants, ran between January 2021 and March 2022, featuring collections from North and South America (Peabody, Yale University; Yerkes Observatory, Williams Bay, Wisconsin; and the Physics Museum of La Plata National University, Argentina), Europe (Palermo Observatory Museum, Italy), Eastern Asia (Electronic Science and Technology Museum, Chengdu), and Africa (South African Astronomical Observatory). Conciseness was requested from speakers and audience alike, with succinctly formulated questions and comments being posted to the video conference chat and read aloud by the moderator. With the duration of each programme being strictly limited to one hour, this approach provided for a lively exchange that proved adequate to the format and reduced the risk, always significant during the pandemic, of screen exhaustion.

Similarly, one of the authors of this chapter (Pedro Raposo) teamed up with one of the volume editors (Sara Schechner) in a live virtual talk about sundials,[22] taking advantage of the remarkable sundial collection of the Adler Planetarium in Chicago (where Raposo was employed at the time). The goal was to promote the recently published Adler sundial catalogue (part 1) authored by Schechner.[23] The hour-long programme was part of Adler Astronomy Live, a series of live YouTube presentations about astronomy and its history that ran in 2021 as part of the Adler's efforts to stay connected with its existing audiences and garner new ones during the COVID lockdown. Consisting of a dialogue around selected sundials facilitated by an Adler content manager and producer, the programme featured pre-recorded video segments showing the essential design and functions of the highlighted instruments. Commentary and explanations were provided live by the speakers. The audience could use the chat function to post questions and comments in real time, some of which were fed into the conversation. These live chats revealed a global, engaged audience with many of its participants showing previous knowledge or at least a keen interest in the subject. Although it is likely that the event attracted mainly sundial and history-of-astronomy enthusiasts, the resulting video, currently still available on YouTube, retains its value as a general introduction to sundials and their history, illustrated with hands-on demonstrations of historical devices from a major collection. The focus on history and cultural

22 Adler Planetarium, "Adler Astronomy Live: Sundials," posted 8 October 2020, YouTube, 55:50, https://www.youtube.com/watch?v=Xfo9ifTixIg.

23 Sara J. Schechner, *Time of Our Lives: Sundials of the Adler Planetarium* (Chicago: Adler Planetarium, 2019).

significance is important since other YouTube videos about sundials (some of which, it must be noted, boast much higher view numbers) tend to focus on the design and construction of sundials, without much cultural commentary or historical contextualisation.

Digital video platforms are, of course, not static. In recent years the platform TikTok has advanced the very short video as a popular form of user-generated audiovisual content. Although the format is not favourable to extended demonstrations or detailed explanations, it nevertheless offers opportunities for engaging a wider audience with scientific instruments. After experimenting with the format during the pandemic, the Adler Planetarium carried out a pilot project to connect actual instruments in galleries with related short videos that were made available on the Adler's social media channels. The videos featured very brief explanations and demonstrations of essential features. Museum visitors could access this additional content while in the museum galleries by scanning a QR code placed next to the respective label. The goal was not to duplicate or extend the content of the label but rather to provide the visitor with an audiovisual complement that counteracted the static appearance of the instrument as presented in the display case. This initiative expanded on a previous experience with three-dimensional models of selected instruments that could be accessed in a similar manner by registered users of the astronomy app *Night Sky*. In this case, users were presented with simple three-dimensional models that could be rotated, but which add little to the experience of observing the actual instrument in the gallery. In contrast, the Adler videos featured a collections staff member handling the actual instrument while providing succinct explanation and commentary. This human element offers an actual experience of communication as opposed to a simple instance of interaction with a digital surrogate through an electronic device.

Notwithstanding their own limitations and challenges, digital video platforms and apps can play an important role in consolidating and expanding existing communities of instrument students, enthusiasts, and people with a casual interest in the subject, and at least to some extent, make it more accessible to new audiences.

4 Instruments, Digital Humanities, and Artificial Intelligence

Current practices for digitally annotating, processing, and archiving cultural heritage align with the principles underlying Linked Open Data (LOD), the Semantic Web, and Digital Humanities (DH), while relying on pre-existing ontologies and structured schemas. The creation of an ontology is a central theme in almost all DH projects, be they academic or institutional. This has led to the development of a plethora of ontologies and classification systems, often with the ambition of becoming a standard – a universal ontology, so to speak. A disproportionate focus on the organisation of knowledge, however, can obscure the information being structured.[24]

The recent rise of Large Language Models (LLMs), with OpenAI's Generative Pre-trained Transformer (GPT) models at the forefront, has the potential to transform these practices because of their strength in working with natural language. The use of natural language to describe scientific instruments can enable researchers and archivists to incorporate pertinent information without adhering to pre-established data structures or vocabularies, and thus to work faster and include a variety of viewpoints. Furthermore, the capabilities of LLMs to translate and transform text enable the provision of information in multiple languages, styles and formats, opening new opportunities for outreach and dissemination.

Judging from the experiences of one of the authors of this chapter (Floor Koeleman), a tool such as ChatGPT can be of great help to process historical sources.[25] For example, Peter Apian's untranslated work *Instrumentum primi mobilis* (Nuremberg, 1534) is now more accessible for studying his *primum mobile* instrument, thanks to an online scan of the book (*Fig. 10.3*).[26] Providing ChatGPT with the digital image of a page, one can instruct it to transcribe the sixteenth-century Latin text and to translate it into English (*Table 10.1*). This allows for democratised access to digitised archival material and to query this material interactively.

In another experiment, ChatGPT proved less useful. Images of early modern scientific instruments were uploaded with a request for a visual interpretation (*Fig. 10.4*). It was clear that this kind of task requires sophisticated knowledge above the capabilities of current LLMs. The more general the kind of knowledge entailed, the better the performance. Koeleman uploaded digital images of three instruments to ChatGPT along with the question: *what scientific instrument is this*? The first was the "*Potenzmaschine*" (inv. no. 1457) in the collection

24 For example, in the case of SIMEOn (see note 7). Sarah Middle and Alex Butterworth, "Towards an Ontology of Pre-20th Century Scientific Instrument Types," in *Metadata and Semantic Research: 16th Research Conference, MTSR 2022, London, UK, November 7–11, 2022, Revised Selected Papers*, ed. E. Garoufallou and A. Vlachidis (Cham, Switzerland: Springer Nature, 2023), 89–94.

25 ChatGPT, accessed 3 July 2024, https://chatgpt.com/. At the time of writing, the model GPT-4o is the latest version.

26 "Instrumentum primi mobilis," Internet Archive, accessed 3 July 2024, https://archive.org/details/instrumentumprimooapia/page/n7/mode/2up.

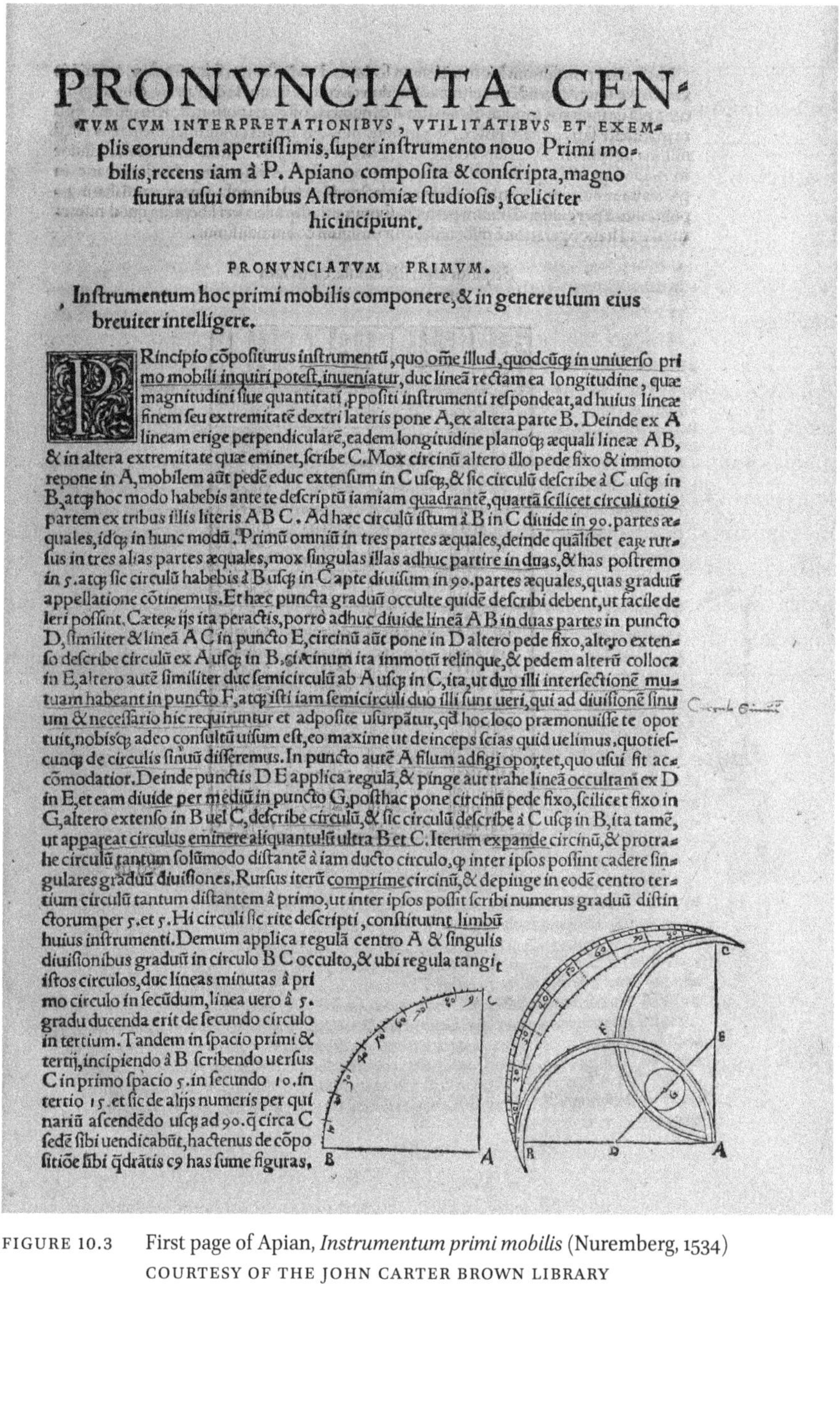

PRONVNCIATA CENTVM CVM INTERPRETATIONIBVS, VTILITATIBVS ET EXEMPLIS eorundem apertiſſimis, ſuper inſtrumento nouo Primi mobilis, recens iam à P. Apiano compoſita & conſcripta, magno futura uſui omnibus Aſtronomiæ ſtudioſis, fœliciter hic incipiunt.

PRONVNCIATVM PRIMVM.

Inſtrumentum hoc primi mobilis componere, & in genere uſum eius breuiter intelligere.

PRincipio cōpoſiturus inſtrumentū, quo om̄e illud, quodcūq; in uniuerſo primo mobili inquiri poteſt, inueniatur, duc lineā rectam ea longitudine, quæ magnitudini ſiue quantitati ꝓpoſiti inſtrumenti reſpondeat, ad huius lineæ finem ſeu extremitatē dextri lateris pone A, ex altera parte B. Deinde ex A lineam erige perpendicularē, eadem longitudine planoq; æquali lineæ A B, & in altera extremitate quæ eminet, ſcribe C. Mox circinū altero illo pede fixo & immoto repone in A, mobilem aūt pedē educ extenſum in C uſq;, & ſic circulū deſcribe à C uſq; in B, atq; hoc modo habebis ante te deſcriptū iamiam quadrantē, quartā ſcilicet circuli totiꝰ partem ex tribus illis literis A B C. Ad hæc circulū iſtum à B in C diuide in 90. partes æquales, idq; in hunc modū. Primū omniū in tres partes æquales, deinde quālibet earū rurſus in tres alias partes æquales, mox ſingulas illas adhuc partire in duas, & has poſtremo in 5. atq; ſic circulū habebis à B uſq; in C apte diuiſum in 90. partes æquales, quas graduū appellatione cōtinemus. Et hæc puncta graduū occulte quidē deſcribi debent, ut facile deleri poſſint. Cæterū ijs ita peractis, porrò adhuc diuide lineā A B in duas partes in puncto D, ſimiliter & lineā A C in puncto E, circinū aūt pone in D altero pede fixo, altero extenſo deſcribe circulū ex A uſq; in B, circinum ita immotū relinque, & pedem alterū colloca in E, altero autē ſimiliter duc ſemicirculū ab A uſq; in C, ita, ut duo illi interſectionē mutuam habeant in puncto F, atq; iſti iam ſemicirculi duo illi ſunt ueri, qui ad diuiſionē ſinuum & neceſſario hic requiruntur et adpoſite uſurpātur, qd̄ hoc loco præmonuiſſe te oportuit, nobiſq; adeo conſultū uiſum eſt, eo maxime ut deinceps ſcias quid uelimus, quotieſcunq; de circulis ſinuū diſſeremus. In puncto autē A filum adfigi oportet, quo uſui ſit accōmodatior. Deinde punctis D E applica regulā, & pinge aut trahe lineā occultam ex D in E, et eam diuide per mediū in puncto G, poſthac pone circinū pede fixo, ſcilicet fixo in G, altero extenſo in B uel C, deſcribe circulū, & ſic circulū deſcribe à C uſq; in B, ita tamē, ut appareat circulus eminere aliquantulū ultra B et C. Iterum expande circinū, & protrahe circulū tantum ſolūmodo diſtantē à iam ducto circulo, ꝙ inter ipſos poſſint cadere ſingulares graduū diuiſiones. Rurſus iterū comprime circinū, & depinge in eodē centro tertium circulū tantum diſtantem à primo, ut inter ipſos poſſit ſcribi numerus graduū diſtinctorum per 5. et 5. Hi circuli ſic rite deſcripti, conſtituunt limbū huius inſtrumenti. Demum applica regulā centro A & ſingulis diuiſionibus graduū in circulo B C occulto, & ubi regula tangit iſtos circulos, duc lineas minutas à primo circulo in ſecūdum, linea uero à 5. gradu ducenda erit de ſecundo circulo in tertium. Tandem in ſpacio primi & tertij, incipiendo à B ſcribendo uerſus C in primo ſpacio 5. in ſecundo 10. in tertio 15. et ſic de alijs numeris per quinariū aſcendēdo uſq; ad 90. q̄ circa C ſedē ſibi uendicabūt, hactenus de cōpoſitiōe ſibi q̄drātis cꝰ has ſume figuras.

FIGURE 10.3 First page of Apian, *Instrumentum primi mobilis* (Nuremberg, 1534)
COURTESY OF THE JOHN CARTER BROWN LIBRARY

TABLE 10.1 ChatGPT's transcription and translation of the header and first sentence of *Figure 10.3*

Transcription	Translation
PRONVNCIATVM PRIMVM.	FIRST PRONOUNCEMENT.
Instrumentum hoc primi mobilis componere, & in genere usum eius breviter intelligere.	To compose this instrument of the first mobile, and briefly understand its general use.
Principio copositurus instrumenti, quo omne illud, quodcúg; in universo primi mobilis inquiri potest, inuenietur, duae lineae rectae una longitudine, quae magnitudini stue quantitati, possit instrumenti respondet, ad huius lineae finem seu extremitate dextri lateris pone A, ex altera parte B.	First, to begin composing the instrument, whereby whatever can be inquired in the universe of the first mobile will be found, two straight lines of equal length, which correspond to the magnitude of the instrument, should be drawn, place A at the end or extremity of this line on the right side, B on the other side.

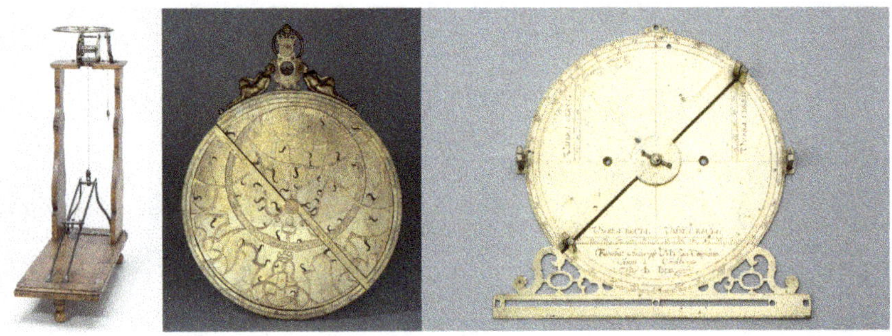

FIGURE 10.4 Three images uploaded to ChatGPT. *From left to right*: unknown maker, *Potenzmaschine*, 1770–1805, metal, wood, brass, textile, and lead, 62.4 × 19.9 × 47.5 cm. DEUTSCHES MUSEUM, 1457. Gualterus Arsenius, astrolabe, 1572, brass, 34.1 cm (diameter). MUSEO GALILEO, 1103. Michel Coignet, simple theodolite and plane table rule, 1602, brass, 21.7 × 20.6 cm COLLECTION OF HISTORICAL SCIENTIFIC INSTRUMENTS, HARVARD UNIVERSITY, DW0659

of the Deutsches Museum in Munich.[27] Very little is known about this instrument, and what it is exactly remains unknown.[28] ChatGPT could not provide a convincing answer either; it identified the image as a Foucault pendulum designed in 1851. Secondly, Koeleman uploaded an image of one of the astrolabes by Gualterus Arsenius (inv. no. 1103) in the collection of Museo Galileo.[29] In this case, ChatGPT described the instrument as an astrolabe, adding that it was "possibly made by a renowned maker such as Arsenius or another prominent instrument maker of the period." Thirdly, ChatGPT was given an image of Michel Coignet's simple theodolite and plane table rule (inv. no. DW0659) from the Collection of Historical Scientific Instruments at Harvard University.[30] The relative obscurity of this instrument once again confounded the LLM, resulting in its identification as a quadrant.

ChatGPT can provide initial guidance by referencing "the principles of dynamics and uniformly accelerated motion" in the case of the *Potenzmaschine* and alluding to instruments "used in astronomy, navigation, and surveying" with regard to the simple theodolite. But the general-purpose LLM currently is not sophisticated enough to perform historically contextualised accurate visual analysis of these historical and uncommon instruments. Furthermore, repeating the experiment does not produce consistent results. In subsequent chats, the instruments are identified as the Atwood machine invented in 1784, an astrolabe – this time without any reference to Arsenius – and a quadrant inscribed with the maker's name and the date. The inscription is transcribed as "Michael Coignetus" and "Anno a Christo nato MDCVI," which is impressive, even though the date is slightly incorrect (it should be MDCII).

---

27 "Potenzmaschine," Deutsches Museum Digital, accessed 10 July 2024, https://digital.deutsches-museum.de/item/1457/.

28 This object was studied by Koeleman during the International Seminar "Material Culture in the History of Physics," funded by the Wilhelm and Else Heraeus-Foundation and held at the Deutsches Museum from 26 February to 2 March 2018. The lecturers were Christian Forstner, Johannes-Geert Hagmann, Peter Heering, and Richard Kremer.

29 "Astrolabe," Virtual Museum, Museo Galileo, accessed 12 August 2024, https://catalogue.museogalileo.it/object/Astrolabe_n12.html.

30 "Simple theodolite and plane table," Waywiser, Collection of Historical Scientific Instruments, Harvard University, accessed 12 August 2024, http://waywiser.fas.harvard.edu/objects/14657/simple-theodolite-and-plane-table-rule. Such an instrument can be found in several *constcamer* paintings, as discussed in Floor Koeleman, "Kijken en Kennis: een Zeventiende-Eeuws Constcamer-Schilderij," *Wonderkamer: Magazine voor Wetenschapsgeschiedenis* 4 (2021): 8–15. Ad Maas is the editor of the magazine published by Gewina, the Belgian-Dutch Society for History of Science and Universities.

On a fundamental level, LLMs predict what the next word in a sentence will be. The output is therefore linguistically correct but not necessarily logical or factually correct. The human-like quality of the responses are by no means indicative of reasoning or perception. As a result, LLMs such as ChatGPT are prone to produce persuasive assertions with a total disregard for veracity.[31] One of the options available to overcome ChatGPT's current shortcomings is to customise a version for a specific purpose, by uploading files as a knowledge base, augmenting the model's capabilities with external information.

Scholars wishing to make use of ChatGPT and similar tools may need to take the time to get used to it, experiment with its features, approach its outputs critically, and have the flexibility to adjust their research questions and methodologies.

Chatbots rely on interaction and iterative exchanges, which requires the user to gain a certain proficiency in prompt design. The process is similar to learning how to effectively use a library catalogue, for instance, where carefully constructing the queries and understanding the system is also essential to obtain meaningful results. The key distinction, however, lies in the use of natural language. While library catalogues require precise search terms, chatbots allow for more conversational exchange, making the interaction intuitive and flexible, yet still demanding a careful approach to crafting prompts for optimal performance.

As such tools continue to evolve and their capacity scales up, it will be possible to increasingly integrate them into workflows. Training LLMs specifically on data related to scientific instruments, for example, effectively transforms them into domain specialists. Consequently, they increasingly will be able to function as personal assistants, not merely answering questions but also aiding the whole research process.

The inclusion of scientific instruments in early modern visual culture is of special interest to Koeleman. Her research using DH methods has made it clear that high-resolution, deep-zoom images still require a keen eye to detect previously unnoticed elements. A case in point is the gigapixel campaign of the Musée des Beaux-Arts de Lyon, that has been applied to the *Four Elements* series by Jan Brueghel the Elder and Hendrik van Balen.[32] In the "ultra-high

31 Michael Townsen Hicks, James Humphries, and Joe Slater, "ChatGPT Is Bullshit," *Ethics and Information Technology* 26, no. 2 (June 2024): 38, https://doi.org/10.1007/s10676-024-09775-5.

32 Discussion of gigapixel campaign for *Le Feu* by Jan Brueghel the Elder and other paintings in the series, Musée des Beaux-Arts de Lyon, accessed 3 July 2024, https://www.mba-lyon.fr/en/node/162. For a full view of *l'Air* discussed in the text, with detail shown in *Figure 5*, see: https://www.mba-lyon.fr/sites/mba/files/gigapixel/lair/lair.html.

resolution" image of *Air* (1611) a telescope is revealed under the left hand of the figure, the earliest dated representation of this instrument in a painting. This had gone unnoticed for hundreds of years until Koeleman discovered it on her screen in 2021 (*Fig. 10.5*).[33]

FIGURE 10.5 Jan Brueghel I, *Air*, 1611, oil on canvas, 46 × 83 cm, detail revealing telescope under figure's left hand
MUSEE DES BEAUX-ARTS DE LYON, A 77 © LYON MBA. PHOTO: ALAIN BASSET

33 Floor Koeleman, "Visualizing Visions: Re-Viewing the Seventeenth-Century Genre of Constcamer Paintings" (PhD diss., University of Luxembourg, 2021), 373. For more information on Coignet's simple theodolite, see 358.

5 Conclusion

Bringing instruments into the digital realm will never result in a homogeneous, consistent, and universal body of data and digital assets; rather, it will form a multifarious digital conglomerate to which collections professionals, researchers interested in instruments, and instrument enthusiasts will continue to add, and which they must continue to navigate as best as they can. However, the digital space opens up innumerable opportunities for sharing and disseminating information about instruments, for promoting engagement with the topic beyond a relatively small community of dedicated experts, and for advancing collaborative work beyond geographical constraints. This chapter is itself an example, having been researched and written by two authors, one in the United States and the other in Switzerland, and discussed with the volume editors entirely through the use of digital tools (shared documents, email, and videoconferencing).

Digital undertakings can also introduce fallacies such as the "digital as replacement" and "digital as forever." Although it is easy to complain about what it takes to properly manage and steward a physical collection in a traditional brick-and-mortar setting, it is equally easy to overlook the level of investment and resources necessary to keep digital assets alive. Institutions embarking on digitisation campaigns and increasing their digital presence must possess adequate computational infrastructures, train their employees in the skills required for the relevant systems and platforms, and ensure their long-term maintenance. The fact that SIC social media is currently halted due to a lack of volunteers is another reminder of the importance of capacity and labour considerations.

All these elements are vital for the sustainability of online resources. Another case in point is the short lifespan of the *Astrolabe Explorer*, a prototype developed by Oxford's Bodleian Libraries in 2018. The project was meant "to bring together data from across the web about astrolabes in the widest sense, allowing a large community of volunteers to update and maintain."[34] Five years later the link is no longer active on the Bodleian Digital Library.[35]

34 Quote taken from an email sent by Silke Ackermann to the Rete mailing list on 21 September 2018. See also: "A global collection of astrolabes in linked open data," Bodleian Digital Library, accessed 3 July 2024, http://blogs.bodleian.ox.ac.uk/digital/2018/08/08/global-collection-astrolabes-linked-open-data/.

35 The Internet site was archived by the Wayback Machine and can still be viewed at the following link: https://wayback.archive-it.org/org-467/20190925143314/http://glam-discovery.bodleian.ox.ac.uk/astrolabes/, accessed 22 August 2024.

Similarly, nowadays museum exhibitions are often accompanied by an online counterpart. However, once the physical exhibit closes, the future of the digital component becomes uncertain. Here the old-fashioned printed catalogue appears to offer a more stable and long-lasting solution, even though they often become difficult-to-find and expensive items.

The backlash on Facebook following the Cambridge Analytica scandal; dramatic changes to Twitter/X and its reputation after the highly publicised and expensive acquisition of the social network by technology billionaire Elon Musk; the copyright lawsuits faced by the Internet Archive; debates about the prospective interdiction of TikTok in the United States; and the ongoing debates about ChatGPT and whether regulations and limits should be imposed on the development of this and similar AI technologies – these are all signs of the frailties and risks surrounding such digital platforms, their stability, credibility, and continuity. Another issue that should also not be overlooked is the significant environmental impact and carbon footprint of these technologies, especially the high levels of energy consumed by AI servers.

Every new technology, method, or process entails gains and losses. Although we should not forget that collections of artefacts and documents have often been destroyed and lost and that all objects are subject to processes of decay and degradation, physical collections and documentation in traditional media usually entail a greater degree of stability and permanence. However, they pose a plethora of access and preservation challenges. Digital reproductions and models and digital-born materials, which rely on complex, energy-gobbling infrastructures and devices that tend to get outdated quickly, seem much more perishable in comparison.

Still, many possibilities remain open to the scientific instrument community and instrument enthusiasts from around the globe for using these technologies in research and outreach, and in broadening engagement with this important part of material culture. The digital opens up possibilities for access, collaboration, and engagement at a scale and rhythm that would be simply impossible to attain with more traditional media and processes. Thoughtful and reflexive experimentation plus a well-balanced co-existence between the old and the new is possibly what can ensure that scientific instruments and their study will continue to thrive in a digital world dominated by AI, without sacrificing their materiality and the crucial importance of human skill, ingenuity, sensibility, interaction, and communication.

What will remain, not only about the instruments of today and yesterday, but of all the digital assets that are being created around them? What about digital instruments, a concept that deserves more attention from instrument scholars? Will ChatGPT and similar digital tools be regarded as the instruments of the future as they are certain to play an increasing role in the making and dissemination of knowledge? And what will we gain and lose as digital research, collaboration, and outreach continue to evolve and gain prominence, while traditional forms of collecting, preserving, and displaying material culture are increasingly under scrutiny, be it for their historical roots and entanglements in networks of empire and colonialism, for the investment and resources they require, or for the shifting interests and expectations of the audiences they serve? These are just some of the questions facing those who pursue the study of scientific instruments in their multiple epistemic and cultural dimensions and engage with preserving and sharing this material (and increasingly digital) culture with broader audiences. It is very likely that new answers and approaches will increasingly rely on the assistance of the kinds of technologies addressed in this chapter – and possibly new ones yet to emerge.

Acknowledgments

This work was supported by the Swiss National Science Foundation [105211-204305].

Bibliography

BrodeFrank, Jessica, Samantha Blickhan, Adriana Guzman Diaz, and Nick Lake. "A Place for Everything: Adapting Content across Platforms and Audiences – A Series of Case Studies at the Adler Planetarium." In "Revisiting Museums." Special issue, *Museological Review* 25 (2021): 26–37.

Bud, Robert, and Deborah Jean Warner, eds. *Instruments of Science: An Historical Encyclopedia*. Garland Encyclopedias in the History of Science, vol. 2. New York: Science Museum, London, and National Museum of American History, Smithsonian Institution, in association with Garland Publishing, 1998.

Dupré, Sven. "Introduction: Science and Practices of Translation." *Isis* 109, no. 2 (2018): 302–307.

Grier, David Allne, and Mary Campbell. "A Social History of Bitnet and Listserv, 1985–1991." *IEEE Annals of the History of Computing* 22 (2020): 32–41.

Hicks, Michael Townsen, James Humphries, and Joe Slater. "ChatGPT Is Bullshit." *Ethics and Information Technology* 26, no. 2 (June 2024): 38. https://doi.org/10.1007/s10676-024-09775-5.

Koeleman, Floor. "Kijken en Kennis: Een zeventiende-eeuws constcamer-schilderij." *Wonderkamer: Magazine voor Wetenschapsgeschiedenis* 4 (2021): 8–15.

Koeleman, Floor. "Visualizing Visions: Re-Viewing the Seventeenth-Century Genre of Constcamer Paintings." PhD dissertation, University of Luxembourg, 2021. http://hdl.handle.net/10993/48720.

Metropolitan Museum of Art. *Computers and Their Potential Applications in Museums: A Conference Sponsored by the Metropolitan Museum of Art, April 15, 16, 17, 1968.* New York: Arno Press, 1968.

Middle, Sarah, and Alex Butterworth. "Towards an Ontology of Pre-20th Century Scientific Instrument Types." Pp. 89–94 in *Metadata and Semantic Research: 16th Research Conference, MTSR 2022, London, UK, November 7–11, 2022, Revised Selected Papers*, edited by Emmanouel Garoufallou and Andreas Vlachidis. Communications in Computer and Information Science. Cham: Springer Nature Switzerland, 2023.

Mosley, Adam. "'Sundials and Other Cosmographical Instruments': Historical Categories and Historians' Categories in the Study of Mathematical Instruments and Disciplines." Pp. 55–82 in *The Whipple Museum of the History of Science: Objects and Investigations, to Celebrate the 75th Anniversary of R.S. Whipple's Gift to the University of Cambridge*, edited by Joshua Nall, Liba Taub, and Frances Willmoth. Cambridge University Press, 2019.

Schechner, Sara J. *Time of Our Lives: Sundials of the Adler Planetarium*. Chicago: Adler Planetarium, 2019.

Fingerposts

A

From Russia with Love

Zhao Ke

FIGURE A Multimeters, *from left to right*: M104, Soviet Union,1959; C4, China, 1965; and C4-2, China, 1970
ELECTRONIC SCIENCE AND TECHNOLOGY MUSEUM, 20200319M002, 20200319M001, 20200319M003. UNIVERSITY OF ELECTRONIC SCIENCE AND TECHNOLOGY OF CHINA, CHENGDU. PHOTOS: YE GUILAN, ESTM ARCHIVES

The Electronic Science and Technology Museum (ESTM), located at the University of Electronic Science and Technology of China (UESTC) in Chengdu,[1] preserves three electronic meters. Although these "black boxes" look like triplets, they were manufactured in two different countries. The M104 meter was produced in the Soviet Union in 1959 and bears the inscription "CCCP" (in Cyrillic font) on its dial. The C4 millivolt-milliamp meter and the C4-2 ammeter were made in China by the Harbin Meter and Instrument Factory in 1965 and 1970, respectively.[2] Produced in two nations within the same decade, these three instruments raise several questions. Why do they look so similar? And how did they make their way to Chengdu?

1 Chengdu is a city in western China, where the electronics industry base was established during the First Five-Year Plan of China (1953–1957). China built four electronics factories in Chengdu that manufactured components, radar, and communication equipment. UESTC, formerly called the Chengdu Institute of Telecommunications Engineering, was established in 1956.

2 Zhao Ke, *Electronic Tracks of Human Civilization: Highlights from the Collection of the Museum of Electronic Science and Technology* (Chengdu: University of Electronic Science and Technology of China Press, 2021), 61–63 [赵轲. 人类文明的电子轨迹：电子科技博物馆馆藏精品. 成都：电子科技大学出版社, 2021, 61–63].

 | DOI:10.1163/9789004737587_013

In the 1950s, the Soviet Union aided China in establishing its industrial system through the "156 Projects," which were implemented in three ways – through direct equipment supply or factory construction, expert support, and education.[3] The M104 meter was given to the Guoguang Factory (a major microwave tube factory) by the Soviet Union.[4] The Harbin Meter and Instrument Factory, established in 1953, was also part of the "156 Projects."[5] It produced C4 and C4-2 meters, which were supplied to the Chengdu Institute of Telecommunications Engineering, created in 1956. In 2015, these three meters were donated to the ESTM.

Upon opening the external black boxes of these three instruments, we found that the manufacturing technique of C4 and C4-2 was superior, having more regular and standardised wiring. Over a ten-year period, from 1959 to 1970, each branch of the electronics industry in China had gradually developed to an independent level, with improved manufacturing techniques for components. Although the 1959 break between China and the Soviet Union had led to the withdrawal of experts, technology, and equipment from Chinese institutions, the established "factory-personnel-technology" system, developed by the Soviet Union, had laid a strong foundation for sustainable Chinese development. Meanwhile, the level of Chinese industrial manufacturing and management improved its efficiency.

The transfer of meter technology from the Soviet Union to China proved to be successful and was sustained despite the political break in the late 1950s. After 1959, China was able to innovate and develop independently within a comprehensive social-cultural environment, driven by national defence and economic growth.

Here are some lessons from these three meters: (1) Technology transfers from stronger to weaker districts can have long-term beneficial effects. Imitation and learning with the support of the state can lead to innovation and the formation of the concept of "independent development." (2) In the early stage of scientific, technical, and industrial development, external support plays a catalytic

3 "Conclusion of the 'Sino-Soviet Treaty of Friendship, Alliance and Mutual Assistance,'" Ministry of Foreign Affairs, the People's Republic of China, published 17 November 2000, https://www.mfa.gov.cn/eng/zy/wjls/3604_665547/202405/t20240531_11367548.html.

4 The Guoguang Factory was China's earliest, large-scale comprehensive microwave tube factory, operating from 1956 to 1963. It was called the "776 Factory."

5 Chen Xi, *Laying the Foundation Stone: The Beginning of the 156 Projects of Soviet Aid to China I* (Chengdu: Tiandi Press, 2020), 75–76 [陈夕. 奠基：苏联援华156项工程始末 I. 成都：天地出版社, 2020, 75–76].

role. A country's development strategy, such as concentrating resources, can provide strong initial momentum. (3) National and political forces can guide the direction of science and technology development, but they are not the only factors. Economic needs and broader social and cultural contexts offer fertile and necessary ground for sustainable development.

B

If Brass Could Talk

Ewa Wyka

FIGURE B It's me, a torquetum! Attributed to Hans Dorn, Buda, circa 1487 JAGIELLONIAN UNIVERSITY MUSEUM, CRACOW, MUJ 1571-I. PHOTO: GRZEGORZ ZYGIER

What could this enigmatic instrument, constructed of sheets of brass, tell a visitor to the Jagiellonian University Museum today?[1] Who would people its story?

Its tale would feature surviving manuscripts, scientific treatises, familiar characters, and a certain scientist who needs no introduction today. It would not speak of its construction, use, nor name (torquetum), since modern historians of astronomy have given excellent accounts of these matters. It would probably boast about its wealthy sponsor, King Matthias Corvinus of Hungary (r. 1458–1490), whose prestige endures. A patron of science and art, as well as

1 Torquetum, attributed to Hans Dorn, Buda, circa 1487 (42 × 56 × 71 cm). Jagiellonian University Museum, Cracow, inv. No. MUJ 1571-I.

 | DOI:10.1163/9789004737587_014

a vigorous proponent of astrology, Matthias founded a short-lived academy at Pozsony (now Bratislava). When it closed in 1471, Marcin Bylica (1433–1493), the torquetum's future owner and hitherto a lecturer at the academy, joined Corvinus's court as a royal adviser and court astrologer. Bylica, from Olkusz, a small town approximately forty kilometres from Cracow, served the king from 1472 until the ruler's death.

In 1476, Hans Dorn (d. 1509), a widely-known scientific instrument maker, came to Buda at the invitation of the king, who intended to build an astronomical observatory. Entrusting his plans to Dorn and Bylica, he sent them to Nuremberg in 1478 on a mission to buy Regiomontanus's library and instruments. The enterprise was, however, unsuccessful. It was probably during this period that Bylica commissioned Dorn to make three instruments: an astrolabe, a celestial globe, and … our torquetum, all intended to equip the observatory. Dorn completed the globe in 1480, and the torquetum is dated to about 1487. Were it only able, the torquetum might tell us why the globe, planned for the observatory, bears the coat of arms of Marcin Bylica and not those of Dorn or the king. Dorn spent fourteen years at the court in Buda and returned to Vienna only after the king's death in 1490. Three years later, Marcin Bylica also died. In his will, he bequeathed the instruments to the Cracow Academy.

By 1494, the young Nicolaus Copernicus and his brother Andrew had been in Cracow for three years, studying liberal sciences. An important event took place on September 10 of that year; an extraordinary meeting of the academic community, called by the then-rector, Jan Sakran, presented Bylica's remarkable and exceptionally valuable instruments, which had just arrived from Buda, to the Cracow Academy. Surely the young Nicolaus Copernicus was present to witness the event.

If the torquetum could talk, it would tell a story of its experiences over more than five hundred years at the Cracow Academy, of its work in the university observatory, of the turmoil of war, of its health problems – in the form of alterations and repairs – and of its honourable return, finally, to the Jagiellonian University Museum. And, of course, of its meeting with the young Nicolaus.[2]

2 For further reading, see Ludwik A. Birkenmajer, *Marcin Bylica z Olkusza oraz narzędzia astronomiczne które zapisał Uniwersytetowi Jagiellońskiemu w roku 1493* (Cracow: Akademia Umiejętności, 1892); Darin Hayton, "Martin Bylica at the Court of Matthias Corvinus: Astrology and Politics in Renaissance Hungary," *Centaurus* 49 (2007): 185–198; Richard L. Kremer, "How Did the Turketum (or Torquetum) Get Its Name?" in *Scientific Instruments between East and West*, ed. Neil Brown, Silke Ackermann, and Feza Günergun (Leiden: Brill, 2019), 80–107; Jarosław Włodarczyk, "The Torquetum (or Turketum): Was It an Observing Instrument?" *Journal for the History of Astronomy* 53 (2022): 266–299.

C

William Brownrigg's Frontiniac Bottles

Simon Werrett

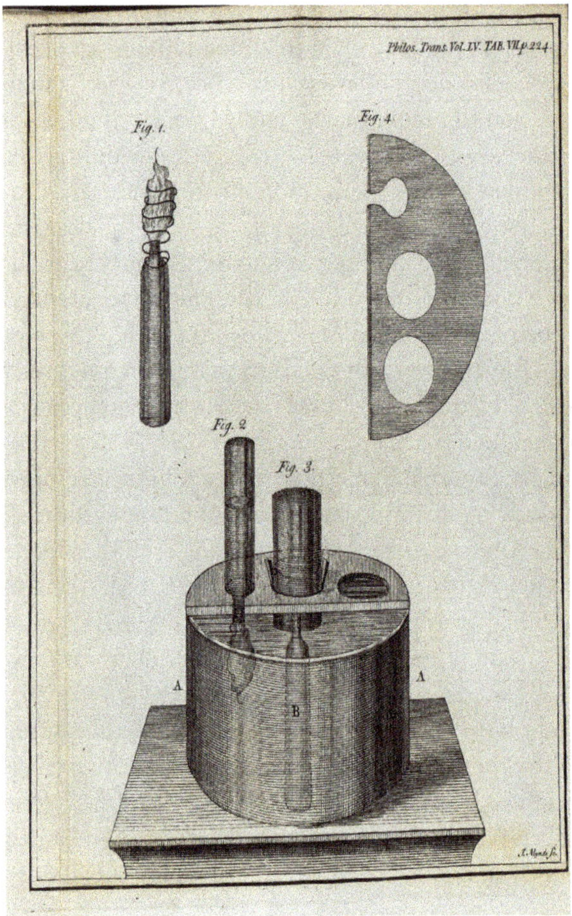

FIGURE C
Brownrigg's apparatus for capturing airs emitted by spa water. *Fig. 1* in the image shows a Frontiniac bottle filled with Pouhon water, with a calf's bladder and wire spiral over the top. *Fig. 2* shows a Frontiniac bottle inverted in a cistern. *Fig. 4* is a wooden board cut with holes to keep the bottles in place in the cistern. From Brownrigg, "An Experimental Enquiry," *Philosophical Transactions* 55 (1765): 224
PHOTO: THE ROYAL SOCIETY

In 1765 the Royal Society's *Philosophical Transactions* included a paper by William Brownrigg on the nature of spa water.[1] Brownrigg had trained as a physician under Boerhaave at Leiden before establishing a medical practice in

1 William Brownrigg, "An Experimental Enquiry into the Mineral Elastic Spirit, or Air, Contained in Spa Water," *Philosophical Transactions* 55 (1765): 218–243.

 | DOI:10.1163/9789004737587_015

Whitehaven in northern England.[2] In the 1740s to 1760s Brownrigg pioneered studies of explosive "damps" found in coal mines and the gases emitted by spa water. In 1764 he took advantage of a visit to the Pouhon spring in Spa, Belgium, to experiment on the waters. He fitted dried calf's bladders over a number of "common quart bottles" and bottles "in which Frontiniac wine is usually kept," expecting to collect emitted air in them but found none.[3] Heating the filled Frontiniac bottles in a copper basin did generate an elastic fluid, into which Brownrigg inserted mice to observe the effects (they died). He concluded that Pouhon mineral water contained a large quantity of air whose suffocating effects might be connected to the damps found in mines.

The Royal Society awarded Brownrigg the Copley medal for his inquiries in 1766, and Joseph Priestley later derided his own chemical apparatus as "nothing more than the apparatus of Dr. Hales, Dr. Brownrigg, and Mr. Cavendish, diversified." To modern eyes, Brownrigg's apparatus hinged on a quaint collection of repurposed wine bottles. Frontiniac (Frontignac) was "a luscious kind of rich Wine, made at Frontiniac, near Montpelier [sic], in France."[4] Why Frontiniac? Brownrigg noted that the long, slender bottles for this sweet muscadine could contain 21 ounces avoirdupois, so their shape and capacity may have been valuable to him. He was also operating in an environment away from home, and wrote that in Spa, he experimented in a manner "such as my engagements, and the apparatus that I could readily there procure, would admit of."[5] No doubt the consumption of wine was common in leisurely resorts like Spa. When Laurence Sterne invited his wife to the Spa in 1762, he reminded her to bring a "bottle-skrew [...] to uncork us our Frontiniac."[6]

Chymical experiments in the eighteenth century often involved expediency, substitution, and making use of what was to hand. Priestley collected airs in "cylindrical jars ... about ten inches long, and 2½ wide, being such as I have generally used for electrical batteries."[7] To move vessels of air from one place to another he made use of "common tea-dishes, which will hold water enough for that time." A "tall beer-glass" served to contain mice when testing the vivifying

2 J.V. Beckett, "Dr William Brownrigg, F.R.S.: Physician, Chemist and Country Gentleman," *Notes and Records of the Royal Society* 31 (1977): 255–271.

3 Brownrigg, "An Experimental Enquiry," 221.

4 Abel Boyer, *Dictionaire royal, françois-anglois, et anglois-françois; tiré des meilleurs auteurs qui ont écrit dans ces deux langues, par Mr. Boyer*, 2 vols. (Amsterdam, 1727), 2: n.p.

5 Brownrigg, "An Experimental Enquiry," 220.

6 Laurence Sterne to Mrs Sterne, June 14, 1762, in Laurence Sterne, *The Works*, 8 vols. (London, 1794), 6: 76.

7 Joseph Priestley, *Experiments and Observations on Different Kinds of Air*, 6 vols. (London, 1774–1786), 1: 6, 8.

capacity of some air.[8] This practice of "making use," what today might be called repurposing, was not accidental, but fitted early modern ideals of "thrift," the need to balance buying new with making use of what one already owned.[9] A Christian moral, thrift applied to rich and poor alike, all being expected to make the most of Earthly things, considered as gifts from God. Brownrigg was a wealthy landowner by the 1760s, so he did not need to repurpose old bottles for lack of resources. If something did the job, it should be made to serve the purpose. Chemists in the nineteenth century recognised this approach as widespread and challenged it, insisting only specialised apparatus would lead to scientific progress. Some chemists resisted. Richard Dennis Hoblyn wrote in 1841, "It is not by the variety and costliness of his instruments that the philosopher is known, but by the knowledge and skill with which he uses them."[10] One imagines that Brownrigg and many other eighteenth-century experimenters would have agreed.

8 Priestley, *Experiments and Observations*, 8.

9 Simon Werrett, *Thrifty Science: Making the Most of Materials in the History of Experiment* (Chicago: University of Chicago Press, 2019).

10 Richard Dennis Hoblyn, *A Manual of Chemistry* (London, 1841), 2.

D

Smoke and Mirrors

Robert W. Smith

FIGURE D The Hubble Space Telescope seen in orbit around the earth in 2009 at an altitude of about 320 miles. The Telescope's aperture door is open so light from astronomical objects can reach its optical system
NASA PHOTOGRAPH

In 1977, plans for the Hubble Space Telescope (then named the Space Telescope) were in serious trouble. Although intended as a joint venture of the US space agency, NASA, and the European Space Agency, no funding was included in NASA's budget for Hubble, not even for feasibility studies, let alone to begin detailed design work, building, and operation. The goal was to launch, with the space shuttle, a reflecting telescope with a primary mirror of 2.4 metres diameter into orbit around the Earth. The shuttle would transport Hubble to a place beyond the atmosphere and so where the stars do not twinkle. It would thereby provide new, and potentially transformative, observing capabilities. Hubble's designers intended it to be the first astronomical spacecraft to be regularly serviced and repaired in orbit. Astronauts could therefore upgrade its

 | DOI:10.1163/9789004737587_016

array of scientific instruments over time. However, the telescope's estimated cost of well over a billion dollars (current year) had led to significant opposition from US Congressional representatives who preferred to support missions with more practical pay-offs. In 1977 it seemed as if NASA had given up on Hubble.

But all was not what it seemed. The senior NASA manager responsible for space science had deliberately left Hubble out of the budget, and so to outsiders it seemed that NASA was in effect cancelling Hubble. The manager's aim was to provoke a reaction from those astronomers and industrial contractors eager for Hubble's construction to get underway. The NASA manager wanted Hubble's supporters to advocate more energetically for the project and so drive it forward. The *actual* cancellation of the project was the last thing the senior NASA manager wanted. And this apparent cancellation provoked an angry reaction as astronomers organised on a scale never before seen in support of an astronomical project. What was coming into being, in fact, was a formidable coalition of supporters of Hubble that included not just astronomers, but a wide range of institutions from universities to industrial contractors, White House and Congressional staffers, the European Space Agency, sympathetic journalists and various officials in US security agencies. Without such a coalition, the planned Hubble Space Telescope would have been technically and scientifically, but not politically, feasible.

In this process of coalition building, the organisation of the astronomical enterprise in the United States was itself reconstituted by its necessary institutional, financial, political, and ideological ties to society.[1] For much of the twentieth century, a few leading astronomers had interacted with patrons to secure funds for new telescopes. For astronomers, the "selling" of the Hubble Space Telescope was radically different. Here was a grassroots effort with astronomers playing explicitly political roles. An important shift was the willingness of astronomers to engage in lobbying and for astronomers to act more directly as a community.

In the study of scientific instruments that are larger in various ways than the traditional "big science," it is not enough to examine their scientific and technical aspects and to limit our attention to those people explicitly working

1 James H. Capshew and Karen A. Rader, "Big Science: Price to the Present," *Osiris* 7 (1992), 3–25, see 16.

on that instrument. We must also scrutinise the wider politics associated with these projects. We need to examine the coalitions of supporters (and groups of opponents) as these are crucial to whether or not a proposed megascience project gets off the drawing board and into operation. This means at times peering into a world of smoke and mirrors where a project's apparent cancellation is nothing of the sort; rather, it is instead a tactic to win support.[2]

---

2 On the making of the Hubble Space Telescope, see Robert W. Smith (with contributions by Paul Hanle, Robert Kargon, and Joseph Tatarewicz), *The Space Telescope: A Study of NASA, Science, Technology and Politics* (New York: Cambridge University Press, 1993). For Hubble's time in orbit, see David DeVorkin and Robert W. Smith, *Hubble Cosmos: 25 Years of New Vistas in Space* (Washington, DC: National Geographic Society, 2014); Christopher Gainor, *Not Yet Imagined: A Study of Hubble Space Telescope Operations* (Washington, DC: NASA, 2021). An example of the actual cancellation of a megaproject is provided by Michael Riordan, *Tunnel Visions* (Chicago: Chicago University Press, 2015).

E

A Catalyst for Experiences and Memories

Elisabetta Rossi

FIGURE E Cockcroft-Walton particle generator and accelerator of the Centro Informazioni Studi ed Esperienze (CISE), Milan, circa 1952–1965, on display at the Museo Nazionale della Scienza e della Tecnologia, Milan. To protect the large object sustainably and without breaking the budget, the museum's exhibition designer chose a wire mesh cage rather than glass showcase; the design also recalls the safety devices employed during the accelerator's operational life
MUSEO NAZIONALE SCIENZA E TECNOLOGIA LEONARDO DA VINCI, 11800. PHOTO: ALESSANDRO NASSIRI

When I first visited the Museo Nazionale Scienza e Tecnologia Leonardo da Vinci in Milan, a huge apparatus, visually impressive in size and shape, caught my eye. Its label read, "Cockcroft-Walton particle generator and accelerator – CISE (Milan) – 1952–1965 ca."[1] The machine is a descendant of the original

1 Cockcroft-Walton particle generator and accelerator, deployed by the Centro Informazioni Studi ed Esperienze (CISE), Milan, 1952–1965; now in the Museo Nazionale Scienza e Tecnologia Leonardo da Vinci, Milan, inv. no. 11800. Luca Reduzzi, "L'acceleratore Cockcroft-Wilson del Museo della Scienza di Milano e il suo ruolo nello sviluppo del nucleare civile in Italia," *Giornale di Fisica* 65, suppl. no. 2 (2024): 135–149.

 | DOI:10.1163/9789004737587_017

so-called "Cockcroft-Walton" particle accelerator system, devised in the early 1930s by physicists John Cockcroft and Ernest Walton, winners of the 1951 Nobel Prize for achieving the first accelerator-induced nuclear reaction.[2] For many years, devices of this type were the main investigative tools available to nuclear physicists, and although being surpassed by today's technologies, they remain a milestone in the history of physics.

As the museum exhibit reveals, the "Cockcroft-Walton" in the Museo is linked to a specific, local experimental framework in the field of particle physics. It comes from the Centro Informazioni Studi Esperienze or CISE in Milan, the first pioneering research centre to be inaugurated in Italy despite the material difficulties in immediate post-war times.[3] The machine was built between 1948 and 1952, and when it became obsolete in 1956, CISE donated the voltage multiplier and the accelerating column to the Museo, which displayed the components for a few decades but then dismantled the instrument and put it into storage.

In 2015–2016 the Museo activated an interdisciplinary restoration project, on the occasion of inaugurating a permanent gallery, "Extreme," dedicated to particle physics.[4] Displayed were the two main parts of the apparatus,

---

2 John Douglas Cockcroft and E.T.S. Walton, "Experiments with High Velocity Positive Ions. – (I) Further Developments in the Method of Obtaining High Velocity Positive Ions," *Proceedings of the Royal Society A* 136, no. 830 (1932): 619–630.

3 On the CISE, see Carlo Lombardi, "La questione dell'energia nucleare," in *Storia dell'industria elettrica in Italia*, vol. 5, ed. Giorgio Zanetti (Rome-Bari: Laterza, 1992), 589–644; Giovanni Battimelli, Michelangelo de Maria, and Giovanni Paoloni, "Dal 'disastro della fisica' all'elettrosincrotrone di Frascati: Le origini dell'Istituto Nazionale di Fisica Nucleare," *Analysis. Rivista di cultura e politica scientifica*, nos. 3–4 (2002); Giovanni Paoloni, "Nuclear Energy and Science Policy in Post-War Italy," in *Nuclear Italy: An International History of Italian Nuclear Policies during the Cold War*, ed. Elisabetta Bini and Igor Londero (Trieste: EUT, 2017), 229–243.

4 The restoration of the object was conducted by STRATI S.N.C. (a restoration workshop that deals with preventive conservation and restoration of cultural heritage in various fields) in the person of Marianna Cappellina, assisted by Giulia Basilissi and Maria Baruffetti, under the supervision of the museum's director of collections Laura Ronzon, curator Luca Reduzzi, and restoration manager, Claudio Giorgione with Claudia Porta, in collaboration with Simona Casonato, who collected oral memories of CISE employees. Historian of physics Leonardo Gariboldi (University of Milan) and his students were also involved, collaborating with former CISE employees (physicists, engineers, chemists) to recover the research centre's archive now preserved at the Fondazione ISEC – Istituto per la Storia dell'Età Contemporanea in Sesto San Giovanni (Milan). See "Restauro del generatore e acceleratore di particelle Cockcroft-Walton," Museo Nazionale della Scienza e della Tecnologia Leonardo da Vinci, accessed 11 January 2025, https://www.museoscienza.org/it/collezioni/restauri/cockcroft-walton.

originally connected with cables: the generator (240 × 65 × 215 cm, h × w × d) and the accelerator (200 × 110 × 230 cm). As a graduate student, I got the chance to learn more about this restoration project, which was conducted parallel to research on the history of the apparatus. The first aim of the intervention was to stop the material degradation and restore the original appearance of the instrument by correcting errors due to previous displacements and reassembly. The second aim was to recover the material object's cultural context by collecting testimonies of people who had worked directly with the accelerator, such as former CISE employees. These oral accounts enabled the restorers to understand some particularities of the instrument's technical configuration and thereby make the restoration more historically accurate.[5]

The project enabled the "Cockcroft-Walton" to be a catalyst for curators, conservators, students, and previous users to explore history, memory, and experience within a public-history and museum context.[6] This peculiarity led the accelerator to be featured in an award-winning documentary film, *Handle with Care*, directed by Francesco Clerici and developed in collaboration with the Museo. The documentary presents behind-the-scenes restoration practices and methods, and it emphasises the importance of manual labour as intangible heritage.[7]

This comprehensive project carried out by the Museo demonstrates the potential of its collections to be used for both research and public engagement. Although the Milan "Cockcroft-Walton" is just one of the many accelerators preserved internationally, the project has managed to tell a unique story about

5 Audiovisual documents and oral histories, Museo Nazionale della Scienza e della Tecnologia Leonardo da Vinci Archive, Milan: A. Ascoli, interviewed by S. Casonato, L. Gariboldi, L. Reduzzi, and G. Zorzi, 9 June 2016 (video); G. Marcazzan, interviewed by S. Casonato, L. Gariboldi, L. Reduzzi, and G. Zorzi, 9 June 2016 (video); T. Rossini, interviewed by S. Casonato and L. Reduzzi, 25 September 2016 (video).

6 On the history of the instrument and related bibliography, see Simona Casonato, "Intangible Heritage, Science, and Identity: National Narratives and the Documentation of Science in Practice," in *Understanding Use: Objects in Museums of Science and Technology*, ed. Tim Boon, Elizabeth Haines, Arnaud Dubois, and Klaus Staubermann (Washington, DC: Smithsonian Institution Scholarly Press, 2024), 166–187.

7 *Maneggiare con cura* [*Handle with care*] (78 minutes, color), directed by Francesco Clerici, produced by Jon Barrenechea, Gaby Ramsperger, and Francesco Clerici, written by Francesco Clerici and Simona Casonato. (Milan, 2017). See also, "Maneggiare con cura," Museo Nazionale della Scienza e della Tecnologia Leonardo da Vinci, accessed 11 January 2025, https://www.museoscienza.org/it/collezioni/maneggiare-con-cura.

particle physics in the context of post-war reconstruction in Italy. Despite being "old" apparatus, the "Cockcroft-Walton" continues to capture the attention of historians of science, physicists, and museum visitors. And through video documentation, visitors can witness the interdisciplinary research behind the restoration.[8]

8 I would like to thank Laura Ronzon (Director of Collections) and Ida Morisetti (Project Manager – Collections and Curators) at the Museo Nazionale Scienza e Tecnologia Leonardo da Vinci for their valuable feedback on the draft of this essay and for their continued support.

F

Trusting the Black Box

Richard Paselk

FIGURE F
Beckman Model G pH Meter, circa 1955 with open sample compartment. This instrument appears identical to the original 1937 catalogue description except for a round meter replacing a square one, and a knob replacing a screw for the zero adjust
PHOTO: BY AUTHOR

The twentieth century saw scientific and laboratory instruments go from devices where the physics or chemistry is obvious and operation is intuitive to "black boxes" making remarkable measurements with little understanding needed by the operator. Davis Baird has argued that the research, development, marketing, and use of these instruments enabled I.B. Cohen's fourth big scientific revolution.[1]

1 Davis Baird, "Analytical Chemistry and the 'Big' Scientific Instrumentation Revolution," in *From Classical to Modern Chemistry: The Instrumental Revolution*, ed. Peter J.T. Morris (Cambridge: Royal Society of Chemistry, 2002), 52.

 | DOI:10.1163/9789004737587_018

For a scientist and educator, black-box instruments present a problem: how do you know the outputs represent the phenomena correctly, and how do you convince students to be properly sceptical of these results? In my own experience teaching chemistry laboratories, this transition requires changes in behaviour and greater awareness of the phenomena measured. For example, with vacuum systems I kept a mercury manometer available (where the physics is visible) to check the electronic vacuum gauges. You often need deeper knowledge of measurements to know when black boxes fail.

The half-decade from 1936–1941 transformed the nature of scientific instruments with the introduction of two key "black-box" instruments into the chemistry laboratory: Arnold Beckman's Model G pH meter[2] and the Beckman DU UV-Vis spectrophotometer.[3]

The pH meter invented by Beckman is an exemplary black box instrument, familiar to any chemist or laboratory biologist working in the second half of the twentieth century. Beckman's pH meter was the first integrated laboratory instrument and the first chemical instrument with electronics at its heart.[4] Before his invention, measuring pH required multiple independent components connected together in a very specific manner, as well as significant interpretation and calculation to get a result. Determining pH requires the measurement of voltage differences generated between two chemical electrodes (essentially batteries). For pH, these electrodes consist of a reference electrode and an electrode sensitive to hydrogen ions, either a hydrogen electrode or a glass electrode. In early setups, these electrodes had to be large or complex to generate the current needed to measure low voltages with a potentiometer, sensitive galvanometer, switches, standard cell, and other components.

Beckman revolutionised this measurement by combining a number of innovations. First, he designed an electronic amplifier that was compact, robust, and highly sensitive to the smallest electrical currents because of its two triode-tube design. This enabled the size of the electrodes to be reduced. (In 1941 he used the same amplifier for the Beckman DU UV-Vis spectrophotometer.)[5] Second, he built an integrated, compact instrument around the amplifier, fitting all the components into a walnut veneer case. Chemists could now purchase a

2 "Development of the Beckman pH Meter: National Historic Chemical Landmark," American Chemical Society, accessed 12 April 2023, http://www.acs.org/content/acs/en/education/whatischemistry/landmarks/beckman.html.

3 Arnold Thackray and Minor Meyers, Jr., *Arnold O. Beckman: One Hundred Years of Excellence* (Philadelphia: Chemical Heritage Foundation, 2000). "DU spectrophotometer," Wikipedia, accessed 19 April 2023, https://en.wikipedia.org/wiki/DU_spectrophotometer.

4 "Development of the Beckman pH Meter."

5 "DU spectrophotometer," Wikipedia.

pH meter and immediately begin making simple, precise, and repeatable measurements of the pH of a solution in both laboratory and industrial situations. Little training or skill is required of the operator; today, students are taught to use successor instruments in an hour or two in introductory laboratory classes. Beckman's pH meter eliminated the need for specialised training and knowledge … but required trust in the black box.[6]

6 Additional description and illustrations of the Model G can be seen at Richard Paselk, "Early North Coast Scientific Instruments, c. 1850–c. 1950: Laboratory Instruments for Health and Industry," accessed 4 September 2024, https://sites.google.com/humboldt.edu/paselkr1/scientific-instrument-museum-exhibits/early-north-coast-scientific-instruments/laboratory-instruments-for-health-and-industry.

G

Forgers, Connoisseurs, and Collectors

Joshua Nall

FIGURE G *On left*: Astrolabe by Johannes Bos, 1597, as illustrated in the sales catalogue for the Anton Mensing collection – *Sammlung Mensing: Altwissenschaftliche Instrumente, Katalog* (Amsterdam: F. Muller & Co., 1924), vol. 2, plate 10. This instrument is now in the collection of the Adler Planetarium, Chicago, USA. *On right*: The Whipple Museum's forged copy of the Bos astrolabe, likely made in Amsterdam around 1925
IMAGES © WHIPPLE MUSEUM AND LIBRARY, CAMBRIDGE, UK, CR 13:22, WH.0355

Few instruments can have motivated as much scholarship as this curious little astrolabe. Signed "IOHANNES BOS DIE 24 MARTII 1597," it first caught the attention of instrument enthusiasts in the late 1920s when the collector Robert Stewart Whipple purchased it from the dealer Gertrude Hamilton for the considerable sum of £70 10s. But it was only once the astrolabe arrived at the University of Cambridge – to whom Whipple donated his entire collection in 1944 – that it began to garner more critical attention. Here, the Bos astrolabe underwent forms of scrutiny entirely unprecedented in the study of instruments. Rather than admiring instruments as individual specimens of past scientific technique, the historian of science Derek Price began

 | DOI:10.1163/9789004737587_019

to assess instruments *en masse*, as part of a wider study of the scientific instrument-making trade. Aided by new resources from public museums – above all the printed catalogue – Price could start to compile extensive datasets of instruments, against which any one object could be compared.[1]

And what Price found was immediately troubling. Consulting the catalogues revealed at least two more instruments that were apparently identical to the Whipple Museum's Bos astrolabe, each, implausibly, signed with the very same date. Price suspected forgery, a heretofore almost unknown feature of historical instruments and a potentially damaging revelation for collectors, the trade, and now the museums that hoped to pioneer their use as historical sources. Price soon found evidence of four further fakes in the Whipple Museum, each of a piece with similar forged instruments in collections across Europe and the United States and each traceable back to a single source: the dealership Frederik Muller & Co., then under the direction of the collector and dealer Anton Mensing. The forgers, whomever they were, appear to have been copying pieces illustrated in two of the company's sales, one held in 1911 (parts of the Strozzi collection), the other in 1924 (Mensing's own collection).[2]

Price's revelation opened up a wide range of questions and challenges for curators and scholars who have hoped to understand historical instruments as sources. On the one hand, he drew particular attention to Mensing, his company, and his collection (a great deal of which ended up at the Adler Planetarium in Chicago).[3] After Price, it remained an open question as to how extensive the spread of "Mensing fakes" might be. It was only after the completion of the comprehensive "Anton Mensing Scientific Instrument Project" (1992–2003) that we received a clear answer: Price had identified the great

1 Boris Jardine, "Like a Bos: The Discovery of Fake Antique Scientific Instruments at the Whipple Museum," in *The Whipple Museum of the History of Science: Objects and Investigations, to Celebrate the 75th Anniversary of R.S. Whipple's Gift to the University of Cambridge*, ed. Joshua Nall, Liba Taub, and Frances Willmoth (Cambridge: Cambridge University Press, 2019), 201–221.

2 Derek J. Price, "Fake Antique Scientific Instruments," *Actes du VIIIe Congrès International d'Histoire des Sciences: Florence-Milan 3–9 Septembre 1956* (Florence: Vinci, 1958), 380–394.

3 Liba Taub, "'Canned Astronomy' versus Cultural Credibility: The Acquisition of the Mensing Collection by the Adler Planetarium," *Journal of the History of Collections* 7, no. 2 (1995): 243–250. Mensing himself was exonerated from personal involvement in the making of the forgeries in W.F.J. Mörzer Bruyns, "Frederik Muller & Co. and Anton Mensing," *Quaerendo* 34, nos. 3–4 (2004): 211–239. Further aspects of Mensing's career and influence are explored in William Cumiford, "A Mensing Link? The Henze Collection of European Scientific Instruments," *Bulletin of the Scientific Instrument Society*, no. 28 (March 1991): 14–18; and W.F.J. Mörzer Bruyns, "The Amsterdam Scheepvaartmuseum and Anton Mensing: The Scientific Instruments," *Journal of the History of Collections* 7, no. 2 (1995): 235–241.

majority of the "Mensing fakes," even if he had only scratched the surface of the wider topic of forged instruments.[4] By opening up this area of enquiry, Price can be credited with inspiring a growing body of scholarship that encompasses, at the least, the tricks and ruses of a sometimes unscrupulous trade;[5] details of the unmasking of master forgers;[6] comparative laboratory analysis of genuine and fake instruments;[7] and the exposure of more forgeries from beyond the Mensing paradigm – including within the collection that started it all, the Whipple Museum in Cambridge.[8]

---

4 S. Johnston, W.F.J. Mörzer Bruyns, J.C. Deiman, and H. Hooijmaijers, "The Anton Mensing Scientific Instrument Project: Final Report," *Bulletin of Scientific Instrument Society*, no. 79 (2003): 28–32. The fruits of the project's principal symposium, held in Leiden in 1999, are published in Peter de Clerq, ed., *Scientific Instruments: Originals and Imitations, The Mensing Connection* (Leiden: Museum Boerhaave, 2000).

5 David Weston, "All that Glitters Is Not Gold," *Bulletin of the Scientific Instrument Society*, no. 21 (1989): 10–12.

6 Harriet Wynter, *The Price of Every Thinge* (Glasgow: The Grimsay Press, 2010), 232–259.

7 G.B. Stephenson, B. Stephenson, and D.R. Haeffner, "Investigations of Astrolabe Metallurgy Using Synchrotron Radiation," *Materials Research Bulletin* 26 (2001): 19–23.

8 Boris Jardine, Joshua Nall, and James Hyslop, "More Than Mensing? Revisiting the Question of Fake Scientific Instruments," *Bulletin of the Scientific Instrument Society*, no. 132 (March 2017): 22–29.

H

Anatomical Models at a Historic Women's College

Jacqueline Marie Musacchio

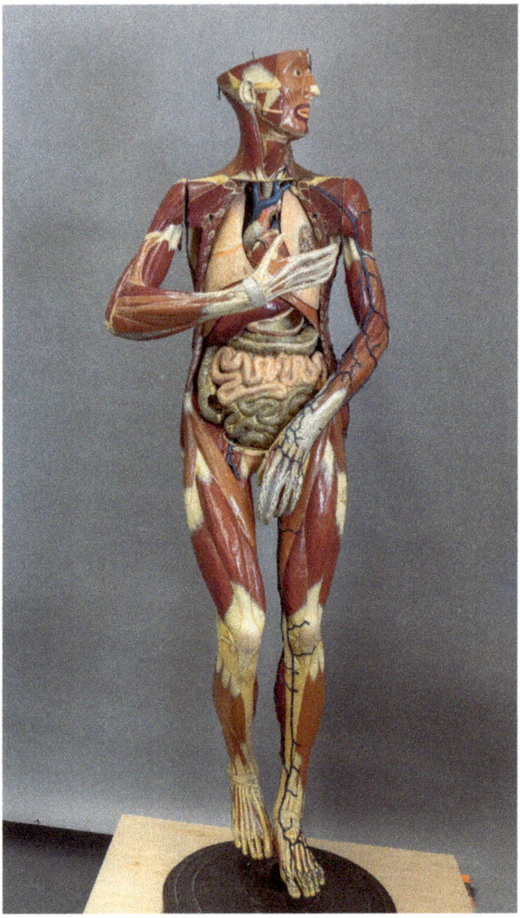

FIGURE H
Anatomical model of a woman, Maison Auzoux, Saint Aubin d'Ecrosville, 1928
SCIENCE CENTER, WELLESLEY COLLEGE. © KATHY TARANTOLA PHOTOGRAPHY

Wellesley College's Sage Hall was built in 1927 to house the popular Botany department; over the years it was renovated to include additional science departments, a library, classrooms, and office and administrative spaces. When it was emptied in anticipation of its demolition in 2019, we uncovered hundreds of objects – instruments, models of all sorts in a variety of materials, slides, posters, and more – that had been used to teach science in the decades following the American college's opening in 1875. Wellesley was at the

 | DOI:10.1163/9789004737587_020

forefront of women's education in the sciences and indeed in other fields, too, and the objects used to teach our early students reveal a great deal about this history. Fearing, with reason, that these objects would be discarded, I was part of a small number of faculty determined to find them new homes. Included in this group of objects was an almost life-size papier-mâché anatomical model of a woman made by Maison Auzoux, forgotten and abandoned in a closet for at least half a century.

Founded in 1827 by Louis Thomas Jérôme Auzoux in Saint Aubin d'Ecrosville, about fifty kilometres west of Paris, Maison Auzoux led the market for botanical, zoological, and anatomical models for more than a century, selling to schools and museums around the world from its Parisian shop, printed trade catalogues, and displays at international exhibitions.[1] Auzoux described the models as "clastique" or "clastic," from the Greek word "to break," because they were designed to be disassembled and reassembled at points marked by tiny manicules on their surfaces; the numbered parts, identified in booklets that accompanied the models, aided in this process. Auzoux's anatomical models were easier and safer to dissect than a cadaver, and his combination of papier-mâché with cork and clay proved particularly durable. Despite the regular dissections she must have endured, Wellesley's woman is in near-perfect condition and, unlike the few other surviving examples I can locate, she retains all of her parts.

We know that Wellesley's founders, Pauline and Henry Fowle Durant, were eager to provide the latest pedagogical tools to their students. Interactive models like those by Auzoux responded to what was known in art history as the "Wellesley Method" of experiential learning; faculty taught in this manner across the curriculum. The Durants purchased a collection of Auzoux's botanical models at the 1878 Exposition Universelle in Paris. Around that time, Durant tried to purchase one of Auzoux's male models, but was told orders were backlogged a year. Fortuitously, Harvard had one to sell, so Durant sent the Harvard chemist and Wellesley early benefactor Eben Norton Horsford to examine it, to ensure that it could be "introduced to polite society;" in other words, Durant wanted a male model without reproductive organs for his female students.[2] Archival photographs indicate that he did acquire one, but it is lost today.

1 See especially Bart Grob, *The World of Auzoux* (Leiden: Museum Boerhaave, 2000); and Bart Grob, Elizabet Nijhoff Asser, and E. Manú Giaccone, *Papieren anatomie: De wonderschone papier-machémodellen van dokter Auzoux* (Leiden: Museum Boerhaave, 2008).

2 Henry Fowle Durant to Eben Norton Horsford, undated (between 1878–1881), Sylvester Manor Archive MSS 208, box 71, folder 59, Fales Library and Special Collections, New York University Libraries, New York.

As the biology curriculum evolved to include reproductive biology, Wellesley acquired Auzoux's models of a foetus in the womb and both male and female torsos, as well as, in 1928, this female model. The pose deliberately echoes that of the celebrated Medici Venus (first century BCE, now in the Uffizi Gallery, Florence), but of course this is an écorché figure; the muscles, fat, nerves, and blood vessels are vividly painted and, when her skull and abdominal cavity are opened, her brain and organs – including reproductive organs – are visible. Unfortunately, the decision was made to display the model in our newly renovated Science Center with her abdominal cavity closed, hiding all of her organs and rendering her much less remarkable and useful to students today.

1

The Frog Drawing Machine

Phil Loring

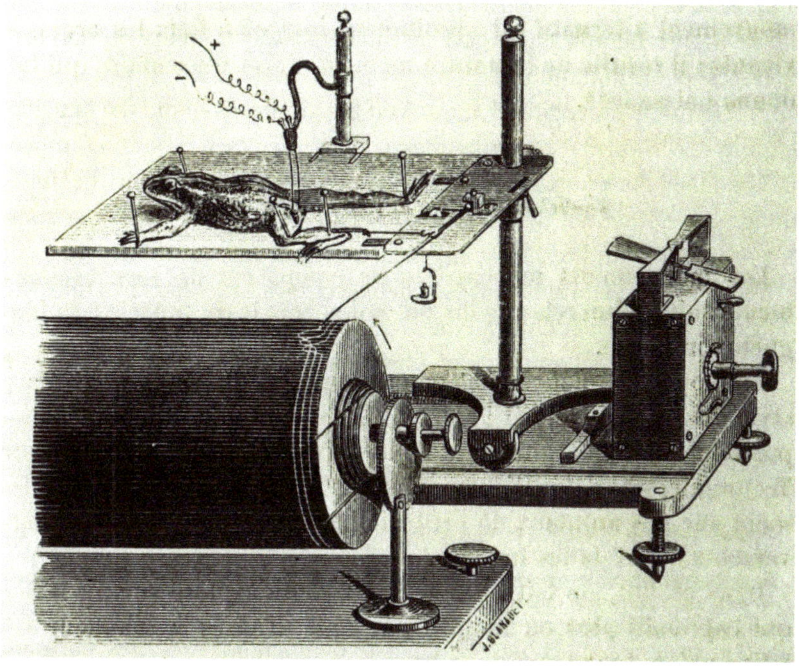

FIGURE 1 A frog inscribing traces on a blackened cylinder. E.J. Marey, *La méthode graphique en sciences expérimentales* (Paris, 1878)
GETTY RESEARCH INSTITUTE

In 1851, physiologist and physicist Hermann von Helmholtz created what he called his "frog drawing machine" by first coating a champagne glass with soot and rotating the glass lengthwise. He then attached a frog's leg to a stylus to scratch a thin line onto the glass. One rotation of the glass traced a full contraction of the leg. Helmholtz transferred the traces to a flat surface and sent them to the Paris Academy of Sciences to illustrate his arguments for the speed of nerve conduction.[1]

1 Henning Schmidgen, *The Helmholtz Curves: Tracing Lost Time* (New York: Fordham University Press, 2014).

 | DOI:10.1163/9789004737587_021

Kymographs, invented by German physiologist Carl Ludwig in 1847 and mass-produced from 1883, worked on the same principle. A stylus cleverly attached to a still-living animal's body (or part thereof) scratched lines on a smoked sheet of paper wrapped around a cylindrical drum. The drum rotated at variable speeds, driven by clockwork or by an electric motor. It was a messy process, what with all the sooty paper and bodily fluids, but it was transformative. Rather than counting the twitches of a frog leg, or the frantic beats of a rabbit's heart, kymograph traces gave researchers a way to freeze motion into visible traces for examination.

And, as Helmholtz recognised, it was as if the animal itself did the drawing, making the marks on the paper by means of its own "self-recording" movements. Gradually the kymograph became "*the* device for visualizing the pulse, breathing, muscle action, nervous activity, and other physiological actions, present in research and teaching alike," as historian Alistair Kwan wrote.[2] It would not be replaced until the 1960s, when electronic amplifiers like the EKG, EMG and EEG became widely available.

In the burgeoning laboratories of physiology of the late nineteenth and early twentieth centuries, an ever-wider range of animals were disassembled, their living muscles and nerves integrated with kymographs. At the same time, particularly in the Anglo-American world, antivivisectionist opponents targeted those laboratories as stages for grotesque displays of cruelty. Answering the question of extrapolation became crucial, particularly in medical research: could results obtained from experiments with frogs, turtles, mice, rats, guinea pigs, dogs, cats, pigeons, primates, and zebrafish (just to name a few), be used to draw conclusions applicable to humans?[3] Kymographs and their partners, the experimental animals, stand at the foundation of what historian and philosopher Donna Haraway called "the technoscientific pharmaceutical research-industrial complex," which profoundly shapes our world today.[4]

The flexible and ubiquitous kymograph also performed work for the behavioural psychologist B.F. Skinner. In the 1930s, beginning with rats and pigeons, Skinner ingeniously adapted kymographs to track the behaviour of

2 Alistair Marcus Kwan, "'Do Not Kill Guinea Pig Before Setting Up Apparatus': The Kymograph's Lost Educational Context," *Teorie Vědy / Theory of Science* 38, no. 3 (2016): 301–335.

3 Frederic L. Holmes, "The Old Martyr of Science: The Frog in Experimental Physiology," *Journal of the History of Biology* 26, no. 2 (1993): 311–328; Karen Rader, *Making Mice: Standardizing Animals for American Biomedical Research, 1900–1955* (Princeton: Princeton University Press, 2004); Daniel P. Todes, *Pavlov's Physiology Factory: Experiment, Interpretation, Laboratory Enterprise* (Baltimore: Johns Hopkins University Press, 2002).

4 Donna Jeanne Haraway, *When Species Meet* (Minneapolis: University of Minnesota Press, 2008).

"whole organisms," meaning animals that were intact and free to move about (although caged). By the 1940s, Skinner's kymograph had evolved into a more specialised device called a cumulative recorder, specifically designed to record the actions of an animal over a period of hours. He sought to understand how animals learn to learn.[5]

Today, Skinner's tradition of research into animal behaviour informs the way animal laboratories operate; it helps reduce stress and distress in both the animals and their human caretakers. "What happens," asked Haraway, "if experimental animals are not mechanical substitutes but significantly unfree partners, whose differences and similarities to human beings, to one another, and to other organisms are crucial to the work of the lab and, indeed, are partly constructed by the work of the lab?"[6] From isolated tissues to working partners, the organism as experimental subject is evolving; still, however, the kymograph ("wave-writer") remains an icon of the biomedical laboratory.

5 Alexandra Rutherford, *Beyond the Box: B.F. Skinner's Technology of Behavior from Laboratory to Life, 1950s–1970s* (Toronto: University of Toronto Press, 2009); Kennon A. Lattal, "Steps and Pips in the History of the Cumulative Recorder," *Journal of the Experimental Analysis of Behavior* 82, no. 3 (November 2004): 329–355.

6 Haraway, *When Species Meet*.

J

Solid Gold

Neil Handley

FIGURE J Gold nose spectacles, English, pre-1723, modified to wig style circa 1780
BRITISH OPTICAL ASSOCIATION MUSEUM, LDBOA1999.1923, THE COLLEGE OF OPTOMETRISTS, LONDON

Many scholars look through spectacles to see clearly, but sometimes looking at them is also worthwhile.[1] Spectacles are an applied technology of obvious interest to historians of optics, but their study has often been comparatively neglected. Spectacles were made by some of history's most important opticians: Adams, Ayscough, Bate, Dollond, Marshall, and Martin, to name but a few from London. They made, sold, and stocked far more of them than the microscopes and telescopes featured on their trade cards, which were

1 Recent treatments include: Travis Elborough, *Through the Looking Glasses: The Spectacular Life of Spectacles* (London: Little, Brown Book Group, 2021); Stefana Sabin, *In the Blink of an Eye: A Cultural History of Spectacles* (London: Reaktion, 2021); and Neil Handley, *Cult Eyewear* (London: Merrell, 2011).

 | DOI:10.1163/9789004737587_022

produced less extensively, but have been afforded greater academic attention.[2] The way customers selected their spectacles was often decidedly non-scientific (trial and error, or as I'll describe, inheritance), and their perceived ordinariness has been one barrier to their serious consideration, dismissed as mere (disposable) commodities. Another has been their fragility; made from organic materials, bone, wood, and leather, or cheap copper wire, most examples have not survived.

Perhaps a valuable solid gold pair, not thrown away, but cherished, will halt this disinterest?[3] Its untypical material nature may nonetheless reveal a wider context, that of spectacles as personal keepsakes rather than practical implements.

Solid gold is a totally unsuitable material from which to fashion spectacle frames, being generally too soft for what need to be robust items that will be handled constantly in daily use and may require regular reglazing. The rarity of gold spectacles is explained not because they were luxurious, but impractical.

Purchased by the British Optical Association Museum in 1945 from a private individual, these spectacles are of the eighteenth century "wig" style (worn over powdered wigs that covered the ears) and feature an unusually ornamental bridge and unprecedented spoked-ring terminals. They were mistaken, at the time of acquisition, as a Victorian jeweller's distraction. Stylistically, we would now date them to about 1780, a conclusion made possible by the ongoing research of mostly amateur collectors.[4] Genealogical research, made easier by online access to the National Archives, has provided another route to reassess the emotional value assigned to this item. Researching the engraved names ("ELIZA VICKRIS" on the right and "ANNE HYAM" on the left) has offered up a fascinating possibility that they are an even earlier pair, altered as a *memento mori*.

The limited space permitted me, though more than the fifty-eight words of the object's current display caption, enables only a summary of our conclusions, rather than the justificatory details.[5] Suffice to say, the will of one Elizabeth

---

2 For example: John Millburn, *Retailer of the Sciences: Benjamin Martin's Scientific Instrument Catalogues, 1756–1782* (London: Vade-Mecum, 1986); Anita McConnell, *R.B. Bate of the Poultry 1782–1847* (London: Scientific Instrument Society, 1993); John Millburn, *Adams of Fleet Street: Instrument Makers to King George III* (Aldershot: Ashgate, 2000). For trade cards see: H.R. Calvert, *Scientific Trade Cards in the Science Museum Collection* (London: HMSO, 1971).

3 British Optical Association Museum, LDBOA1999.1923, The College of Optometrists, London. MusEYEum Online Catalogue, accessed 11 January 2025, www.musEYEum.org.

4 I refer readers to the journal *Ophthalmic Antiques* (1982–present).

5 I acknowledge the contribution of Sandra Adams of Ottawa, Canada, family historian and Vickris family descendant.

Vickris written in 1723, the year before her death, leaves "Gold Spectacles" to a daughter (another Elizabeth, the sister of Anne). These could only have been nose spectacles as spectacles with sides would not be advertised, by Edward Scarlett Sr, until later that decade. We surmise that the recipient of this inherently valuable family heirloom, if it is indeed the pair mentioned in the will, had them updated and then engraved in honour of her mother and sister (who died in 1750). This rare instance of contextualising dates permits us to declare them a unique transitional design.

K

Better Together

Richard Dunn

FIGURE K Trade card, H.G. Blair & Co., Cardiff, about 1860
SCIENCE MUSEUM, LONDON, 1934-121/13. © SCIENCE MUSEUM/SCIENCE & SOCIETY PICTURE LIBRARY

A trade card from 1860 is a good way to think about the ways in which the public history of technological innovation has treated two iconic navigational instruments.

The sextant and chronometer were developed from the mid-eighteenth century as instrumental aids for determining a ship's longitude (east-west position) at sea. As has been well told in bestselling books, exhibitions, and on television, both innovations were inspired by the large rewards available under the Longitude Act of 1714.[1] Importantly, the detailed study of instruments and related archives has guided this historiography thanks to significant contributions from university academics, museum curators, professional practitioners, and collectors — a rich mix characteristic of Scientific Instrument Commission

1 Published works include Dava Sobel, *Longitude* (New York: Walker & Co., 1995); Richard Dunn and Rebekah Higgitt, *Finding Longitude* (Glasgow: Collins, 2014).

 | DOI:10.1163/9789004737587_023

meetings.[2] It is, nevertheless, a peculiarity of navigation that its different tools fit into often distinct niches as regards collecting, classification, and study: instrument studies; horology; cartography; history of mathematics. Keeping hold of the bigger picture can require nimble footwork.

When the National Maritime Museum put on a display in 1962 to celebrate 200 years since the first sea trial of John Harrison's fourth marine timekeeper (H4), its curators presented longitude determination by lunar distance as "the rival method" to that represented by Harrison's machine.[3] Rivalry, which has become a hallmark of longitude historiography, naturally makes for compelling narratives, and some sense of it did emerge in the disputes between Harrison and the Commissioners of Longitude. Yet competition was not characteristic of mariners' experiences at sea. Navigators did not expect that one method would prove to be *the* solution. Chronometers and sextants were fine additions to the equipment available, but there was no need to choose between the methods they represented. The well-trained navigator drew on old and new instruments and practices alike to gather as much data as possible, then weighed up the evidence to decide the right course.[4] Besides, the new instruments were co-dependent: astronomical observations needed good timekeepers; local time had to be determined astronomically before being compared with chronometer time; timekeepers were checked by reference to the skies.[5] Co-dependence endured with the introduction of new methods

2 Notable examples include William J.H. Andrewes, ed., *The Quest for Longitude* (Cambridge MA: Harvard University Collection of Historical Scientific Instruments, 1996); J.A. Bennett, *The Divided Circle* (Oxford: Phaidon, 1987); Jonathan Betts, *Marine Chronometers at Greenwich* (Oxford: Oxford University Press/Royal Museums Greenwich, 2017); Charles H. Cotter, *A History of the Navigator's Sextant* (Glasgow: Brown, Son & Ferguson, 1983); Rupert T. Gould, *The Marine Chronometer* (London: J.D. Potter, 1923); Peter Ifland, *Taking the Stars* (Newport News: Mariners' Museum, 1998); W.E. May, "How the Chronometer Went to Sea," *Antiquarian Horology* 9 (1974): 638–663; Willem Mörzer Bruyns, *Sextants at Greenwich* (Oxford: Oxford University Press/National Maritime Museum, 2009); Saul Moskowitz, "The World's First Sextants," *Navigation: Journal of The Institute of Navigation* 34 (1987): 22–42; Alan Stimson, "Some Board of Longitude Instruments in the Nineteenth Century," in *Nineteenth Century Scientific Instruments and their Makers*, ed. Peter de Clercq (Leiden: Museum Boerhaave, 1985), 93–115.

3 National Maritime Museum, *4 Steps to Longitude* (London: National Maritime Museum, 1962).

4 W.E. May, *A History of Marine Navigation* (Henley-on-Thames: G.T. Foulis, 1973); Margaret Schotte, *Sailing School: Navigating Science and Skill, 1550–1800* (Baltimore: Johns Hopkins University Press, 2019).

5 Emily Akkermans, "Chronometers and Chronometry on British Voyages of Exploration, 1819–1836" (PhD thesis, University of Edinburgh, 2020).

such as position-line navigation in the nineteenth century, and the sextant outlived the method for which it was originally intended.

Attempting to tell the story afresh at Greenwich in 2014, the temporary exhibition *Ships, Clocks & Stars* tried instead to present the development of new instruments and methods as complementary. The exhibition's conclusion came not with the trial and initial adoption of the sextant and chronometer in the eighteenth century, but with their combined application in the nineteenth using commercially available instruments for routine navigation and, crucially, the extensive charting activities on which better navigation relied. Thus, one of the final exhibits was a case with chronometers, sextants, octants, charts and nautical almanacs displayed together as if in a nineteenth-century shop window.

Blair and Co.'s trade card depicts this relationship perfectly: chronometers and sextants offered together with charts, compasses, and the other instrumental accoutrements of the well-equipped navigator.

L

Paper Circles

Gaye Danışan

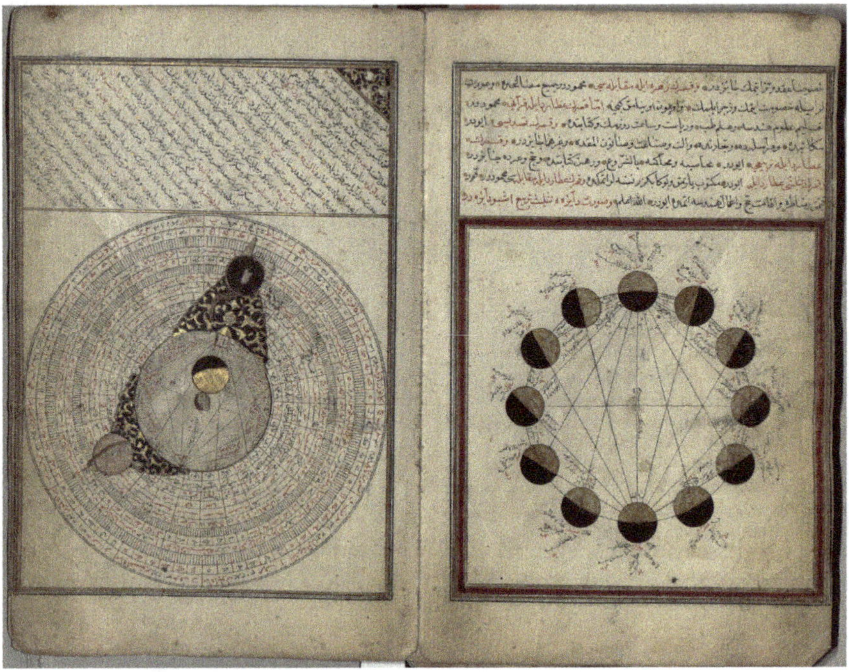

FIGURE L Ottoman volvelle featured within a perpetual calendar attributed to *Şeyh Vefâ*, seventeenth century. This corpus includes various perpetual calendars, featuring two world maps
TALAT ÖNCÜ PRIVATE COLLECTİON, ANKARA, TÜRKİYE, F. NP

While studying Ottoman timekeeping and calendars, I encountered a remarkable example: A perpetual calendar that includes a paper instrument with two moving circles from 1676.[1] This unexpected discovery sparked a 2017 research

1 Sheïkh Vafa, Ibrahim Shahidi ibn Khouda Dede, *Rûznâme-i Şeyh Vefâ* (*Le calendrier du sheïkh Vafa, dans une rédaction différente de celle des n°s 186 et 194 de l'Ancien fonds, avec, toutefois, la même préface*), Bibliothèque nationale de France, MS Supplément turc 537, copy date: H.1087 (1676 CE), f. 6v.

 | DOI:10.1163/9789004737587_024

project that later expanded into a broader exploration of the role of paper instruments in Ottoman scientific and practical activities.[2]

Embarking on this research journey presented several challenges. The physical fragility of paper instruments often limits their preservation and accessibility, complicating comprehensive study and classification. Traditionally, paper instruments have been overshadowed by more durable metal counterparts, contributing to their under appreciation in the history of science. Additionally, the role of paper instruments in Ottoman astronomy had not been thoroughly explored before, making it challenging to trace them in archives, especially in digital catalogues and museum collections due to the scarce information and uncertainty about their integration into manuscripts. To navigate these obstacles, I focused on identifying Ottoman paper instruments in astronomical literature without imposing chronological limitations, categorising them based on type, use, and historical period, while excluding drawn-on paper affixed to astronomical devices.

Building on this foundation, I embarked on a study of Ottoman calendars between 1550 and 1710, focusing on the relationship between volvelles and calendars.[3] I found significant differences between Ottoman and European volvelles in their historical contexts. Although European volvelle-making traditions may have inspired Ottoman scholars, the ultimate designs reflected distinctive Ottoman cultural and scientific characteristics. Significantly, these instruments are often to be found within calendars, which underscores their practical applications and suggests their preservation has been linked to their everyday utility. Volvelles allowed users to visualise interconnected astronomical concepts efficiently, enhancing the understanding of complex information in Ottoman society. Therefore, tracing the history of each example has been crucial in order to understand it as a product of a specific maker and user.

After recognising the historical significance of these instruments, we organised interactive workshops that brought together researchers, librarians, and curators to facilitate the discovery of hidden materials before they might deteriorate or become lost.[4] As a result, researchers from various fields began

---

2 The project entitled *Paper Instruments in the History of Ottoman Astronomy* was funded by a grant from the Scientific Instrument Society from 2017 to 2020.

3 The project entitled *A Comparative Study on Theoretical and Practical Aspects of Scientific Activity in the Ottoman Empire: Annual and Perpetual Calendars (1550–1710)* received funding from the Scientific and Technological Research Council of Turkey from 2020 to 2022 (Project No: 119K827).

4 For more detail see Gaye Danışan, "Avrupa ve Osmanlı Volvelleri: Tarihi, Yapısı ve Kullanımı Üzerine Karşılaştırmalı Bir Çalışma," paper presented in the workshop "Bilimsel Etkinliğin Teorik ve Pratik Yönleri: Osmanlı Takvimleri (1550–1710) Çalıştayı," Istanbul, 21 March 2022,

to share information about the volvelles they encountered, and we received insights from private collectors about their presence in their collections.[5]

To date, I have identified nineteen specimens from the sixteenth to the nineteenth centuries and divided them into four categories: calendar-related instruments (including volvelles, gurre-nâme, and date converters), qibla indicators, celestial planispheres, and interdisciplinary instruments for mathematics and navigation. Most of these instruments were found in perpetual calendars or manuscripts related to timekeeping. The earliest known example is an incomplete Ottoman volvelle from a corpus of calendars in the sixteenth century, while most surviving examples date from the seventeenth century.[6] Considering the small production of copies due to the absence of printing technology and the challenges of preserving paper instruments, these extant specimens are noteworthy. But it is reasonable to expect that more examples are likely to be identified within calendars or other manuscripts.

I am presently investigating the adaptation and diffusion of portable astronomical instruments within Ottoman geographical regions from 1500 to 1700.[7] Despite mysteries that continue to surround Ottoman paper instruments, I am hopeful that further studies will uncover their significant role in the history of science.

---

book of abstracts, 23, https://osmanlitakvimleri.com/workshop_calendars/; "Tracking Astronomical Instruments Through Ottoman World and Beyond," workshop, Istanbul, 28 November 2023, Osmanli Takvimleri, accessed 30 September 2024, https://osmanlitakvimleri.com/project_works_instruments/.

5 Following the first workshop, I was informed about three distinct volvelles, one of which was brought to my attention by my colleague, Meltem Kocaman, from the History of Science Department at Istanbul University. I would like to extend my sincere gratitude to Meltem Kocaman for informing me about the volvelle in Talat Öncü's collection, and to Talat Öncü for his generosity in sharing the manuscript from his private collection in Ankara, Türkiye, for the purposes of this research.

6 For more detail, see: Gaye Danışan, "The Paper Instruments in the History of Ottoman Astronomy," *Bulletin of the Scientific Instrument Society*, no. 149 (2021): 34–43.

7 The project entitled *Portable Astronomical Instruments: The Processes of Adaptation and Diffusion of Medieval Islamic and European Examples in the Ottoman Geography* (1500–1700) is funded by the Turkish Academy of Sciences (TÜBA) through the Outstanding Young Scientists Awards (GEBİP) from 2022 to 2025.

M

Fingerprints and Forensic Methods

Tiemen Cocquyt

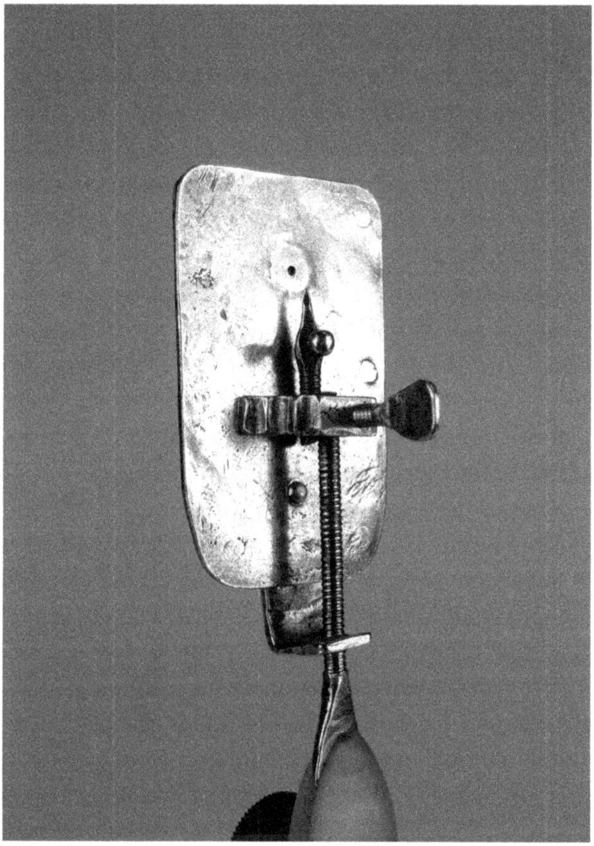

FIGURE M
Silver, single-lens microscope by Antoni van Leeuwenhoek, with a magnification of 248 times, produced circa 1673–1723
COURTESY ZUYLENBURGH COLLECTION, THE NETHERLANDS

In the 1670s Antoni van Leeuwenhoek (1632–1723) opened up the microworld with his self-produced, single-lens microscopes. Of the hundreds of microscopes Leeuwenhoek made, only a handful have been preserved. Each of these testifies to the Delft microscopist's handiwork. They all are very similar, but no two are identical.

This variability makes these instruments challenging for an instrument historian. Traditionally, we study scientific instruments by classifying them by style or by design. Leeuwenhoek's microscope design, however, is so straightforward that there is not much by which we can classify his microscopes. A "unique

 | DOI:10.1163/9789004737587_025

fingerprint" of Leeuwenhoek's microscopes is difficult to define. Consequently, assessing and studying these microscopes requires a lot of connoisseurship.

This reliance on connoisseurship is evident in how Leeuwenhoek's preserved microscopes have traditionally been assessed.[1] Production traces give Leeuwenhoek microscopes their distinctive "look and feel." The hallmarks that occur on silver copies may demonstrate the microscopes' antique status, as does wear and tear of the metal plates, screws or lens surfaces. In addition, non-functional details – production features that do not make sense for the instrument's operation – are often considered as indicators of authenticity. Exactly because these criteria are difficult to objectify, researchers have held on to the few aspects of Leeuwenhoek microscopes that can easily be quantified, such as the pitch of the screw threads. But what do variations in threading really tell for a microscope maker who single-handedly produced hundreds of microscopes over a period of forty years?

When in 2015 a new Leeuwenhoek microscope turned up, we felt that the method of investigation would benefit from including novel, forensic techniques.[2] We put our hopes on XRF analysis – a forensic method by means of which the atomic composition of metal alloys can accurately be determined.[3] Apart from the silver content, or the copper and zinc that make up brass, historic metals contain many more trace elements that are leftovers or contaminations from the smelting process. By turning to XRF, it was hoped that the alloy composition of Leeuwenhoek's preserved microscopes would offer us an objective, "unique fingerprint" of the metals that Leeuwenhoek used. Newly discovered microscopes could then subsequently be evaluated against this fingerprint.

Much to our surprise, the metal composition of the handful of genuine microscopes we had available was found to show enormous diversity. Even individual parts of the same, original, microscopes differed substantially in silver or brass composition. In hindsight, this outcome should not have been surprising for a handyman like Leeuwenhoek, who sometimes refined his own

---

1 Jacob van Zuylen, "On the Microscopes of Antoni van Leeuwenhoek," *Janus* 68 (1981): 159–198; Brian J. Ford, *The Leeuwenhoek Legacy* (Bristol: Biopress, 1991), 141–181.

2 Tiemen Cocquyt, "The Identification of a Silver Microscope of Antoni van Leeuwenhoek (1632–1723)," *Bulletin of the Scientific Instrument Society*, no. 130 (2016): 30–39.

3 This analysis method has incidentally been deployed in instrument studies before, on astrolabes for instance. See for example: B. Gratuze and J.-N. Barrandon, "Nouvelles analyses de l'astrolabe Latin AI. 86–31," *Physis: Rivista internazionale di storia della scienza* 32 (1995): 433–438; Jan C. Deiman, "Imitations among the Mensing Instruments," in *Scientific Instruments: Originals and Imitations*, ed. Peter de Clercq (Leiden: Museum Boerhaave, 1999), 98–112. In recent years, hand-held XRF devices have made analysis much more accessible.

metal from ore.[4] In a sense, the "objective" XRF data mimicked the stylistic features of his microscopes: superficially similar, but never identical. Eventually we learned to direct our attention to the trace elements in the silver composition, and the newly turned-up microscope signed off convincingly as a product compatible with seventeenth-century production after all. In addition, we found striking similarities between the new microscope's lens and the lens in another, authentic Leeuwenhoek microscope.[5]

If XRF analysis had failed to provide us with a "unique fingerprint," perhaps we had been asking the wrong question. The real power of XRF lies in systematic measurements, in the power of numbers. Only then do broader developments come to the surface. The significance of an XRF evaluation stands or falls with the availability of a sound reference data set to which individual results can be matched. In addition, our evaluation of the newly discovered Leeuwenhoek microscope taught us that seemingly "objective" forensic methods also require careful interpretation. The data needs to be put in context. These methods are valuable as a complement to existing research methodologies, but it does not seem likely that they will ever take out the role of connoisseurship in instrument studies.

The possibilities that forensic methods offer for instrument studies are very promising. Due to their sheer number of instruments, our collections can rightfully be seen as goldmines for building reference data sets for future forensic studies.[6] Yet, for a balanced assessment of an instrument's authenticity, a holistic approach that takes into account both subjective and objective views, both connoisseurship and forensic insights, remains paramount.

4 Huib J. Zuidervaart and Douglas Anderson, "Antony van Leeuwenhoek's Microscopes and Other Scientific Instruments: New Information from the Delft Archives," *Annals of Science* 73 (2016), 1–32.

5 On this lens, also see: Tiemen Cocquyt, Zhou Zhou, Jeroen Plomp, and Lambert van Eijck, "Neutron Tomography of Van Leeuwenhoek's Microscopes," *Science Advances* 7, no. 20 (2021), https://doi.org/10.1126/sciadv.abf2402.

6 This is currently pursued through XRF on British instrument collections in the research project *Tools of Knowledge: Modelling the Creative Communities of the Scientific Instrument Trade, 1550–1914*, accessed 10 January 2025, https://toolsofknowledge.org/.

N

Archimedes and Spectakles in S^t. Pauls Church Yard

Gloria Clifton

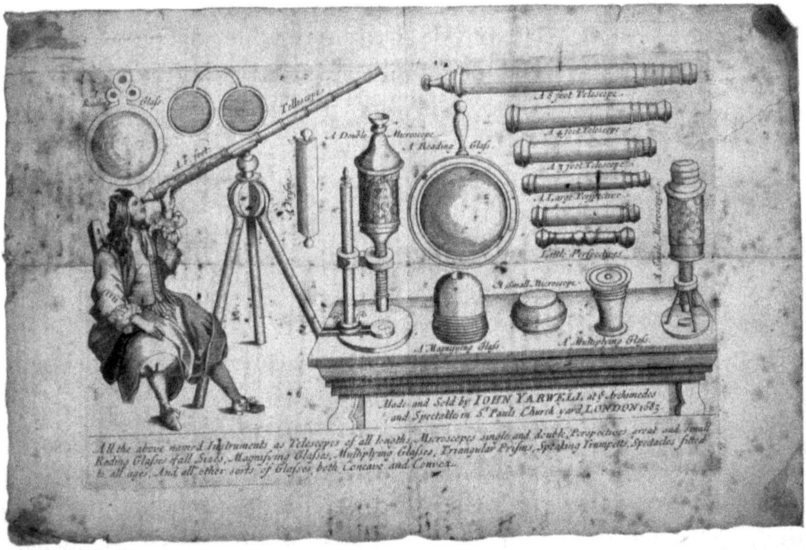

FIGURE N John Yarwell's trade card of 1683
IMAGE © WHIPPLE MUSEUM OF THE HISTORY OF SCIENCE, CAMBRIDGE, WH.3806

My fingerpost instrument is not actually an instrument but a trade card or advertisement issued by John Yarwell, the London optical instrument maker, the first surviving example being dated 1683.[1] Earlier advertisements by mathematical and optical instrument makers usually took the form of statements in books on mathematical and technical subjects that the instruments described or illustrated could be obtained from certain named makers.[2] By contrast Yarwell's trade card was clearly intended to stand by itself, being copiously

1 Trade card of John Yarwell, Wh.3806, Whipple Museum of the History of Science, Cambridge, UK.

2 Numerous examples are given by David Bryden in his article, "Evidence from Advertising for Mathematical Instrument Making in London, 1556–1714," *Annals of Science* 49, no. 4 (1992): 301–336.

© GLORIA CLIFTON, 2026 | DOI:10.1163/9789004737587_026

illustrated with engravings of telescopes, microscopes, and other instruments sold by optical instrument makers, including spectacles. It is an important signpost in a number of respects. As far as I have been able to discover, it is the first illustrated trade card to be issued by an instrument maker depicting a range of instruments. Other trade cards from the late seventeenth century, such as that of John Coggs, showed the maker's shop sign rather than the instruments sold.[3] However, by around 1700 a few other leading instrument makers issued trade cards which included numerous illustrations of instruments, for example Thomas Tuttell,[4] so it seems likely that Yarwell helped to initiate a trend as far as instrument makers were concerned. Other London makers to have elaborately illustrated trade cards in the first half of the eighteenth century were Fisher Combes and Nathaniel Hill, and the practice became more common thereafter.[5]

Yarwell later updated his card, probably in the 1690s, with a new address. The 1683 card had the address "Archimedes and Spectakles in S^t^. Pauls Church yard, LONDON." The address on the later undated card was "Archimedes and three Golden Prospects in S^t^. Pauls Church Yard London," and he also added the note "With all Sorts of Weather Glasses," but did not change the illustrations on the card, presumably because of the cost of engraving copper plates.[6] This points to the way in which a series of trade cards by the same maker help to illuminate the development of a maker's work, suggesting that "prospects" or telescopes were becoming a particularly important part of his output, and that he had added weather glasses to his range.

---

3 Michael A. Crawforth, "Evidence from Trade Cards for the Scientific Instrument Industry," *Annals of Science* 42, no. 5 (1985): 453–554, see 501.

4 Crawforth, "Evidence," 535. Thomas Tuttell, trade card, 1934–123, Science Museum, London, accessed 10 January 2025, https://collection.sciencemuseumgroup.org.uk/objects/co8015956/thomas-tuttell.

5 Nathaniel Hill, trade card, 1934–102, Science Museum, London, accessed 10 January 2025, https://collection.sciencemuseumgroup.org.uk/objects/co65688/nathaniel-hill; Fisher Combes, trade card, 1951–685/17, Science Museum, London, accessed 10 January 2025, https://collection.sciencemuseumgroup.org.uk/objects/co8002720/trade-card-fisher-combes. See also many examples in H.R. Calvert, *Scientific Trade Cards in the Science Museum Collection* (London: Science Museum, 1971).

6 John Yarwell, trade card, Y,5.310, British Museum, London, accessed 10 January 2025, www.britishmuseum.org/collection/search?museum_number=Y,5.310.

The Yarwell trade card is also a fingerpost in being the first to be discussed in any detail by a historian of science, featuring in an article in 1951 by Robert S. Whipple in *Annals of Science*.[7] Whipple was especially interested by the fact that the back of the example now in the Museum bearing his name was used by Sir Isaac Newton to make notes, but he did also comment on what could be learned about an instrument maker from such ephemera.

7 Robert S. Whipple, "John Yarwell or the Story of a Trade Card," *Annals of Science* 7, no. 1 (1951): 62–69.

O

"If in Doubt, Leave It Well Enough Alone"

Ileana Chinnici

FIGURE O
The restored Secchi meteorograph on display at the entrance to the Rome Astronomical Observatory at Monte Porzio Catone, near Rome
INAF OSSERVATORIO ASTRONOMICO DI ROMA

Father Secchi[1] would have been proud to see his spectacular meteorograph on display at the INAF astronomical observatory in Monte Porzio Catone near Rome. Of course, he would also have been a bit confused to see it in this new location, far from the Collegio Romano observatory where it had operated for

1 Ileana Chinnici, *Decoding the Stars: A Biography of Angelo Secchi, Jesuit and Scientist* (Leiden: Brill, 2019).

 | DOI:10.1163/9789004737587_027

decades. In fact, that observatory does not exist anymore: it was merged with the Campidoglio observatory and its equipment, including the meteorograph, was relocated to Monte Mario. Later, the glorious instrument was moved to a new site in Monte Porzio Catone, where it is now preserved.

The meteorograph is considered a prototype of modern self-recording meteorological stations. Designed by Angelo Secchi from 1859 to 1867, the instrument was made by the brothers Brassart and equipped with a French clock. It simultaneously measured temperature (with a bimetallic thermometer), pressure (with a balance barometer), humidity (with a wet-bulb and dry-bulb thermometer), rainfall (with a special pluviometer), and wind speed and direction (with a windvane and a cup anemometer), and continuously recorded the results.[2]

The idea of a continuous recording of meteorological parameters was not new – but Secchi developed it in the best possible way. In this complex and complete machine, the measurements were recorded with twelve pens moved by electro-mechanical arms on two clockwork-driven wooden boards, each covered with a large paper sheet and travelling vertically within the frame of the instrument. One board descended completely in two days; the other in ten days.[3]

The meteorograph was a marvel in operation. "The indication of the different meteorological sensors (placed on the roof of the observatory) were transmitted to the writing pens with a sophisticated electromechanical system including 11 telegraphs, 6 counters, a pendulum clock, and an incredible arrangement of wheels, counterweights, levers, chains, pulleys and Watt parallelogram linkages."[4] In 1867 it was awarded the Grand Prix at the *Exposition Universelle* in Paris. Standing 2.7 m tall, 1.5 m wide, and 0.6 m deep, it must have wowed the public while in operation during that exhibition, especially every time the hammer regularly struck the big brass bell on top of the instrument.

The advantage of having the data output by graphs instead of numbers was multiple: it freed scientists and assistants from the need to visually record measurements at fixed times; it offered a general overview of simultaneous trends in the meteorological parameters; and it enabled visual comparison.

2 Angelo Secchi, *Descrizione del meteorografo dell'Osservatorio del Collegio Romano* (Rome: Tipografia delle Belle Arti, 1870).

3 Paolo Brenni, "Secchi's Meteorograph: Rediscovering and Restoring a Forgotten Scientific Instrument," *Science and Technology for Cultural Heritage* 1 (1992): 227–235.

4 Paolo Brenni, "Secchi's Meteorograph," *Bulletin of the Scientific Instrument Society*, no. 34 (1992): 12.

The cost of the meteorograph, however, was unaffordable for most observatories at that time. Simpler and cheaper versions[5] were later sold at greatly discounted rates and installed in Jesuit colleges in Manila, Havana, and Shanghai. None of these originals, however, has survived.[6] The meteorograph in Rome is thus the only one extant.

In the 1990s the meteorograph was in poor condition when Paolo Brenni considered the challenges of restoring, what he considered, one of "the most impressive, complicated and expensive recording instruments ever made."[7] He proposed a conservative and yet functional restoration, and his thoughtful work proved to be a model for future conservators and restorers.

Restoration of the meteorograph was a sort of "summa" of typical restoration work on scientific instruments, usually made of mixed materials, because the instrument had wooden, iron, silver, and brass parts.[8] To deal with such variety at the same time was greatly exciting for Paolo who spoke of *his* Secchi meteorograph.[9]

By this work, Paolo also established a restoration philosophy for the SIC community. It had these principles: no excessive reconstruction should be done; there should be no loss of original information; and "if in doubt, leave it well enough alone."[10] Today these criteria are widely accepted in the restoration of scientific instruments.

---

5 Emilio Brassart, *Descrizione del meteorografo Secchi modificato* (Roma: Tipografia delle Belle Arti, 1872).

6 Paolo Brenni, "Secchi's Meteorographs Outside of Italy," in *Angelo Secchi and Nineteenth Century Science: The Multidisciplinary Contributions of a Pioneer and Innovator*, ed. Ileana Chinnici and Guy Consolmagno (Cham: Springer, 2021), 175–194.

7 Brenni, "Secchi's Meteorograph," *Bulletin of the Scientific Instrument Society.*

8 Paolo Brenni, "Strumenti scientifici: il restauro del meteorografo Secchi," *OPD Restauro* 5 (1992): 99–107.

9 Paolo Brenni, "Il meteorografo di padre Angelo Secchi," *Nuncius* 8, no. 1 (1993): 197–247.

10 Brenni, "Secchi's Meteorograph," *Science and Technology for Cultural Heritage*, 234.

P

The Air-Pump and *Leviathan*

Terje Brundtland and Rebekah Higgitt

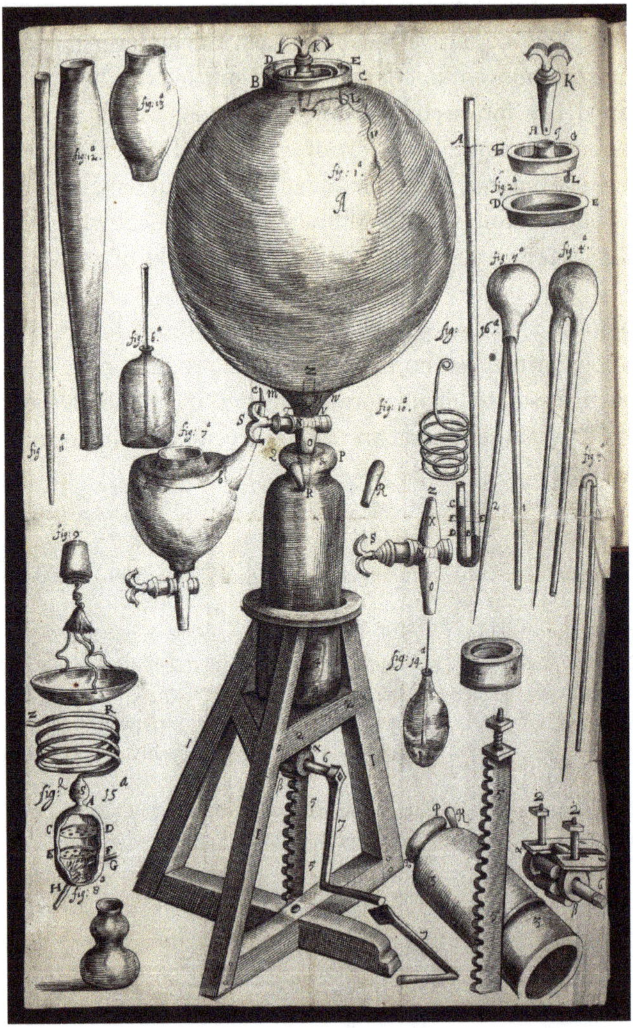

FIGURE P The air pump depicted in this plate from Robert Boyle's *New Experiments* (1660) was reproduced in various ways on the front covers of different editions of *Leviathan and the Air-Pump*
LINDA HALL LIBRARY OF SCIENCE, ENGINEERING & TECHNOLOGY

 | DOI:10.1163/9789004737587_028

A conversation between Terje Brundtland and Rebekah Higgitt about instrument studies and Shapin's and Schaffer's *Leviathan and the Air-Pump*, published forty years ago.

TB: My first encounter with this book was in 1997, when I started the MSc course on scientific instruments at MHS in Oxford.

RH: Mine was at a very similar time – in 1996–1997, when I was doing an interdisciplinary Masters in Seventeenth-Century Studies at Durham. My first degree was in history, so this was part of my introduction to HPS and SSK.[1] My focus was on the political and intellectual history it supplied – I was slower to grasp the insights into experimental practice. The book's claim that "solutions to the problem of knowledge are solutions to the problem of social order" – from the idea that facts are produced through social processes to the relationship between political and epistemological authority – had become axiomatic.[2]

TB: My background was as a laboratory technician, specialising in modern vacuum technology, yet this book was almost inaccessible at first. Social and political topics were not my thing.

RH: Looking back, I find I didn't cite the book in my dissertation. By the time I was doing my PhD in the early 2000s, my sense of the book's ubiquity meant that I almost avoided it. At that time, I hadn't really begun to think about artefacts and instruments, and it didn't prompt me to.

TB: For technically inclined instrument historians and museum curators, the book can be problematic and off-putting. It has sections containing series of words, concepts, problems, discussions and so on, which might seem unfamiliar, irrelevant, and unnecessary for those working hands-on with old scientific instruments, as well as for modern technicians.

RH: Once unpacked, I think the book does contain important insights for scientists and technicians, revealing origins of assumptions and practices they make take for granted. I didn't fully appreciate this until *after* I'd had a curatorial post, working with objects and audiences, and had then gone back to core STS texts for teaching.[3] Then, I quite naturally turned to the air

1 History and Philosophy of Science (HPS) and Sociology of Scientific Knowledge (SSK).
2 Steven Shapin and Simon Schaffer, *Leviathan and the Air-Pump: Hobbes, Boyle and the Experimental Life* (Princeton: Princeton University Press, 2011), 332.
3 Science and Technology Studies (STS).

pump to introduce students to seventeenth-century experimental philosophy, replication, and witnessing, and the social and political nature of fact making and communication – though I got them to read Shapin's "Pump and Circumstance" rather than *Leviathan*.[4]

TB: For me, it was when I focused on the technical and practical issues that the message of the book became gradually clearer: the accounts of experiments and work undertaken, using metal, glass, sealing wax, mercury, etc., instead of the discussions of social, political, and philosophical issues. It became an interesting source about seventeenth-century technology, but also what the dispute between Boyle and Hobbes was in fact all about. It's not written for technicians, although it's very much about them, their work, their use of materials, and their routines.

RH: I am not sure that the authors spent a lot of time with air pumps as artefacts, though they pay close attention to the written accounts of them and their use. How do you feel the book stands up as a technical account?

TB: *Leviathan* is maybe not the first text I would recommend to someone interested in seventeenth-century air pump practice. It contains many technical details about Boyle's pumps, but these are often hidden inside SSK issues. It requires a lot of effort for a technician to penetrate down to the hard details.

I did cite *Leviathan* in my DPhil dissertation and some articles, but I do not always agree with the authors, especially when it comes to the importance of Papin's third, twin-barrel air pump for the further development of pneumatics in England. I found no smooth line connecting Boyle's and Hauksbee's pumps.

RH: The book is undoubtedly a touchstone for historians of science and STS scholars. Although the original reviews were variously "critical, approving, mixed, bland, and point-missing," it's now a classic.[5] Less so in instrument studies,[6] though I wonder if this is changing – many science curators are

4 Steven Shapin, "Pump and Circumstance: Robert Boyle's Literary Technology," *Social Studies of Science* 14, no. 4. (1984): 481–520.

5 Simon Schaffer and Steven Shapin, "Final Comment: Author's Response," *Isis* 108, no. 1 (2017): 143–144, see 144. See retrospectives such as Anna Marie Roos on H-Net in 2000 https://www.h-net.org/reviews/showrev.php?id=3996; the "Second Look" in *Isis* 108, no. 1 (2017); and the introduction to the 2011 edition.

6 Shapin and Schaffer credit Jim Bennett rather than themselves for inspiring studies of instruments and instrumental techniques: "Introduction to the 2011 edition," *Leviathan and the Air-Pump*, xxxiv.

PhD-trained in a field shaped by *Leviathan*, where much of what was once provocative has become common ground. Scientific controversies, failed or superseded programmes, hidden technicians, and virtual witnesses have all featured in museum displays.

But what about air pumps? Do they have an exemplary place for instrument historians and curators? I was fascinated to see from your article that Shapin and Schaffer drew on the account of early air pumps by George Wilson, the first director of the Industrial Museum in Edinburgh (forerunner of my institution), written back in 1849.[7] Has our subsequent attention to the air pump been because of, or in spite of, *Leviathan*?

TB: The air pump has been described as one of the six most important instruments developed in the seventeenth century, together with the barometer, the thermometer, the telescope, the microscope, and the pendulum clock.[8] While there is a rich literature about these five, there are few texts in English diving deep into seventeenth-century air-pump technology. I think the attention given to the air pump by sociologists and historians of science is largely because of *Leviathan*, though its wide range of applications – ranging from status symbol to research instrument or demonstration device, performing phenomena for a paying public – means that it could hardly be ignored. No seventeenth-century air pumps have survived which explains, in part, why *Leviathan* is focused on textual rather than material culture.

Approaching the story of Boyle's air-pumps from the instrument perspective, there remain many opportunities for instrument historians and curators. Among others, one challenging task could be to write a text like *Leviathan and the Air-Pump for Technicians*.

RH: Nice idea! Or perhaps an annotated edition, in which technical content could add to or sit in tension with the original text. A future project for the SIC?

7 Terje Brundtland, "After Boyle and the Leviathan: the Second Generation of British Air Pumps," *Annals of Science* 68, no. 1 (2011): 93–124.

8 W.E.K. Middleton, *The History of the Barometer* (Baltimore: Johns Hopkins University Press, 1964), 3.

Q

The Universal Screw

Randall C. Brooks

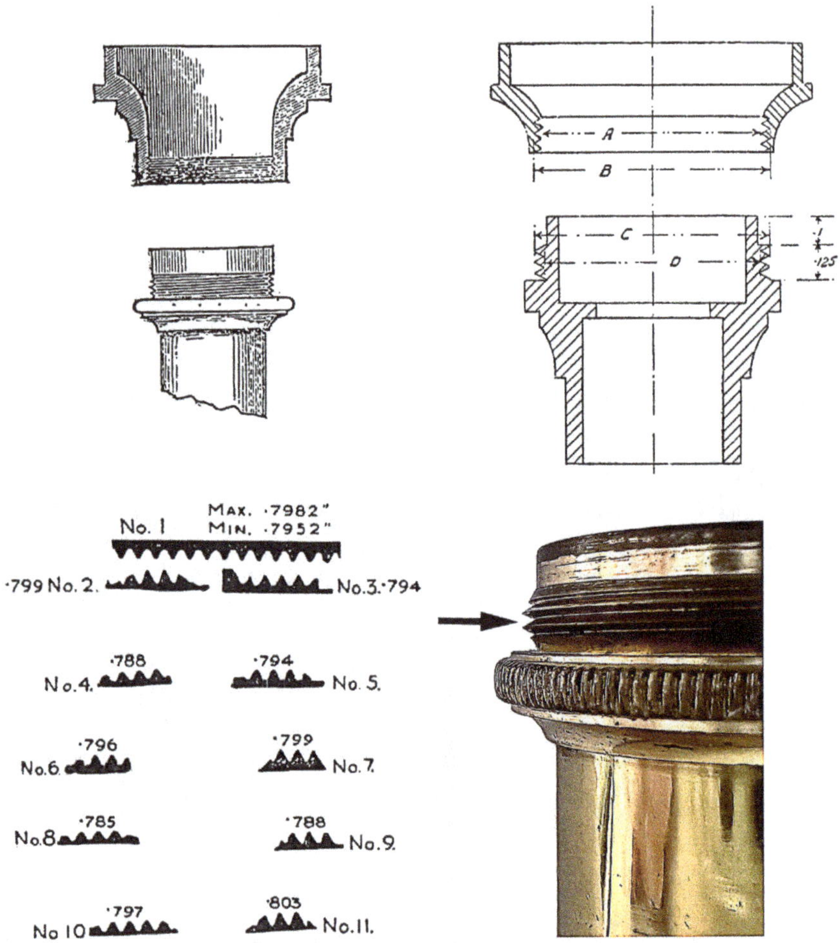

FIGURE Q *Clockwise from upper left*: the 1859 Microscopical Society standard; the improved RMS screw thread in 1896; an example of the RMS thread on a Leitz objective circa 1890, showing that the thread peaks are not as rounded as called for by the RMS standard; and profiles of RMS screw threads measured in 1911, showing anomalies and offering evidence that some makers were not meeting the standards

IMAGE COLLAGE: SARA SCHECHNER

 | DOI:10.1163/9789004737587_029

One would be forgiven for wondering why I was asked to discuss the Royal Microscopical Society screw thread in the context of the SIC anniversary – myself among you. In the early 1980s while searching for a PhD topic, I came across Dr Allan Mills's name (past editor of the *Bulletin of the Scientific Instrument Society*). I sent him a note proposing a topic; he countered that I should study screw threads as a means of identifying fraudulent or reproduction instruments such as copies of Newton's first reflecting telescope. Having dealt with old instruments with "marked" screws, I was intrigued and took up the topic with Allan since the screw is the most ubiquitous "complex device."

For nineteenth-century microscopists, including those who practiced their genteel hobby on Sundays (then a religiously acceptable Sunday activity), lenses for their instruments were expensive. During a period of active innovation in optical design, microscopists wanted the best and latest lenses for improved definition. J.J. Lister's 1830 proposal of changing the objective lens without having to replace the entire instrument was one practical solution.[1] During the leadership of George Shadbolt (1856–1857) as President of the Microscopical Society of London (MSL), a sub-committee was set up to report on the best form of universal attachment of the object glass to the body of a compound microscope.[2] The preliminary report and the gauge specification for what came to be called "the universal screw" were to be James Whitworth's standard (making him the father of the first standard screw). He agreed to fashion the hardened gauges and provide examples to the Society at cost.[3] Conrad Beck's report delivered in 1859, recommended the "Universal Thread" and was illustrated with Whitworth's form (55° angle, 36 TPI, and 0.780-inch diameter) along with the tools and gauges to be employed. Trials and exchanges confirmed that it worked as desired, and it was formally adopted.[4] The Microscopical Society screw thread became the first "standard" applicable to scientific instruments (other than standards required for commercial expedience, such as those for length, mass, or temperature). After 1866, the thread became known as the Royal Microscopical Society screw thread.

1 Gerard L'E. Turner, *God Bless the Microscope! A History of the Royal Microscopical Society over 150 Years* (Oxford: Royal Microscopical Society, 1989), 28.

2 Turner, *God Bless the Microscope*, 32.

3 George Jackson, Charles Brooke, and H. Perigal, Jr, "Report of the Sub-committee of the Microscopical Society on the Best Form of Universal Attachment of the Object-glass to the Body of a Compound Microscope," read 12 November 1857, *Transactions of the Microscopical Society* 6 (1857): 39–41.

4 Richard Beck, "Remarks on the 'Universal Screw,'" read 26 May 1859, *Transactions of the Microscopical Society* 8 (1859): 92–97.

It is not surprising that prominent microscope and lens makers Ross, Smith & Beck, and Powell immediately adopted the Society Thread and were soon followed by others. With the success of the RMS screw thread, other societies (e.g., Royal Astronomical Society and the British Association for the Advancement of Science) developed their own threads for specific requirements relevant to their members' needs.[5]

Subsequent reviews and improvements to the RMS thread were made and illustrated in 1896. A 1911 review illustrated many inconsistencies in the thread profile and pitch, which were addressed in a later review of the standard in 1915. The tools were then deposited with the National Physical Laboratory with new specifications for the tapping tools and with metric equivalents (to no less than four decimal places, which were then impossible to achieve).[6] With only three or four threads, the RMS thread was far from a precision screw but served very adequately for the purpose – accurately positioning the objective on a microscope's optical axis.

Once the RMS screw thread was adopted, it was almost universally accepted as the standard for microscope objective mountings by many European makers including, for example, Leitz. There were occasional subsequent reviews and updated standard definitions (e.g., British Standards Inst., BS 3569: 1962) but no significant changes. However, on 10 January 2022, the "ASTM Standard Specification for Microscope Objective Thread" ended the RMS screw thread's dominance after 163 years. It was the longest surviving instrumental standard! What other aspect of any scientific instrument has remained essentially the same for even a fraction of that time? The current ISO standard is physically the same but now with metric measurements. The penultimate image of the RMS screw thread may be purchased at the ASTM International website.[7]

5 Randall C. Brooks, "Towards the Perfect Screw Thread," *Transactions of the Newcomen Society* 64 (1992–1993): 101–120.

6 The various drawings and images of the screw threads of the Microscopical Society appear in the *Transactions of the Microscopical Society* for 1857 and 1859 (cited above), and subsequently in the *Journal of the Royal Microscopical Society* (*JRMS*). See Conrad Beck, "The Royal Microscopical Society's Standard Screw-Thread for Nose-Piece and Object-Glasses of Microscopes," *JRMS*, ser. 2, 16 (1896): 389–392; F.W. Watson Baker, "Anomalies in Objective Screw Threads," *JRMS*, ser. 2, 31 (1911): 175–177; and [anonymous], "Objective Screw Thread," *JRMS*, ser. 2, 35 (1915): 230–231.

7 A PDF with the details and images for the standard ASTM E210-63 (2016) may be purchased at the ASTM website https://webstore.ansi.org/.

R

A Cheap Berlin Hotel, Midnight

Marvin Bolt and Michael Korey

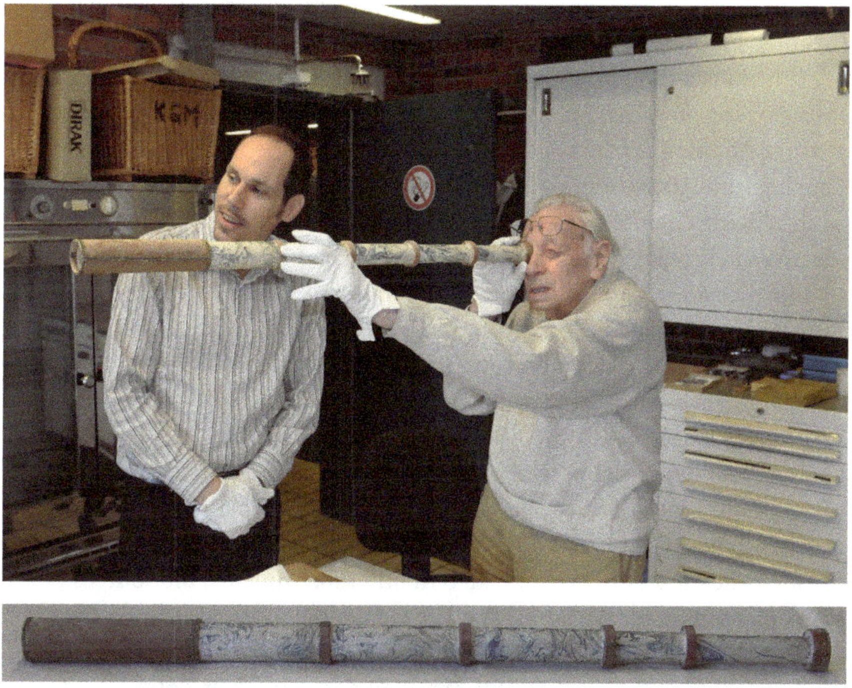

FIGURE R Rolf Riekher (1922–2018) and Michael Korey examining the pre-1617 telescope from the *Pommerscher Kunstschrank* in the conservation studio of the Kunstgewerbemuseum in Berlin on the occasion of Riekher's 85th birthday (2007). The telescope (Inv. P 23) features a planoconvex objective and a planoconcave ocular lens
PHOTOS: MARV BOLT

The (prior) scene. 2015, a cheap Berlin hotel, midnight. A young physicist tries to detect subtle floor vibrations while two aging curators snore.

The instrument to be studied. The world's oldest surviving, securely-dated telescope, at Berlin's Kunstgewerbemuseum (decorative arts museum), never before subjected to close optical analysis. The archives document it as a princely commission, alongside mathematical instruments, games, and other

 | DOI:10.1163/9789004737587_030

components of a filled-to-the-brim curiosity cabinet (*Kunstschrank*) delivered by Philipp Hainhofer in 1617 to Pomerania's Duke Philipp II. Although the cabinet was destroyed in the Second World War, its contents survive, including a painting depicting Hainhofer presenting the *Kunstschrank* to the duke, with its three dozen artisans arrayed in an imaginary group portrait.

The research. Using an optical toolkit incorporating classic techniques (adapted for portability and for use in art museums) alongside a groundbreaking (previously-thought-to-be-impossible) portable interferometer developed by the young physicist, Tiemen Cocquyt, our trio substantiated a striking conjecture: stopping down a spectacle lens triggered the telescope's invention.[1] Looking through the telescope confirmed limited value for military purposes, while the absence of prismatic tinges increased doubt about the historical severity of chromatic aberration. The painting and its courtly context pointed to a further *Kunstschrank*, made by Hainhofer's circle and now at Dresden's Kunstgewerbemuseum.[2] An inventory listed that this once contained another telescope, seemingly lost; storage-room foraging located it. Disassembled, both telescopes revealed a pattern previously unmentioned anywhere: objective and ocular lenses of distinctly different colours. Subsequent compositional analysis pointed to deliberately varied glass composition and provided links to contemporary glass artwork with comparable composition and, thus, associations between their artisans. Whereas borescope examination has yielded the identification of texts embedded in other instruments, the sources of the blank paper tubes of the Berlin and Dresden telescopes remain unknown.

The moral. Advancements in research techniques (such as optical metrology or material analysis) enable us to tackle previously unsolvable problems and to pose innovative new questions requiring direct investigation of objects. The integration of archival documentation, object-handling and usage, and material and iconographic analyses, made possible by wide-ranging collaborative networks, is becoming standard SIC research procedure.[3] Cocquyt's late-hour

1 See Rolf Willach, "The Long Route to the Invention of the Telescope," *Transactions of the American Philosophical Society*, n.s., 98, no. 5 (2008): 1–116.

2 Marvin Bolt and Michael Korey, "The World's Oldest Surviving Telescopes," in *The Origins of the Telescope*, ed. Albert Van Helden, Sven Dupré, Rob van Gent, and Huib Zuidervaart (Amsterdam: Koninklijke Nederlandse Akademie van Wetenschappen, 2010), 231–255.

3 An outstanding example of bringing together these approaches is Huib Zuidervaart and Tiemen Cocquyt, "The Early Development of the Achromatic Telescope Revisited," *Nuncius* 33, no. 2 (2018): 265–344.

technical tweaks supplemented the methods of the earlier toolkit, whose oversized suitcase crushed a curatorial toe and yielded a German emergency room bill less than five percent of its American counterpart; the old toolkit and toe still work, just not in the same way or quite as well. Whether the topic is science, sleeping, or systems of medicine, an abstract and indirect knowledge of performance or practice is one thing; direct experience is quite another.

S

Dowsing for Evidence

Julia Bloemer

FIGURE S
Automatic water finder, built circa 1910
PHOTO: DEUTSCHES MUSEUM, MUNICH

At first glance, the object looks unremarkable. It is a wooden box with a coil of iron wire inside, and above this, a rotating magnetic needle. Maybe it is a magnetometer? At least, nothing spectacular. The story becomes interesting with the related patent: it calls this instrument an "apparatus for detecting underground water courses" by responding "to the natural currents of the Earth."[1]

1 Patent No. 178457, issued by the Imperial Patent Office of the German Empire on July 30, 1906, patented on August 3, 1904.

 | DOI:10.1163/9789004737587_031

Although it does not look like a standard dowsing rod, it was constructed for the same purpose – water finding in the field. How can we deal with objects like this in exhibitions, at the crossroads of pseudoscience and science?

The instrument was invented by Adolf Schmid from Switzerland and got its imperial patent in 1904. In 1912, the freethinker Eduard Aigner donated the water finder to the Deutsches Museum. A year later, the water finder was loaned to the International Construction Trade Fair in Leipzig. It was displayed in the "agriculture" section, and accompanied by images of dowsers, not only from the Middle Ages but also, interestingly, from the "the present age."[2] The water finder and conventional dowsing rods were presented as normal detection instruments that were still in routine use.

Nevertheless, how the instrument functioned was subject to discussion. The inventor Schmid himself attributed the functioning to the influence of electromagnetic Earth currents, and Aigner wrote in his letters to the museum about the needle that followed the Earth's magnetic field. Surprisingly, the patent itself gave no reason that the instrument worked, stating that "for the effect of the device, so far a scientific-physical explanation with certainty could not be given." Instead, positive experiences served as empirical evidence for the functioning of the water finder. Schmid stressed, "The inventor carried out tests over several years with the help of engineers and physicists according to a strictly scientific method and they have convincingly shown that the needle of the apparatus reacts to the fluctuations of the … electrical Earth currents … caused by underground flowing water."[3] Aigner added even more evidence: "The Soil Department of the Government of India in Bombay reports very encouraging results with an English apparatus manufactured on the same principle."[4] Other conventional dowsing rods that Aigner donated to the Deutsches Museum also had prominent users. One dowsing rod was directly connected to the water supply office in Munich, another with the municipal building office. Even the German Emperor at that time supported water dowsing in German Southwest Africa.

Today the water finder stays forgotten in the museum storage. But it should return to public view. It could be an opportunity to talk explicitly about the role of empirical evidence, standardised tests, and criteria for scientificity,

---

2 List of the objects that were provided from the Deutsches Museum to the Construction Trade Fair in Leipzig 1913, 2189, Verwaltungsarchiv, Archiv des Deutschen Museums, Munich.

3 Adolf Schmid, *Patentierter Automatischer Quellenfinder. Gebrauchsanweisung* (n.d.).

4 Eduard Aigner to Mr. Trautwein, 4 June 1912, 1201/6, Verwaltungsarchiv, Archiv des Deutschen Museums, Munich.

as well as how the scientific community adjusts its boundaries repeatedly. It would be a story about people who tried to find evidence for and against dowsing, and by doing so, explored the boundary between science and pseudoscience. Discussion of the water finder might help the public understand better how scientific knowledge is generated, that change is part of the story along with stability, and that scientific consensus plays a role.

T

On Displaying Very Large Globes

Jim Bennett and Sylvia Sumira

FIGURE T The large globes made by Vincenzo Coronelli for Louis XIV, 1683
BIBLIOTHÈQUE NATIONALE DE FRANCE. PHOTO: SYLVIA SUMIRA

One of the most spectacular projects in the field of historic scientific instruments in recent years was the conservation, restoration, and redisplay of the pair of very large globes of 1683 – celestial and terrestrial – by the Venetian cosmographer and globe-maker Vincenzo Coronelli. Each globe measures almost four metres in diameter. The project was completed in 2006, when the restored and remounted globes were installed in a public gallery at the Bibliothèque nationale de France in Paris. This was a major involvement in our discipline by a national institution, which addressed in a very public setting many of the critical issues and dilemmas of our subject and our professional practice; the extent of restoration, choice of materials used, method of display, and interpretation. Remarkably the project was complemented in 2016

 | DOI:10.1163/9789004737587_032

by a parallel scholarly achievement of substance, the publication of *Vincenzo Coronelli Cosmographer (1650–1718)* by Marica Milanesi.[1] Issues raised within these complementary projects have enriched our understanding of perhaps the greatest globe-maker in history.

Apprenticed as a carpenter, Coronelli's origins were humble, but we might suggest that his education by the Franciscans in his native Venice and especially his entry into the order as a friar served to mitigate his modest social position. He rose to the role of Father General, albeit briefly, and established a large workshop in the Gran Casa dei Frari for producing globes, maps, prints, and books. It is remarkable that the large globes were only the second pair he made, having been commissioned by the courtier Cardinal d'Estrées in about 1680 to create them as a gift to the king, Louis XIV, to be installed at the palace of Versailles, which was then under construction. The story of the globes thus concerns not only the social standing of their maker but the protocols of royal symbolism and courtly patronage in its most exacting setting – that of the Sun King, whose every appurtenance was an evocation of a unique statecraft.

It was announced that the globes were finished in November 1683, but in fact much remained to be done, not least the challenge of providing a mounting. It would be twenty years before all was complete, with Coronelli taking no further part in the work. In 1703 the globes found a home in two pavilions in the park of the castle in Marly, in a very large version of the traditional globe mount, with a meridian ring supported by pillars rising to a horizon ring. The globes remained at Marly until 1714 but were then removed to the Bibliothèque du Roi at the Louvre.

Since Marly, other venues have faced the display challenge posed by the king's globes; most recently, it was the Bibliothèque nationale. A chequered history had created a legacy of display solutions,[2] which must have presented the designers with interesting choices and a variety of legitimate possibilities. The success of the result is widely acknowledged. Suspension above head-height rather than supporting stands has created an open, accessible display which, after viewers recover from the astonishment of the initial impression, draws them into the installation.

1 Marica Milanese, *Vincenzo Coronelli Cosmographer (1650–1718)*, Terrarum Orbis 13 (Turnhout: Brepols, 2016).

2 Hélène Richard, *Les Globes de Coronelli* (Paris: Bibliothèque nationale de France/Seuil, 2006), 24–35.

A parallel challenge to the display has been the major programme of restoration – perhaps the largest ever undertaken on a pair of globes.[3] This is given due acknowledgement in the display with a filmed account of the work.

Finally, historical insights continue to reward students and scholars. The globes were, above all, a major undertaking in the discipline of cosmography, whose role and significance has been increasingly appreciated in recent years. The terrestrial globe, however, offers a very different panorama of the world from the mathematical science of cosmography, its detailed iconography presenting its peoples and their customs and activities, viewed from the particular standpoint of late seventeenth-century Europe.[4]

---

3 Nathalie Buisson, Alain Roger, Daniel Vigears, and Thierry Borel, "Les Globes de Coronelli," *Actualités de la conservation*, no. 27 (January–June 2008): 1–4.

4 Catherine Hofmann, "'Incomparable' Globes or 'Old Devices as Cumbersome as They Are Useless'? The Vicissitudes of the Great Globes of Coronelli, 1683–1915," *Globe Studies*, no. 53/54 (2007): 24–41.

Index

www.ingramcontent.com/pod-product-compliance
Lightning Source LLC
Chambersburg PA
CBHW071334230926
51849CB00002B/393
9789004737570